A Biology of the Algae

A Biology of the Algae

third edition

Philip Sze

Georgetown University

Boston, Massachusetts   Burr Ridge, Illinois   Dubuque, Iowa
Madison, Wisconsin   New York, New York   San Francisco, California   St. Louis, Missouri

About the cover: Enlarged branch tips of the intertidal brown alga, *Fucus spiralis,* containing the reproductive structures.

WCB/McGraw-Hill

A Division of The **McGraw·Hill** *Companies*

A BIOLOGY OF THE ALGAE

Copyright © 1998 by The McGraw-Hill Companies, Inc. All rights reserved. Previous editions © 1986, 1993 by Wm.C. Brown Communications, Inc. Printed in the United States of America. Except as permitted under the United States Copyright Act of 1976, no part of this publication may be reproduced or distributed in any form or by any means, or stored in a data base or retrieval system, without the prior written permission of the publisher.

 This book is printed on recycled, acid-free paper containing 10% postconsumer waste.

1 2 3 4 5 6 7 8 9 0 QPD/QPD 9 0 9 8 7

ISBN 0-697-21910-0

Publisher: Michael D. Lange

Sponsoring Editor: Margaret J. Kemp

Developmental Editor: Kathleen R. Loewenberg

Marketing Manager: Thomas C. Lyon

Project Manager: Vicki Krug

Production Supervisor: Deborah Donner

Cover Designer: John Rokusek, Rokusek Design

Cover Image: Philip Sze

Compositor: Interactive Composition Corporation

Typeface: 10/12 Times Roman

Printer: Quebecor Printing Book Group (DBQ)

Library of Congress Cataloging-in-Publication Data

Sze, Philip, 1945–
    A biology of the algae / Philip Sze. — 3rd ed.
        p.   cm.
    Includes bibliographical references (p.   ) and index.
    ISBN 0-697-21910-0
    1. Algae. 2. Algology.   I. Title.
    QK566.S97   1997
579.8—dc21   97-10511
                                                                                CIP

www.mhhe.com

CONTENTS

Preface vii

1

Introduction to Algal Characteristics and Diversity 1

Diversity of Photosynthetic Pigments 3
Reserve Carbohydrates 7
Diversity of Cellular Organization 7
Molecular Phylogeny 13
Morphologic Diversity 14
Ecologic Diversity 16
Algae and Humans 18
Summary 19
Further Reading 20

2

Cyanobacteria 21

Cellular Structure 22
Diversity 24
  Nonfilamentous Cyanobacteria 24
  Filamentous Cyanobacteria without
    Specialized Cells 25
  Heterocystous Cyanobacteria 25
  Prochlorophytes 29
Ecology 29
Origin of Chloroplasts 35
Summary 37
Further Reading 38

3

Green Algae (Division Chlorophyta) 39

Class Prasinophyceae 43
Class Chlorophyceae 45
  Chlorophycean Diversity 47
  Chlorophycean Ecology and Commercial
    Uses 64

Class Ulvophyceae 67
  Ulvophyte Diversity 68
  Ulvophyte Ecology and Commercial
    Uses 77
Class Charophyceae 79
  Charophyte Diversity 79
  Charophyte Ecology 86
  Evolution of Land Plants 87
Summary 87
Further Reading 88

4

Division Chromophyta 89

Classes Chrysophyceae and Synurophyceae 91
  Representative Genera 91
  Ecology of Chrysophytes and
    Synurophytes 94
Classes Tribophyceae, Eustigmatophyceae, and
  Raphidophyceae 96
  Class Tribophyceae 96
  Class Eustigmatophyceae 98
  Class Raphidophyceae 98
Class Bacillariophyceae—The Diatoms 98
  Reproduction in the Diatoms 101
  Ecology and Diversity of the Diatoms 105
Class Phaeophyceae 108
  Isomorphic Generations 109
  Heteromorphic Generations 114
  Brown Algae without a Free-Living Haploid
    Phase (Orders Fucales, Durvillaeales) 122
  Brown Algal Ecology and Commercial
    Uses 126
Summary 127
Further Reading 128

5

Haptophytes, Dinoflagellates, Cryptomonads, and Euglenophytes 129

Division Haptophyta 129
Representative Haptophyte Genera 132
Haptophyte Ecology 133
Division Dinophyta 134
Typical Dinoflagellate Cell 135
Dinoflagellate Ecology 140
Division Cryptophyta 144
Typical Cryptomonad Cell 144
Cryptomonad Ecology 146
Division Euglenophyta 146
Typical Euglenophyte Cell 146
Euglenophyte Ecology 148
Division Chlorarachniophyta 149
Summary 149
Further Reading 150

6

Red Algae (Division Rhodophyta) 151

Subclass Florideophycidae 153
Thallus Structure 154
Life Cycles 160
Postfertilization Events and Development of the Carposporophyte 169
Pit Plugs 170
Subclass Bangiophycidae 171
Red Algal Ecology and Commercial Uses 174
Summary 177
Further Reading 178

7

Phytoplankton 179

Environmental Factors Influencing Phytoplankton Growth 180
Light 180
Density Stratification 180
Nutrients 184
Flotation and Sinking 190
Grazing 193
Other Influences 194
Adaptive Strategies of the Phytoplankton 195
Primary Production by Phytoplankton 195
Marine Phytoplankton 197
Temperate Oceans 199
Tropical and Subtropical Oceans 199
Polar Oceans 200
Coastal Oceans 203
Coastal Upwelling and Estuaries 204

Freshwater Phytoplankton 207
Temperate Lakes 207
Tropical and Polar Lakes 209
Rivers 210
Summary 211
Further Reading 212

8

Freshwater Benthic and Terrestrial Algae 213

Freshwater Benthic Algae 213
Standing Water 215
Symbiotic Algae 218
Flowing Water 219
Terrestrial Algae 221
Summary 224
Further Reading 224

9

Benthic Marine Algae 225

Introduction to Marine Macroalgae 225
Morphologic Types 225
Life Cycles 228
Production and Food Chains 230
Coral Reefs 231
Soft-Bottom Communities 237
Mangrove Swamps and Seagrass Meadows on Tropical Coasts 237
Temperate Shorelines 238
Atlantic Rocky Shores 239
Spray Zone 240
Intertidal Emergent Surfaces 241
Tidepools 244
Subtidal Region 245
Pacific Rocky Shores 248
Summary 250
Further Reading 250

Appendix: Useful References for Studies with Algae 251

Glossary 253

References 261

Index 271

Contents

P R E F A C E

The need for a third edition of *Biology of the Algae* is gratifying and, I hope, reflects the continuing desire to introduce undergraduate students to this fascinating group of organisms. As in the earlier editions, my overall goal is to provide an overview of the algae that emphasizes their morphologic, evolutionary, and ecologic diversity. The book can serve either as the primary text for a one-term course on the algae, in which the instructor will want to expand on topics of interest with supplementary readings, or as a text for a course that includes the algae as well as other topics, such as an algae-fungi course or an aquatic botany course. I have assumed that the reader is familiar with basic concepts covered in a standard introductory biology course.

Organization

Text organization is the same as in previous editions. Chapter 1 is a general introduction to the algae that emphasizes the aspects of their diversity to be considered in more detail in subsequent chapters. Chapters 2 through 6 cover the different groups of algae, distinguishing among divisions and classes and using representative genera to show morphologic variation. Chapters 7 through 9 describe the roles of algae in different ecosystems. Research on many aspects of algal biology has been prolific since the second edition. A challenge of the revision process was selecting new material for inclusion while trying to keep the text to approximately the same length.

Molecular comparisons using specific genes or ribosomal RNA are an exciting new tool for evaluating relationships among algal groups. I have added a short section on molecular phylogeny to chapter 1, focusing on the use of RNA in the small subunit of ribosomes and raising some interesting questions about how closely the different algal divisions are related. In subsequent chapters, I have indicated where this molecular evidence agrees or disagrees with relationships indicated by the more traditional approaches to classification, which are based on structure and pigmentation.

I treat the evolution of algal photosynthetic systems as a one- or two-step process. The basic system in eukaryotic algae is derived from cyanobacteria either directly as a primary endosymbiotic event, as seen in the green and red algae, or indirectly as a secondary symbiotic event, in which the chloroplasts of some groups are derived from another eukaryotic alga. In the final section in chapter 2 on the cyanobacteria, I introduce the origin of the chloroplasts of eukaryotic algae.

I have made some fairly significant changes from the second edition in the treatment of several groups of algae. Molecular evidence does not support either the separation of the prochlorophytes from the cyanobacteria in chapter 2 or a role for the prochlorophytes as ancestors of green algal chloroplasts. In chapter 3, division of green algae into four classes (Prasinophyceae, Chlorophyceae, Ulvophyceae, Charophyceae) seems warranted based on ultrastructural differences, especially basal body orientation, and molecular comparisons. In chapter 4, the division Chromophyta contains algae with chlorophyll c and heterokontous flagellation, which includes the brown algae (Phaeophyceae) but excludes the haptophytes. The haptophytes (or prymnesiophytes), which were treated as a class of the Chrysophyta in the second edition, have been moved to chapter 5 as a separate division and their coverage expanded, reflecting recognition of their importance in the oceans. In chapter 5, I use division Dinophyta for the dinoflagellates instead of division Pyrrophyta and have added a brief section on *Chlorarachnion*.

The material on algal ecology in the last three chapters, especially information on coastal phytoplankton and the Antarctic region, has been updated. In chapter 7, I added a general introduction to primary production and in chapter 9 replaced the introductory listing of environmental factors affecting benthic algae with a short section on regulation of ecologic cycles of macroalgae.

Each chapter concludes with a short summary and a list of general references, which should serve as guides to the published literature. An appendix at the end of the text lists references that describe methods for studying algae and guides for algal identification. Key terms in the text are defined in the end-of-book glossary.

Acknowledgements

Students in my courses at Georgetown University and at the Shoals Marine Laboratory have provided the inspiration for *Biology of the Algae*, and have also served as a "testing ground" for its contents. I am also indebted to everyone who has taken the time to send comments, and owe special thanks to the publications and authors who have allowed me to reproduce their illustrations. As reviewers, the following people provided valuable suggestions for the preparation of this edition:

Marilyn H. Baker
University of Alaska, Anchorage

Dean W. Blinn
Northern Arizona University

Richard Fralick
Plymouth State College

Ronald W. Hoham
Colgate University

James R. Sears
University of Massachusetts

William J. Wardle
Texas A & M University at Galveston

Peggy A. Winter
University of West Florida

Finally, I would like to thank the staff of the Blommer Library (Georgetown University) for their assistance in locating books and journals in the midst of a major renovation.

Preface

1

Introduction to Algal Characteristics and Diversity

Living cells appeared on earth over 3.5 billion years ago. These first cells lacked a well-defined nucleus and other complex cellular compartments, and obtained energy as either heterotrophs, consuming organic material from the surrounding seawater, or as autotrophs, using inorganic material. Autotrophs included chemosynthetic forms, similar to bacteria found today near geothermal vents in the deep seas, and photosynthetic forms, similar to present-day sulfur bacteria using a single photosystem in photosynthesis.

Cyanobacteria, the first algae, appeared over 3 billion years ago. They introduced photosynthesis with two photosystems, in which water is split and oxygen is given off as a by-product. This oxygen release profoundly affected the earth. Accumulation of oxygen in the atmosphere led to formation of an ozone layer, which protected against high-energy ultraviolet radiation, and to development of aerobic respiration in living cells to break down organic material more effectively. Under these new conditions, another type of cell, with a nucleus and complex cellular compartments (organelles), evolved about 1.5 billion years ago. Among these early eukaryotic cells were the ancestors of today's eukaryotic algae. Today, algae continue to be important producers of oxygen and organic material in ocean and freshwater environments, while other algae have adapted to living on land or in symbiotic associations with other organisms.

Algae have evolved into a diverse group of photosynthetic organisms, ranging in size from microscopic single cells to complex, multicellular seaweeds many meters long. Their cellular structures, their cell arrangements to form multicellular bodies or thalli, and their pigments for photosynthesis vary greatly. Algae span the range of structural complexity from bacteria to plants. Both algae and land plants have photosynthetic systems based on chlorophyll *a*, but algae lack plants' complex reproductive structures. In development, algae do not form embryos within protective coverings the parents produce. Also, algal reproductive structures do not have sterile cells—all cells are potentially fertile. None are formed exclusively to provide protection or nutrition during development. The algae also include a few "colorless" or nonphotosynthetic species that are closely related to photosynthetic species.

Traditionally, the study of algae, called **phycology** (less correctly, algology), has dealt with photosynthetic organisms that are not bryophytes or vascular plants. With the recognition that we can no longer divide living organisms into "plants" and "animals," a five-kingdom system has become popular. In this system, the eukaryotic algae are placed in the kingdom Protista, which also includes fungal protists and protozoa, and the prokaryotic algae in the kingdom Monera with

Table 1.1   *Divisions (=Phyla) and Classes of Algae*

Prokaryotic Algae

Division Cyanophyta (cyanobacteria or blue-green algae)
  Class Cyanophyceae (includes prochlorophytes)

Eukaryotic Algae

Chloroplasts Surrounded by Two Membranes
Division Rhodophyta (red algae)
  Class Rhodophyceae
Division Chlorophyta (green algae)
  Class Prasinophyceae
  Class Chlorophyceae
  Class Ulvophyceae
  Class Charophyceae

Chloroplasts Surrounded by More Than Two Membranes
Division Chromophyta (=Chrysophyta)
  Class Chrysophyceae (golden brown algae)
  Class Synurophyceae
  Class Tribophyceae (=Xanthophyceae) (yellow-green algae)
  Class Eustigmatophyceae
  Class Raphidophyceae (=Chloromonadophyceae)
  Class Bacillariophyceae (=Diatomophyceae) (diatoms)
  Class Phaeophyceae (=Fucophyceae) (brown algae)
Division Haptophyta
  Class Prymnesiophyceae (=Haptophyceae)
Division Dinophyta (=Pyrrophyta) (dinoflagellates)
  Class Dinophyceae
Division Cryptophyta (cryptomonads)
  Class Cryptophyceae
Division Euglenophyta
  Class Euglenophyceae
[Division Chorarachniophyta
  Class Chlorarachniophyceae]

Table 1.2   *Levels of Classification*

	Ulva lactuca L.	*Laminaria saccharina* (L.) Lamouroux	Suffix
Division	Chlorophyta	Chromophyta	-phyta
Class	Ulvophyceae	Phaeophyceae	-phyceae
Order	Ulvales	Laminariales	-ales
Family	Ulvaceae	Laminariaceae	-aceae
Genus	*Ulva*	*Laminaria*	(variable)
Species	*Ulva lactuca*	*Laminaria saccharina*	(variable)

eubacteria. Recent molecular evidence suggests that the ancestors of different groups of eukaryotic algae may have acquired photosynthetic systems from different sources and thus are less closely related to each other than to other groups of protists. This indicates that the algae represent a number of independent evolutionary lines and thus are not one well-defined taxonomic group. This text classifies the algae into eight major divisions (or phyla) with seventeen classes (table 1.1). These divisions, and even some classes within divisions, exhibit great diversity. We will examine differences among algae with special attention to the following features:

1. Light-harvesting pigments for photosynthesis
2. Polysaccharide reserve
3. Cellular organization
4. Molecular phylogeny
5. Morphology
6. Ecology

Variation in these characteristics separates algae at different taxonomic levels. (Table 1.2 reviews the major taxonomic levels.)

Diversity of Photosynthetic Pigments

For photosynthesis, pigments absorb light, which is converted into chemical energy in ATP (adenosine triphosphate) and NADP (nicotinamide adenine dinucleotide phosphate), which are used to synthesize organic compounds from carbon dioxide. Usable light for photosynthesis is in the 400–700 nanometer range of the electromagnetic spectrum, referred to as **photosynthetically active radiation** (PAR). Individual pigments selectively absorb certain wavelengths of PAR. The principal pigment in all algae is chlorophyll *a*, but accessory pigments absorb other wavelengths of PAR and transfer the light energy to chlorophyll *a*.

Photosynthetic pigments associate with proteins in the thylakoid membranes of a cell to form light-harvesting complexes or photosystems (phycobiliproteins are on thylakoid surfaces in cyanobacteria and red algae). Each photosystem consists of several hundred pigments funneling energy into a reaction center consisting of a special chlorophyll-*a*–protein unit. Two types of photosystems operate together to reduce NADP and to convert ADP (adenosine diphosphate) to ATP (fig. 1.1). Photosystem I is composed primarily of chlorophyll *a*, while photosystem II contains chlorophyll *a* and several accessory pigments. Algae may respond to their light environment by increasing or decreasing their overall pigment content (number and/or size of photosystems), changing the composition of accessory pigments (chromatic adaptation), varying the ratio of photosystem I to photosystem II, and adjusting the relative flow of electrons in photosystem I and photosystem II (Chow, Melis, and Anderson 1990).

The Calvin cycle is the principle pathway for the formation of organic compounds from carbon dioxide. Here, the energy and reducing potential of NADP and ATP are used to synthesize sugars, usually represented as glucose molecules. The glucose may be oxidized immediately for energy, combined in short polysaccharides for transport, or stored in long chain polysaccharides.

Reaction center activities and the Calvin cycle are similar in all algae, but algae show considerable diversity in their accessory pigments. Table 1.3 summarizes principal photosynthetic pigments in the different algal divisions. Photosynthetic pigments are divided into three classes: **chlorophylls, carotenoids,** and **phycobilins.** In addition to chlorophyll *a*, many algal groups have another form of

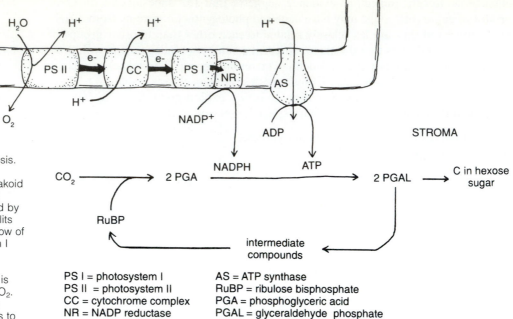

INSIDE OF THYLAKOID

STROMA

PS I = photosystem I
PS II = photosystem II
CC = cytochrome complex
NR = NADP reductase

AS = ATP synthase
RuBP = ribulose bisphosphate
PGA = phosphoglyceric acid
PGAL = glyceraldehyde phosphate

Figure 1.1
Summary of photosynthesis. Pigments organized into photosystems in the thylakoid membranes absorb light energy. Energy absorbed by photosystem II (PS II) splits water and generates a flow of electrons to photosystem I (PSI), where NADP (nicotinamide adenine dinucleotide phosphate) is reduced. PS II gives off O_2. Accumulation of H^+ ions inside the thylakoid leads to the synthesis of ATP (adenosine triphosphate) when the H^+ ions recross the membrane. The Calvin cycle occurs in the stroma of a chloroplast. Carbon dioxide combines with ribulose bisphosphate (RuBP) to produce phosphoglyceric acid (PGA). Ultimately, some of the carbon atoms in PGA are used to synthesize sugar (glucose), and other carbon atoms regenerate RuBP. ATP and NADP are used in the process.

Table 1.3   Photosynthetic Pigments and Carbohydrate Reserves

Division	Principal Photosynthetic Pigments	Carbohydrate Reserve
Cyanophyta	Chlorophyll *a*; phycocyanobilin, phycoerythrobilin	Starch (α-1,4-linked glucan)
Rhodophyta	Chlorophyll *a*; phycoerythrobilin	Starch
Chlorophyta	Chlorophylls *a, b*	Starch
Chromophyta	Chlorophylls *a, c_1, c_2*; fucoxanthin	Chrysolaminarin or laminarin (β-1,3-linked glucan)
Haptophyta	Chlorophylls *a, c_1, c_2*; fucoxanthin	Chrysolaminarin (β-1,3-linked glucan)
Dinophyta	Chlorophylls *a, c_2*; peridinin (some chlorophylls *a, c_1, c_2*; fucoxanthin)	Starch
Cryptophyta	Chlorophylls *a, c_2*; phycocyanobilin or phycoerythrobilin	Starch
Euglenophyta	Chlorophylls *a, b*	Paramylon (β-1,3-linked glucan)

chlorophyll, either chlorophyll *b* or chlorophyll *c* (fig. 1.2). Chlorophyll *c* may be present in two or sometimes three slightly different forms, designated c_1, c_2, and c_3. While chlorophylls are green pigments, carotenoids are brown, yellow, or red. Although approximately sixty carotenoids occur in algae (Rowan 1989), the major carotenoids participating in photosynthesis are fucoxanthin, peridinin, siphonaxanthin, and possibly β-carotene (fig. 1.3). Most algae have β-carotene, but usually, it has secondary importance. In some green algae, siphonaxanthin may

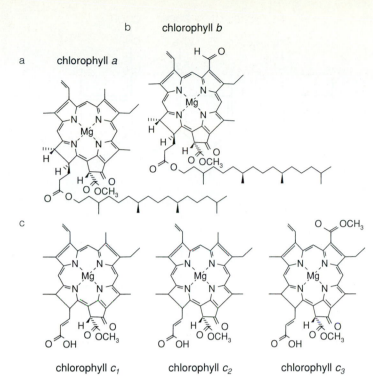

b   chlorophyll *b*

a   chlorophyll *a*

c

chlorophyll *c₁*   chlorophyll *c₂*   chlorophyll *c₃*

Figure 1.2
Chlorophylls. (*a*) Chlorophyll *a*.
(*b*) Chlorophyll *b*. (*c*) Forms
of chlorophyll *c*. (*a–c* from
Egeland et al. 1995, courtesy
Journal of Phycology.)

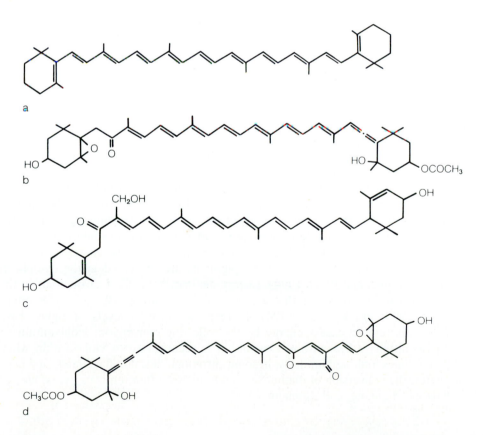

a

b

c

d

Figure 1.3
Carotenoids. (*a*) *β*-carotene.
(*b*) Fucoxanthin.
(*c*) Siphonaxanthin.
(*d*) Peridinin.

Introduction to Algal Characteristics and Diversity   5

Figure 1.4
Phycobilins.
(*a*) Phycocyanobilin.
(*b*) Phycoerythrobilin.

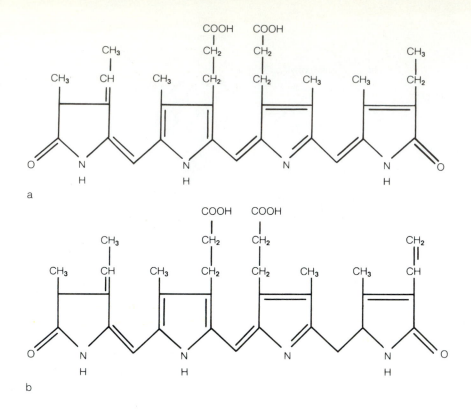

be an adaptation for absorbing light in relatively deep water. Fucoxanthin occurs in some algae with chlorophyll *c*, while peridinin is present in many dinoflagellates. Functions of other carotenoids are poorly understood, but some are important for photoprotection, as discussed later in this section. Three divisions of algae have phycobiliproteins. Each phycobiliprotein consists of several pigment molecules, called phycobilins, tightly bound to a protein. The phycobilins include the red pigment phycoerythrobilin and the blue pigment phycocyanobilin (fig. 1.4).

The dominant photosynthetic pigments often give algae a distinct color and their common names (see table 1.1). However, observed color is not always a reliable indicator of taxonomic position. For instance, not all members of the Rhodophyta (red algae) have an obvious red color; many are tan, dark brown, or another color (compare plates 6*b* and 7*b*).

Exposure to high levels of light may inhibit photosynthesis and even damage the photosynthetic system. To reduce such adverse effects, some carotenoids function in photoprotection by screening the chloroplasts, by deactivating highly reactive forms of oxygen before cellular damage occurs (superoxide dismutase is also important in this role), or by absorbing and dissipating excess energy (Demmig-Adams and Adams 1992). Screening pigments are dispersed in the cytoplasm (Hagen, Braune, and Björn 1994). Energy dissipation involves carotenoids in the photosynthetic photosystems that are part of the "xanthophyll cycle" (Demmig-Adams 1990; Frank et al. 1994). In response to high levels of light, these carotenoids remove excess energy by the following conversion: violaxanthin → antheraxanthin → zeaxanthin. Zeaxanthin is reconverted to violaxanthin when light becomes limiting. Similarly, in some chromophytes, dinoflagellates, and haptophytes, the conversion of diadinoxanthin to diatoxanthin removes excess energy (Olaizola, Bienfang, and Ziemann 1992).

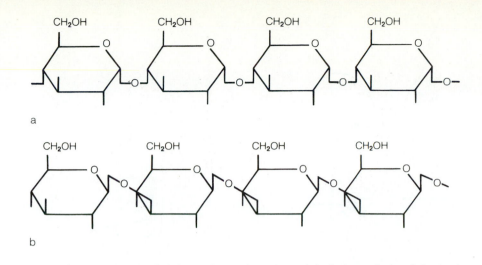

Reserve Carbohydrates

The glucose molecules the Calvin cycle produces may join in long chains (glucans) for storage. Algae have two types of these polysaccharide reserves, depending on how the glucose units link (fig. 1.5). The various forms of **starch** are α-1,4-linked glucans. The β-1,3-linked glucans include **chrysolaminarin** (=leucosin), **laminarin,** and **paramylon.** In both types, side branches may arise from the six-carbon, and the degree of branching and the length of chains vary. The carbohydrate reserve is characteristic for different algal divisions (table 1.3).

Diversity of Cellular Organization

Prokaryotic and eukaryotic forms are included among the algae. All prokaryotic algae belong to the cyanobacteria and lack a nuclear region surrounded by a nuclear envelope and complex organelles, such as chloroplasts, mitochondria, Golgi bodies, and endoplasmic reticula. Prokaryotic algae resemble other eubacteria in their walls, ribosomes, and chromosome structure, but differ from other photosynthetic bacteria by having two photosystems and producing oxygen. The chlorophylls (and carotenoids) of cyanobacteria are on internal membranes of flattened vesicles called **thylakoids,** while phycobiliproteins occur in granular structures called **phycobilisomes** on the outer surfaces of thylakoid membranes.

The other divisions of algae are eukaryotic. Let us consider two hypothetical eukaryotic cells representing primitive and advanced conditions. Primitive eukaryotic algae were probably flagellated (except in Rhodophyta). Figure 1.6*a* is a diagram of a hypothetical primitive phytoflagellate. The beating of two anterior flagella propels the cell through the water. Each flagellum arises from basal bodies and has an axoneme of microtubules in a 9+2 arrangement (nine peripheral pairs and two central microtubules). Sliding of the peripheral pairs of microtubules relative to each other causes flagella to flex. The primitive algal cell lacks a wall but may have a covering of platelike **scales** composed of complex polysaccharides. Scales formed in Golgi vesicles within the cell are deposited on the cell membrane's outer surface, sometimes in overlapping layers (see fig. 4.5*a*). Scales form a protective covering about the cell and represent an intermediate condition between a cell without any external covering and a cell enclosed by a wall.

A conspicuous structure within the hypothetical primitive alga is a single, large chloroplast. It is surrounded by an envelope composed of two closely associated membranes and contains flattened sacs called **thylakoids** (similar to cyanobacterial thylakoids). Photosynthetic pigments are in thylakoid membranes. The chloroplast

Figure 1.6
Flagellated cell representing a primitive eukaryotic cell. (*a*) Structures of vegetative cell. (*b*) Asexual reproduction involving mitosis with a persistent nuclear envelope and cell division by furrowing. ch = chloroplast; cp = centriole pair; cr = chromosomes; ey = eyespot; fl = flagellum; fu = furrow; gb = Golgi body; mi = mitochondrion; mt = spindle of microtubules; nu = nucleus; py = pyrenoid; sc = scale; th = thylakoid

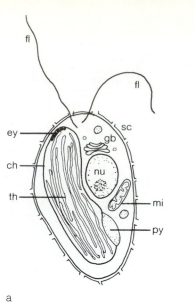

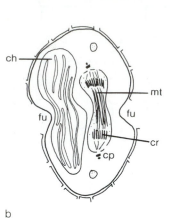

a   b

may contain a distinct region called the **pyrenoid,** whose principal component is the enzyme ribulose bisphosphate carboxylase, which catalyzes the incorporation of carbon dioxide into organic compounds in the Calvin cycle. A pyrenoid also may contain other proteins (Okada 1992). In some algae, carbohydrate reserves accumulate on the pyrenoid. An eyespot, consisting of several layers of carotenoid granules, also may be part of the chloroplast. Most flagellated algae swim toward the sunlight at the water surface and thus show a positive phototaxis. The eyespot helps sense the direction of light but is not the actual photoreceptor, which is either part of the cell membrane or a swelling near the base of one flagellum. The eyespot blocks light from some directions and reflects light from other directions onto the receptor. A chloroplast also contains ribosomes and DNA (Coleman 1985).

Other cellular structures present in the cytosol include mitochondria, Golgi bodies, endoplasmic reticulum, various types of vesicles, ribosomes, and cytoskeletal components. Some eukaryotic algae have mitochondria with flattened cristae, while others have tubular cristae (see table 1.6). In the cell's center is a large nucleus surrounded by a nuclear envelope composed of two membranes and containing chromosomes composed of DNA and proteins. The nucleus may have a distinct nucleolus. Freshwater flagellates often have contractile vacuoles near the bases of their flagella. These organelles regulate the cell's fluid content by slowly filling with excess liquid and expelling it through pores, and also may function in secretion and uptake of material into the cell. Some algal groups store their carbohydrate reserve in cytoplasmic vesicles rather than in the chloroplast. The cell's cytoskeleton is associated with the basal bodies and controls cell shape, position of organelles, movements of materials within the cell, formation of the cell covering, and various aspects of cell division. It consists of connecting fibers between the basal bodies, a fibrous root extending from the basal bodies toward the nucleus, and bundles of microtubules radiating around the cell periphery (see fig. 1.10).

Mitosis and cell division are associated with cellular reproduction. In a primitive alga, the nuclear envelope remains intact, enclosing the mitotic spindle (closed spindle), and centrioles are organizing centers for the microtubules composing the spindle (fig. 1.6*b*). In more advanced algae, the nuclear envelope breaks down, resulting in an open spindle. Other organelles, including the chloroplast,

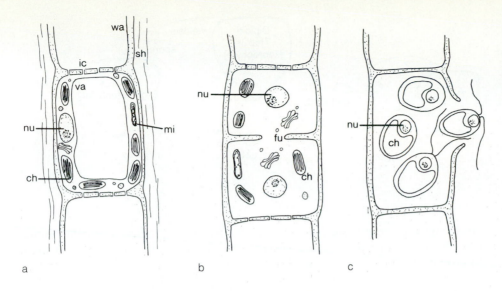

Figure 1.7
Cell of a filament.
(*a*) Structures of a vegetative
cell. (*b*) Vegetative cell
division adding a new cell to
the thallus. (*c*) Formation of
flagellated reproductive cells.
ch = chloroplast;
fu = furrow; ic = intercellular
connection;
mi = mitochondrion;
nu = nucleus; sh = sheath;
va = vacuole; wa = wall.

a   b   c

divide with the nucleus. As mitosis is completed, the cell divides at its equator by **furrowing**—ingrowth of the cell membrane under the control of a contractile ring of actin. Stressful environmental conditions may cause a cell to discard its flagella, form a thick wall, and become a dormant **cyst.** When external conditions are favorable again, the cell becomes metabolically active.

More advanced, multicellular algae are colonies, filaments, blades, or cylinders. The body of an alga is its **thallus.** A filament, with its cells arranged in a linear series, is representative of an advanced alga (fig. 1.7*a*). Filaments appear thread-like and may be branched or unbranched. Each cell is surrounded by a wall and lacks flagella, basal bodies, and an eyespot. Adjacent cells in a filament share a common end wall. Intercellular connections through these end walls may create cytoplasmic continuity between cells. A cell wall consists of a framework of polysaccharide bundles or fibrils surrounded by a mucilaginous material composed primarily of other polysaccharides (mucopolysaccharides). Most algae form fibrils composed of cellulose, but in a few algae, xylans or mannans replace cellulose. Enzyme complexes (cellulose synthase) in the cell membrane control the polymerization of glucose units to form chains of cellulose and the organization of these chains into fibrils. The mucilaginous material filling the space between the fibrils consists of a variety of polysaccharides (sometimes divided into pectic substances and hemicelluloses); proteins also may be present. Mucilage may extend beyond the wall as a gelatinous sheath. Sheaths have a variety of functions, including holding cells together, attaching cells to the substrate, reducing water loss during exposure to air, facilitating nutrient uptake, and producing gliding movement. A sheath also may reduce an alga's susceptibility to epiphytes and herbivores. Mucilaginous polysaccharides are synthesized in Golgi bodies and transported to the cell membrane in vesicles. Some algae deposit a stony layer of calcium carbonate on their walls for protection.

Each cell in a filament has a large central vacuole containing a watery solution of inorganic and organic materials called the cell sap. The cytoplasm, confined to a thin layer adjacent to the wall, contains the nucleus, one or more chloroplasts, and other organelles, as in primitive phytoflagellates. The cytoskeleton consists of actin microfilaments and microtubules. Actin filaments near the cell membrane are involved with cytoplasmic streaming, movement of organelles, and furrowing during cell division. Microtubules composed of tubulin are associated with microtubu-

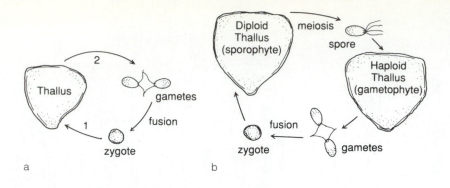

lar-organizing centers, such as centrosomes and basal bodies. The centrosome, near the nucleus, consists of amorphous material surrounding a pair of centrioles. Microtubules control vesicle movements and the arrangement of cellulose fibrils in the wall, and compose the mitotic spindle.

Vegetative cell divisions add new cells to a filament (fig. 1.7*b*), while reproductive divisions form gametes or spores. When a filament reproduces, the contents of one or more cells divide repeatedly to produce a number of small reproductive cells (fig. 1.7*c*), which are released into the surrounding water when the parental wall breaks down. These daughter cells function as either asexual spores or sexual gametes. In **asexual reproduction,** the daughter cells, called **zoospores** when they are flagellated, initiate new filaments. Zoospores, resembling the phytoflagellate described earlier, have chloroplasts, mitochondria, and reserve materials that allow them to meet short-term metabolic demands. After a brief period of motility, zoospores lose their flagella, attach to the bottom, and divide to initiate new filaments. In **sexual reproduction,** the cells the parent releases are **gametes.** Pairs of compatible gametes fuse to form a zygote. If both gametes of a pair are flagellated and similar in size, they are **isogametes.** Gametes that are flagellated but differ in size are **anisogametes.** In **oogamy,** only one gamete is flagellated (sperm), and it fuses with a larger, nonflagellated gamete (egg). Zygotes produced by gametic fusion may develop immediately into new filaments by vegetative growth, or they may undergo meiosis to form spores. Development may be delayed if zygotes become dormant.

The life cycle (or life history) of an alga may involve one or sometimes two vegetative phases, depending on the stage at which meiosis occurs. Algae with a single vegetative phase may be either haploid or diploid (fig. 1.8*a*). Algae with a haploid vegetative phase show zygotic meiosis (zygote undergoes meiosis) and are **haplonts. Diplonts** are algae with a diploid phase and show gametic meiosis. Other algae have life cycles that alternate between haploid and diploid phases. The spore-producing phase is called the **sporophyte** (normally diploid), and the gamete-producing phase is the **gametophyte** (normally haploid) (fig. 1.8*b*). These algae show sporic meiosis and are **diplohaplonts.**

The various groups of eukaryotic algae show distinct ultrastructural features, especially of the chloroplast and flagellar apparatus. The envelope surrounding a chloroplast is usually composed of two closely associated membranes, but many dinoflagellates and euglenophytes have envelopes with three membranes. Other groups have a layer of endoplasmic reticulum surrounding the chloroplast. This **chloroplast ER** (or **CER**) is usually continuous with the nuclear envelope. The thylakoids within a chloroplast may join in groups. In some algal classes, the thylakoids are associated in groups of three, but in other algae, the number may

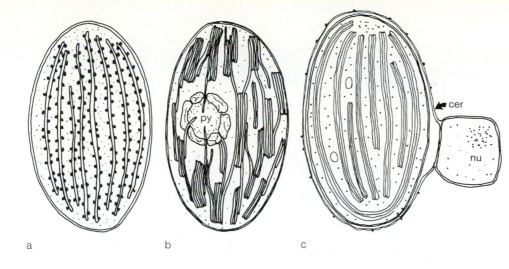

a   b   c

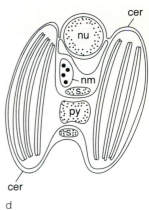

d

e   f

Figure 1.9
Chloroplast types.
(*a*) Rhodophyte chloroplast
with separate thylakoids and
phycobilisomes.
(*b*) Chlorophyte chloroplast
with thylakoids in stacks and
a pyrenoid (py) with starch
on its surface. (*c*) Chromophyte chloroplast with
thylakoids in groups of three
and a chloroplast
endoplasmic reticulum (cer)
with ribosomes on its outer
surface, nu = nucleus.
(*d*) Cryptomonad chloroplast
with periplastidal space
containing nucleomorph (nm)
and starch grains (s).
(*e*) Rhodophyte chloroplast.
(*f*) Chromophyte chloroplast
(brown alga). (*e, f* courtesy
C.N. Pueschel.)

Table 1.4   *Chloroplast Structure of Eukaryotic Algae*

Division	Chloroplast Envelope	Chloroplast ER	Association of Thylakoids
Rhodophyta	Double	Absent	(not associated)
Chlorophyta	Double	Absent	2–6
Chromophyta	Double	Present	3
Haptophyta	Double	Present	3
Dinophyta	Usually triple	Absent	3
Cryptophyta	Double	Present	2
Euglenophyta	Triple	Absent	3

vary or the thylakoids are unassociated. The differences among algal divisions in chloroplast structure are summarized in table 1.4 and illustrated in figure 1.9.

Evidence is strong that the chloroplasts of eukaryotic algae are derived from endosymbionts—cells of one alga living inside cells of a host. The chloroplasts of Rhodophyta and Chlorophyta probably originated from prokaryotic endo-symbionts, as indicated by their envelopes with two membranes. The additional membranes, including CER, around the chloroplasts of other groups may have

Figure 1.10
Flagellar apparatus consisting of emergent flagella (f), transition regions (tr), basal bodies (bb), and cytoskeletal elements associated with the basal bodies. Cytoskeletal elements include connecting fibers (cf) between basal bodies, microtubular roots (mt), and fibrous roots (fr). Tubular hairs or mastigonemes (ms) are on one flagellum; delicate hairs (h) are on the other flagellum, nu = nucleus.

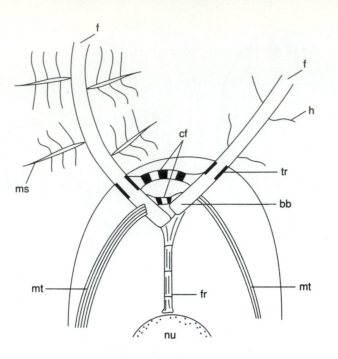

Table 1.5   *Parts of the Flagellar Apparatus*

Flagellum
Flagellar appendages
   Simple hairs (smooth flagella)
   Stiff, tubular hairs (mastigonemes)
   Scales
Transition region (characteristic pattern in different algal groups)
Basal bodies
Flagellar swelling (paraflagellar body) associated with eyespot
Cytoskeleton
   Connecting fibers between basal bodies
   Microtubular roots
   Fibrous roots (striated roots, rhizoplasts)

arisen from eukaryotic endosymbionts. We will discuss the probable chloroplast origin of each algal division in subsequent chapters.

The flagellated stages of different algal groups show characteristic features of the **flagellar apparatus.** The flagellar apparatus consists of the emergent flagella, a transition region where each flagellum joins its basal body, basal bodies, and cytoskeletal elements radiating from the basal body (fig. 1.10 and table 1.5). Flagella may differ in their appendages. Smooth (or "whiplash") flagella have only delicate hairs, while other flagella have stiff hairs (**mastigonemes**) in one or two rows ("tinsel" flagella), or platelike scales on the surface of the flagellar membrane. Table 1.6 summarizes the flagellar types characteristic of different algal divisions. Some algal groups have a swelling (paraflagellar body) near the base of one flagellum that functions as a photoreceptor to allow a cell to respond to the direction of light. The transition region between a flagellum and its basal body, with

Table 1.6   *Type of Flagella and Mitochondrial Cristae*

Division	Flagella	Cristae
Rhodophyta	None	Flattened
Chlorophyta	Smooth flagella	Flattened
Chromophyta	Heterokontous—one smooth flagellum and one flagellum with mastigonemes	Tubular
Haptophyta	Smooth flagella (and haptonema)	Tubular
Dinophyta	Smooth flagella	Tubular
Cryptophyta	Flagella with mastigonemes	Flattened
Euglenophyta	Flagella with nontubular hairs	Flattened

characteristic features for different algal groups, is important when the flagellum is cast off or excised in response to environmental stress or at the time of reproduction. Although basal body structure with nine triplets of microtubules is the same in all algae, the relative orientation of basal bodies varies in different algae and determines flagella position. The cytoskeleton of a flagellated cell is composed of structural proteins radiating from the basal bodies and may consist of three components: connecting fibers between the basal bodies, roots composed of microtubules, and fibrous roots (rhizoplasts). The contractile protein centrin is present in connecting fibers and fibrous roots. The types of cytoskeletal components and their arrangements vary among algal groups and are important in determining evolutionary relationships.

Molecular Phylogeny

Eukaryotic algal cells contain three genetic systems in the nucleus, chloroplasts, and mitochondria. In each, the DNA coding system is based on the sequence of bases (adenine, guanine, cytosine, and thymine) of the nucleotide units. During evolution of a line of organisms, mutations change the base sequences. The most common are point mutations involving the change of a single base, but other mutations may involve base deletion or rearrangement. Most mutations do not significantly affect survivorship and reproductive success (neutral theory of mutations) but accumulate slowly over time. When two evolutionary lines diverge, each independently accumulates its own genetic changes from random mutations. The longer two lines are separate, the more their genetic sequences will differ. Thus, comparison of the sequences of nucleotide bases provides a measure of the evolutionary relationships of contemporary groups.

Evolutionary relationships are commonly estimated by comparing the nucleotide sequences of specific genes or ribosomal RNA (rRNA). Of the different types of RNA in a cell, rRNA is the most common and contributes to the structure of ribosomes. Each ribosome in a eukaryotic cell consists of a large subunit composed of proteins and several chains of RNA, and a small subunit composed of proteins and a single chain of RNA. Nucleotide sequence in the small subunit (SSU) is especially valuable for studying relationships among algal groups. The RNA can be extracted and sequenced or more recently, the nuclear gene coding for SSU rRNA is sequenced. Figure 1.11 shows a phylogenetic tree for eukaryotic algae based on SSU rRNA sequences (simplified from McFadden, Gilson, and Hill 1994; see also Bhattacharya and Medlin 1995). Interpretation of such a diagram requires caution. The relationships are based on sequences determined from a very small

Figure 1.11
Algal phylogeny based on nucleotide sequences of RNA in the small subunit of ribosomes. (After McFadden, Gilson, and Hill 1994.)

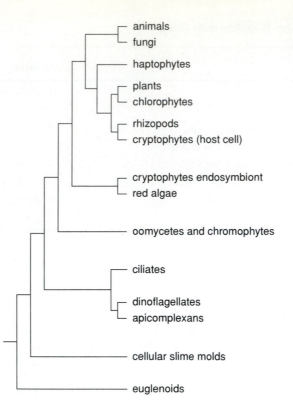

number of species, sequences from different algae must be aligned properly for comparisons (additions and deletions of nucleotides make this difficult), and differences in sequences can be analyzed in several different ways that do not necessarily lead to the same conclusions.

The phylogenetic tree in figure 1.11 suggests some interesting conclusions. The close relationship between the chlorophytes (green algae) and plants is supported, the chromophytes appear to be more closely related to oomycete fungi than to other algae, the removal of the haptophytes from the chromophytes is supported, and the dinoflagellates appear to be related to the ciliates and apicomplexans, which include the malaria-parasite *Plasmodium*. Overall, this analysis suggests that different divisions of eukaryotic algae are not as closely related to each other as to other groups of organisms, and that different algal groups arose independently as several different host cells acquired photosynthetic systems as different endosymbionts.

Morphologic Diversity

Algae range from small, solitary cells to large, complex, multicellular structures. The different types of algal thalli are important in grouping species at intermediate levels of classification. Prokaryotic algae occur as solitary cells, as aggregations and colonies of cells, and as filaments in which cells are arranged in a linear series. Eukaryotic algae show the seven basic morphologic types that follow. In the first two, vegetative cells are flagellated and normally show active motility. In the remaining five types, cells are nonflagellated in the vegetative (nonreproductive) condition. Within each algal class, one or more of the different types is present.

1. **Flagellated solitary cells** (fig. 1.12*a*)   This type is considered primitive in most groups of eukaryotic algae and is believed to have given rise to the other types. Flagellated cells vary in the number and arrangement of their flagella.

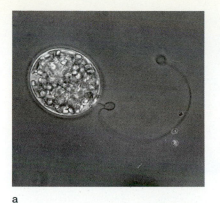

a

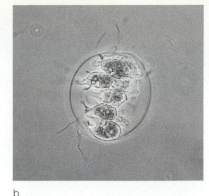

b

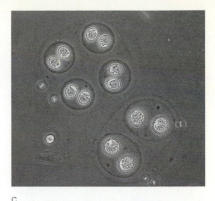

c

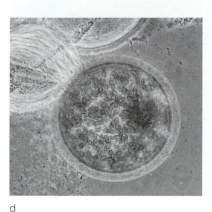

d

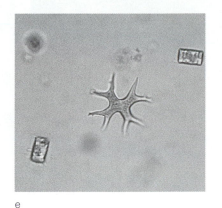

e

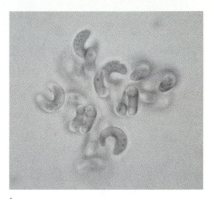

f

Figure 1.12
Morphologic types of algae.
(*a*) Flagellated cell
(*Trachelomonas*). (*b*) Colony
of flagellated cells
(*Stephanosphaera*).
(*c*) Palmelloid aggregations
(*Gloeocystis*). (*d, e*) Non-
flagellated cells (*Eremo-
sphaera, Tetraedron*). (*f*)
Aggregation of cells
(*Kirchneriella*).

2. **Colonies of flagellated cells**   Colonies may vary in cell arrangement and in how cells are held together. Commonly, cells of colonies are arranged either in a flat plate or in a sphere. Colonies may be maintained by direct cellular contact (plate 3*a*), or a mucilaginous framework may hold cells together (fig. 1.12*b*).

3. **Palmelloid aggregations** (fig. 1.12*c*)   Flagellated species may have a temporary palmelloid stage during which cells lose their flagella and secrete extensive mucilage. A palmelloid condition is the normal, persistent condition in some algae. A sheath of mucilaginous polysaccharides, which may have an indefinite shape or a distinctive form, surrounds the cells. The cells are separated from each other in the mucilage and retain features of flagellated cells, such as basal bodies, eyespots, and contractile vacuoles.

4. **Nonflagellated cells and colonies**   Cells may be solitary (fig. 1.12*d, e*), aggregated (fig. 1.12*f*), or associated in regular colonies (plate 2*b*). Mucilage sometimes surrounds colonies. Cell divisions typically result in formation of reproductive cells rather than enlargement of the thallus.

5. **Amoeboid or rhizopodial cells** (fig. 1.13*a*)   This type includes solitary cells and colonies of amoeboid cells. Cells lack walls, although other structures, such as cup-shaped loricae, may enclose them.

6. **Filaments**   In filaments, cells are arranged end-to-end, with adjacent cells sharing a common cross wall. Cytoplasmic connections (plasmodesmata) may extend through the cross walls. In **uniseriate filaments,** the cells are arranged in a single series. **Multiseriate filaments** have more than one series of cells but still

Figure 1.13
Morphologic types
(continued). (*a*) Amoeboid
cell (*Chrysamoeba*).
(*b*) Unbranched filament
(*Erythrotrichia*). (*c*) Branching
filament (*Callithamnion*).
(*d*) Parenchymatous thalli
(*Laminaria*).

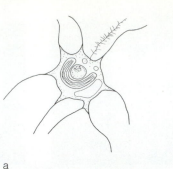

a

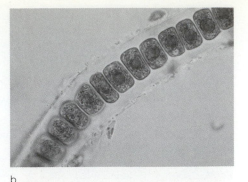

b

c

d

retain a threadlike appearance. Filaments may be unbranched (fig. 1.13*b*) or branched (fig. 1.13*c*). More complex filamentous algae may show differentiation among the branches. **Heterotrichous** filaments have a distinct system of prostrate branches growing attached to the substrate and an erect system of more open branches extending free of the substrate (see fig. 3.26). In some filamentous species, the branches do not spread apart in an open branching pattern but form a compact mass that makes individual branches difficult to see. Such thalli are termed **pseudoparenchymatous** because of their resemblance to the type described in number 7. The relative position of cells within a filament are described by the terms *basal, intercalary,* and *apical* (or *terminal*).

7. **Parenchymatous thalli**   In parenchymatous construction, cell divisions in three dimensions produce a solid mass of cells rather than the threadlike, linear arrangements of a filament. Parenchymatous thalli may be blades (fig. 1.13*d*; plate 2*c*), branching cylinders (terete) (plate 4*a*), or hollow tubes. Parenchymatous construction, which is also characteristic of bryophytes and vascular plants, is the most advanced form.

Ecologic Diversity

Algae grow in both freshwater and marine environments, where they are important photosynthetic producers. They are also found in damp terrestrial environments. As photosynthetic organisms, algae are major producers of oxygen and organic material in aquatic environments. By converting inorganic material into compounds that heterotrophic organisms can use, algae are the first step in many food chains. Some cyanobacteria also convert nitrogen gas (N_2) by nitrogen fixation into forms of nitrogen that other organisms can use. Some algae are important depositors of calcium carbonate. In the ocean, calcareous structures from algae contribute

Table 1.7   Types of Holdfasts of Macroalgae

Thallus appressed to substrate (encrusting algae)

Filament with unicellular holdfast at base forming rhizoidal extensions

Rhizoids from cells near base of thallus

Heterotrichy with modified branches forming base for attachment

Discoid pad of tissue (parenchymatous or pseudoparenchymatous) forming the base of larger macroalgae

Haptera (fingerlike extensions) (characteristic of Laminariales)

to sediments and transport excess carbon from the surface into deep water when they sink.

Algae generally are considered to be autotrophs, requiring only inorganic nutrients, such as carbon dioxide, water, phosphates, and inorganic nitrogen, with light as an energy source. Some algae, however, also need organic growth factors, and others can take up and oxidize organic material as supplemental energy sources. The latter are mixed feeders or mixotrophs and may use organic material dissolved in the surrounding water or ingest particles of organic material (phagotrophy), such as bacteria and detritus. A few algae totally lack photosynthetic systems and depend entirely on heterotrophic metabolism.

Lifestyles divide algae living submerged in either freshwater or marine environments into planktonic (floating) and benthic (bottom-living) forms. Planktonic algae, composing the **phytoplankton** community of a lake or ocean, float freely in the water. Unlike plants on land, members of the phytoplankton are mostly microscopic, occurring as isolated cells or in small colonies. To receive sufficient light for photosynthesis, planktonic algae must remain near the water surface and thus have a variety of adaptations for flotation. Nutrient availability and grazing by herbivorous animals are important in regulating phytoplankton.

Rather than floating with the water, **benthic algae** (phytobenthos) are associated with submerged substrates, and water moves past them. Smaller benthic algae may be associated only loosely with substrates, but larger species are usually securely attached. The entire thallus may be appressed to the substrate as a crust, or a special structure called a **holdfast** may attach the thallus base while the remainder of the thallus extends from the substrate (table 1.7). The type of substrate (sand, rock, plant) determines whether a particular species can grow. Algae grow on a wide range of nonliving substrates and on the surfaces of other organisms as **epiphytes** (or epibionts). Epiphytism, in which one alga grows on another alga or plant, is common. Light availability and herbivores control distributions of benthic algae.

While most planktonic algae are small, benthic algae range in size from microscopic, single cells to large macrophytes. The microbenthos (or periphyton) has algae with thalli too small to be seen without a microscope, while the **macroalgae** (or macrophytes) have individual thalli that are visible without viewing aids. Microbenthic algae may form visible films when abundant. Macroalgae differ from land plants in that they have little or no supporting tissue and take up nutrients over their entire surface. The surrounding water provides buoyancy, reducing the need for supporting tissues, although gas-filled bladders support some large algae. In addition to producing organic material and oxygen, macroalgae also create habitats and shelters for other organisms.

In terrestrial environments, algae are usually confined to relatively damp habitats or are only active when conditions are moist. They sometimes form conspicuous growths on damp soil, tree bark, and snow. Terrestrial or aerial algae must tolerate a wider range of environmental conditions than aquatic algae. On land, algae are exposed to high light levels and must minimize evaporative water loss. Special dormant, resistant stages assist survival between periods favorable for growth.

Symbionts live closely associated with other organisms—either among their tissues (extracellular or intercellular) or inside their cells (intracellular). In many cases, these interactions are mutualistic in that both host and symbiont benefit from the association. The host organism provides protection, a constant environment, and nutrients, such as carbon dioxide from respiration, nitrogenous compounds in wastes, and phosphate, while the endosymbiont produces oxygen and organic material (as sugar alcohols, simple sugars, fatty acids, and amino acids). Common symbiotic algae include the algal components of lichens, green algae (zoochlorellae) in freshwater invertebrates, dinoflagellates (zooxanthellae) in reef-forming corals and other marine invertebrates, and nitrogen-fixing cyanobacteria in a wide range of hosts.

Globally, algae are important as (1) producers of oxygen and organic material, responsible for 30–50% of the photosynthetic production on earth; (2) sources of usable forms of nitrogen from nitrogen-fixing cyanobacteria; (3) steps in nutrient cycles (nitrogen, phosphorus, carbon, oxygen, calcium); (4) habitat creators for other freshwater and marine organisms; and (5) possible weather regulators by releasing compounds that seed clouds and by increasing solar energy absorption.

Algae and Humans

Algae are major producers of organic material and oxygen, but also can affect humans more directly (table 1.8). Sometimes, the negative effects of algae are most obvious, such as when they discolor lakes, clog streams, foul the bottoms of ships, and form slippery films on walkways. Massive growths or **blooms** may severely deplete oxygen when cells decompose, reduce the use of lakes and streams for recreation, and interfere with water purification. Recently, blooms of planktonic algae in coastal waters have increased, including **red tides** of toxic dinoflagellates. Toxins from dinoflagellates and other algae may accumulate in food chains, causing fish kills and contaminating commercially important marine animals. In freshwater, cyanobacterial blooms associated with urban pollution or farm runoff often detract from recreational use of inland waters.

On the positive side, some algae are used for food and in commercial products. Human consumption of macroalgae is more widespread in Pacific and Asian countries, where several red and brown algae are commercially farmed. Some macroalgae have been used as animal feed, fertilizer, and soil conditioners in agriculture. Microalgae are used as feed in the cultivation of commercially important marine animals and in wastewater treatment to remove nutrients and metals.

Important products from algae are **diatomaceous earth** (diatomite) and **phycocolloids** (=hydrocolloids). Diatomaceous earth containing the remains of diatoms is used in filters and polishes. Phycocolloids, such as alginic acid, carrageenan, and agar, are derived from larger marine algae and used as thickeners, gels, and stabilizers in foods, cosmetics, and other commercial products. Relatively few microalgae have commercial use. *Spirulina*, a cyanobacterium in soda lakes, may have potential because of its high protein content. The green algae *Dunaliella* is a good source of β-carotene. The use of algae in biotechnology to synthesize specific compounds of commercial value is likely to increase.

Table 1.8 *Importance of Algae to Humans*

Beneficial Aspects	Examples
Food for humans	*Spirulina, Porphyra, Laminaria, Undaria*
Food for invertebrates and fishes in mariculture	Various phytoplankton
Animal feed	Various macroalgae
Chemical sources	β-carotene from *Dunaliella*
Soil fertilizers and conditioners in agriculture	Soil algae, especially cyanobacteria
Treatment of waste water	Various microalgae
Diatomaceous earth (=diatomite)	Deposits of diatom frustules
Chalk deposits	Remains of coccolithophores (coccoliths)
Phycocolloids used as thickeners and gels	Agar and carrageenan from red algae, alginates from brown algae
Drugs	Several microalgae and macroalgae
Marl	Coralline red algae
Model system for research	Genetic studies with *Chlamydomonas*, developmental studies with *Acetabularia*
Phycobiliproteins for fluorescence microscopy	*Porphyra*

Detrimental Aspects	Causes
Blooms of freshwater algae	Planktonic and benthic algae, especially cyanobacteria
Red tides and marine blooms	Dinoflagellates, coccolithophores
Toxins accumulated in food chains	Cyanobacteria in freshwater, dinoflagellates in oceans (see table 7.2)
Damage to cave paintings, frescoes, and other works of art	Terrestrial algae
Fouling of ships and other submerged surfaces	Marine macroalgae, such as *Enteromorpha*
Fouling of the shells of commercially important bivalves	*Codium*

Summary

1. The algae are a heterogeneous group with few common characteristics. Typically, they have photosynthetic systems with two light-collecting photosystems containing chlorophyll and other pigments, and give off oxygen. The algae include prokaryotic forms, classified with the bacteria, and eukaryotic forms, classified with protists. The algae range in size from solitary cells to complex, multicellular thalli. Their reproductive structures consist of cells that are all potentially fertile, and algae do not form embryos.

2. The algae are divided into eight major groups or divisions, which differ in their photosynthetic pigments, carbohydrate reserves, and cell structures. In addition to chlorophyll *a*, algae may possess other chlorophylls, carotenoids, and phycobilins. Polysaccharide reserves are polymers of glucose that differ in the way the glucose units are linked. Eukaryotic algae differ in chloroplast structure (grouping of thylakoids, presence of chloroplast endoplasmic reticulum, layers of the chloroplast envelope), and in their flagella and associated structures (flagellar appendages, cytoskeleton).

3. Molecular comparisons using nucleotide sequences in ribosomal RNA strongly suggest that the eukaryotic algae are less closely related to each other than to nonphotosynthetic protists. The different groups of algae may have independently acquired their photosynthetic systems as either cyanobacterial or eukaryotic endosymbionts.

4. Each major group of algae may have one or more morphologic types. Seven basic types are flagellated solitary cells, colonies of flagellated cells, palmelloid aggregations, nonflagellated cells and colonies, amoeboid cells, filaments, and parenchymatous thalli.

5. Algae are important producers in marine and freshwater environments as members of planktonic (floating) and benthic (bottom-living) communities. A few algae are terrestrial, growing in moist habitats exposed to the air. Dense algal growths, including blooms of planktonic algae, may interfere with human activities. A few algae are used for food and in commercial products.

Further Reading

Akatsuka, I., ed. 1990. *Introduction to Applied Phycology.* SPB Academic Publishing.

Berner, T., ed. 1993. *Ultrastructure of Microalgae.* CRC Press.

Bhattacharya, D., and L. Medlin. 1995. The phylogeny of plastids: A review based on comparisons of small-subunit ribosomal RNA coding regions. *Journal of Phycology* 31: 489–98.

Bold, H. C., and M. J. Wynne. 1985. *Introduction to the Algae,* 2d ed. Prentice-Hall.

Borowitzka, M. A., and L. J. Borowitzka, eds. 1988. *Micro-Algal Biotechnology.* Cambridge University Press.

Canter-Lund, H., and J. W. G. Lund. 1995. *Freshwater Algae.* Biopress.

Hoek, C. van den, D. G. Mann, and H. M. Jahns. 1995. *Algae.* Cambridge University Press.

Lee, R. E. 1989. *Phycology,* 2d ed. Cambridge University Press.

Lembi, C. A., and J. R. Waaland, eds. 1989. *Algae and Human Affairs.* Cambridge University Press.

Melkonian, M., ed. 1992. *Algal Cell Motility.* Chapman and Hall.

Menzel, D., ed. 1992. *The Cytoskeleton of the Algae.* CRC Press.

Radmer, R. J. 1996. Algal diversity and commercial algal products. *BioScience* 46: 263–70.

Reisser, W., ed. 1992. *Algae and Symbiosis.* Biopress.

Rogers, L. J., and J. R. Gallon, eds. 1988. *Biochemistry of the Algae and Cyanobacteria.* Clarendon.

Round, F. E., and D. J. Chapman, eds. 1982–1995. *Progress in Phycological Research.* Vols. 1–2. Elsevier. Vols. 3–11. Biopress.

Rowan, K. S. 1989. *Photosynthetic Pigments of Algae.* Cambridge University Press.

2

Cyanobacteria

Cyanobacteria, with prokaryotic cells, were the first algae to evolve. Even though this ancient group lacks the cellular complexity of other algae, its representatives continue to be important in many ecosystems. Fossils of what appear to be filamentous cyanobacteria are present in deposits at least 3.5 billion years old (Schopf 1993). Although not the first photosynthetic organisms, cyanobacteria were the first cells to have two photosystems and to give off oxygen as a by-product. They were instrumental in the development of earth's present atmosphere, which preceded the evolution of more complex eukaryotic algae. Today, cyanobacteria are photosynthetic producers in a wide range of freshwater and marine environments, and one of the most common algal groups in terrestrial habitats and in symbiotic associations. Many cyanobacteria also convert nitrogen gas (N_2) into usable forms of nitrogen through nitrogen fixation.

Prokaryotic algae resemble other eubacteria in that they lack a nuclear envelope defining a nuclear region and complex organelles, such as chloroplasts, mitochondria, endoplasmic reticula, and Golgi bodies. However, it is incorrect to say they lack organelles. Their cells contain simple, flattened vesicles called **thylakoids** that contain their photosynthetic systems. No prokaryotic algae are flagellated.

On the basis of their photosynthetic pigments, the cyanobacteria (formerly called blue-green algae) are divided into two groups (table 2.1). Most species have chlorophyll *a* as their only form of chlorophyll and **phycobilins** as accessory pigments. The few known representatives of a second group, the **prochlorophytes,** have two forms of chlorophyll—*a* and *b*—and lack phycobilins (chlorophyll *c* also may be present, Larkum et al. 1994). Molecular evidence does not support separation of prochlorophytes as a distinct group; rather, the three known genera of prochlorophytes are more closely related to other cyanobacteria than to each other

Table 2.1 *Comparison of Typical Cyanobacteria and Prochlorophytes*

	Typical Cyanobacterium	Prochlorophyte
Major photosynthetic pigments	Chlorophyll *a*, phycobiliproteins	Chlorophylls *a*, *b*, (*c*)
Carbohydrate reserve	Starch	Starch
Organization of thylakoids	Separate	Paired
Cell covering	Peptidogylcan wall	Peptidoglycan wall

Table 2.2   *Phycobiliproteins of the Cyanobacteria and Red Algae*

Phycobiliprotein	Phycobilins
Phycocyanin	Phycocyanobilin
Allophycocyanin	Phycocyanobilin
Phycoerythrin	Phycoerythrobilin
Phycoerythrocyanin	Phycocyanobilin, phycobiliviolin

Source: After Hoffmann, Talarico, and Wilmotte 1990.

(Palenik and Haselkorn 1992; Urbach, Robertson, and Chisholm 1992). Thus, prokaryotic algae are treated here as belonging to a single class Cyanophyceae in the division Cyanophyta.

The photosynthetic pigments of cyanobacteria are associated with the thylakoids. Each thylakoid encloses a space distinct from the general cytoplasm. As in eukaryotic algae, the photosystems in the thylakoid membranes are organized around reaction centers containing chlorophyll *a*. When present, phycobiliproteins compose granular structures called **phycobilisomes** on the outer surfaces of thylakoid membranes. Each **phycobiliprotein** consists of several phycobilins tightly bound to a protein (table 2.2). The four types of phycobiliproteins in cyanobacteria are phycoerythrin (PE), phycocyanin (PC), allophycocyanin (APC), and phycoerythrocyanin (PEC). All phycobilin-containing cyanobacteria have allophycocyanin and phycocyanin. Phycocyanin usually gives cells their distinctive blue-green color. Each phycobilisome consists of a core with allophycocyanin and radiating rods with other phycobiliproteins. The phycobilisomes collect light, which is transferred to a reaction center in the thylakoid membrane by the following pathway:

$$\text{PE (or PEC)} \rightarrow \text{PC} \rightarrow \text{APC} \rightarrow \text{chl } a \text{ (in photosystem II)}$$

Cyanobacteria may respond to their light environment by changing the content of their phycobiliproteins. In some species, green light stimulates phycoerythrin synthesis, and red light stimulates phycocyanin synthesis. In others, light quality only affects phycoerythrin synthesis.

Cellular Structure

When viewed through a light microscope, cyanobacterial cells are usually smaller than eukaryotic cells and have few distinct features. Most of their cell structure is only visible with an electron microscope (fig. 2.1). A wall and a sheath surround a typical cyanobacterial cell, and a cell membrane encloses the protoplasm containing the thylakoids, chromosomes, and various granules.

The wall has peptidoglycan as its major component rather than cellulose and is similar to that of gram-negative bacteria in that it has two principal layers: a rigid peptidoglycan layer adjacent to the cell membrane and an outer lipopolysaccharide layer. The peptidoglycan layer consists of polymers of N-acetylglucosamine and N-acetylmuramic acid cross-linked by amino acids. A mucilaginous sheath usually extends beyond the wall. The sheath's consistency may vary from watery and indistinct to firm and well defined, and it may be colorless, or contain a yellow or brown pigment (scytonemin) to protect against exposure to high irradiance (Garcia-Pichel and Castenholz 1991). Some cyanobacteria secrete mucilage to cause a gliding movement when in contact with a substrate or in a dense mat.

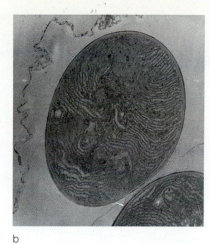

a

b

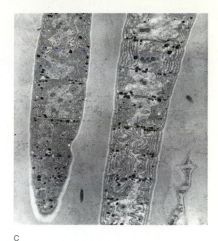

c

Figure 2.1
Typical cyanobacterial cells.
(*a*) Sheath (sh), wall (wa),
thylakoids (th)
 with phycobilisomes,
carboxysomes (pb),
cyanophycin granules (cg),
other granules (gr), and DNA.
(*b*) Electron micrograph of
Gloeocapsa. (*c*) Electron
micrograph of *Oscillatoria.*
(*b* and *c* courtesy of G. B.
Chapman.)

Thylakoids are conspicuous in the outer part of the cytoplasm near the cell membrane. In phycobilin-containing cyanobacteria, the thylakoids with granular phycobilisomes on their outer surfaces are separate from each other, but in prochlorophytes, thylakoids may associate in pairs. Thylakoids are the site of photosynthesis and probably of respiration. Respiratory enzymes are also present in the cell membrane. One cyanobacterium, *Gloeobacter,* lacks thylakoids, and the photosynthetic pigments are in its cell membrane.

Other cytoplasmic structures include carboxysomes and cyanophycin, starch, and polyphosphate granules. Carboxysomes contain the Calvin cycle enzyme ribulose bisphosphate carboxylase and carbonic anhydrase for converting bicarbonate to carbon dioxide. Cyanophycin granules are composed of polypeptides with aspartate and arginine. Starch in cyanobacteria is highly branched and thus is similar to glycogen in animal cells. Starch granules are the principal carbohydrate reserve, while nitrogen is stored in cyanophycin granules (and to a lesser extent in phycobiliproteins), and phosphate in polyphosphate granules.

Ribosomes, the sites of protein synthesis, are common throughout the cytoplasm and are the smaller (70S) type typical of bacteria. In the cell's central region, the genetic material DNA is organized into a circular chromosome (sometimes with multiple copies); smaller DNA loops or plasmids also may be present. Prokaryotic chromosomes differ from those of eukaryotic cells in that they are circular and lack histone proteins, but introns are present (Kuhsel, Strickland, and Palmer 1990). Some planktonic cyanobacteria contain gas vesicles, with membranes enclosing gas-filled spaces. Although the cyanobacteria do not possess the complex cytoskeletons of eukaryotic cells, some contain tubules resembling eukaryotic cytoskeletal elements.

Commonly, cellular reproduction is by ingrowth of the cell membrane and wall at a cell's equator to produce two cells (binary fission) (fig. 2.2*a*). Cell division by unicellular and aggregated cyanobacteria is the principal means of reproduction leading to formation of new individuals. In filamentous cyanobacteria, division as part of vegetative growth adds new cells to an existing filament (fig. 2.2*c*). Filaments usually reproduce by fragmentation. A few cyanobacteria form distinct spores. Exospores result from successive division at one end of a cell (fig. 2.2*b*), while endospores form by multiple divisions within a cell and are released when the parental wall splits.

Figure 2.2
(a) Cell division by *Gloeocapsa*. The cell on the left is undergoing binary fission. Daughter cells from a recent division are on the right. (b) Exospore formation by *Chamaesiphon*. (c) Vegetative divisions (arrows) in a filament of *Lyngbya*.

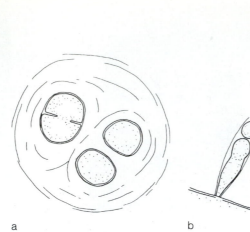

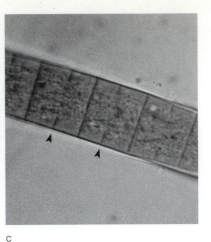

a   b   c

Table 2.3   *Classification of Prokaryotic Algae*

Division Cyanophyta—cyanobacteria (formerly blue-green algae)
    Class Cyanophyceae
        Order Chroococcales—solitary cells, aggregations, and colonies not
            reproducing by spores
        Order Chamaesiphonales—solitary cells, aggregations, and short filaments
            reproducing by spores
        Order Oscillatoriales—unbranched filaments without specialized cells
        Order Nostocales—unbranched filaments with heterocysts and akinetes
        Order Stigonematales—branched filaments (sometimes multiseriate) with
            heterocysts and akinetes

Diversity

The cyanobacteria are divided into nonfilamentous and filamentous types. The presence or absence of specialized cells, called heterocysts and akinetes, further subdivides the filamentous species. Table 2.3 summarizes the classification of the cyanobacteria.

Criteria for distinguishing different species and genera should remain constant under different environmental conditions. However, culture studies have shown that growth conditions may change characteristics traditionally used to distinguish species, such as sheath consistency and cell size. While many of the described cyanobacterial species are not valid and should be combined, no revision of cyanobacterial classification has received universal acceptance. The older, more traditionally recognized names are used here, and the reader is cautioned that the same names are sometimes used in a broader sense.

Nonfilamentous Cyanobacteria

Spherical or elongate cells of nonfilamentous cyanobacteria may be solitary or associated in aggregations and colonies, and often are surrounded by extensive mucilage. *Gloeocapsa* is a relatively simple cyanobacterium. Its spherical cells grow in small aggregations held together by a firm mucilaginous sheath (fig. 2.3a). The sheath is often colored and has distinct layers that indicate each cell's contribution. To reproduce, each cell elongates and divides into two daughter cells (fig. 2.2a).

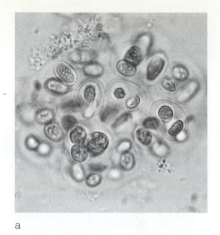
a

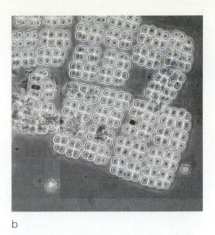
b

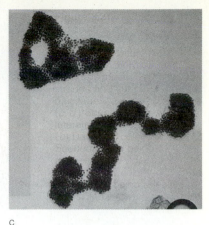
c

Figure 2.3
Representative
nonfilamentous
cyanobacteria.
(*a*) *Gloeocapsa* with
mucilaginous sheath holding
cells together.
(*b*) *Merismopedia* with a
regular arrangement of cells.
(*c*) Colonies of *Microcystis*
composed of a large number
of cells containing gas
vesicles.

Colonial cyanobacteria have a more definite form or larger size, compared to aggregations such as *Gloeocapsa*. However, no absolute criterion distinguishes a cyanobacterial aggregation from a colony. In flat colonies of *Merismopedia,* cells are arranged in regular rows and columns in one plane held together by mucilage (fig. 2.3*b*). Colonies of *Microcystis* consist of hundreds of spherical cells in a mucilaginous sheath (fig. 2.3*c*). *Microcystis* cells appear black when viewed through a microscope because of gas vesicles for buoyancy.

Filamentous Cyanobacteria without Specialized Cells

Oscillatoria is an unbranched filament composed of disk-shaped cells arranged in a single series (fig. 2.4*a*). All of the cells in a filament are alike except for terminal cells with a modified shape. *Oscillatoria* often grows in mats of interwoven filaments. Individual filaments may show active gliding movement. How this motion is produced is not fully understood but probably involves mucilage secretion (Häder and Hoiczyk 1992). This movement may adjust filament density in a mat in response to environmental conditions.

Lyngbya is similar to *Oscillatoria,* but its filaments have a firm mucilaginous sheath that normally extends beyond the terminal cell (fig. 2.4*b*). When a sheath is present, as in *Lyngbya*, the term **trichome** refers to the series of cells, while the filament includes the sheath and cells.

Both *Oscillatoria* and *Lyngbya* grow by adding cells to an existing filament (fig. 2.2*c*). A new filament begins when part of a filament breaks off. This is often a short segment from the end of a filament, called a **hormogonium** (fig. 2.4*c*), that forms when the wall between cells splits or when a cell dies. Hormogonia sometimes differ from vegetative trichomes in cell shape, greater motility, and presence of gas vesicles.

Nonheterocystous cyanobacteria vary in thickness and form of the sheath and in whether filaments are straight or spiraled (fig. 2.4*d, e*).

Heterocystous Cyanobacteria

Other cyanobacteria have specialized cells, called **akinetes** and **heterocysts.** Akinetes are thick-walled cells formed after a period of active growth that survive in a dormant state when conditions are unfavorable for further growth (fig. 2.5*a*). Akinetes are larger than vegetative cells and have thick walls and granular cytoplasm with an abundance of cyanophycin granules. They may remain inactive for

a

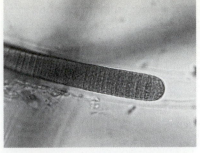

b

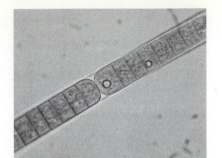

c

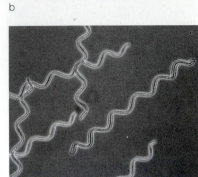

d

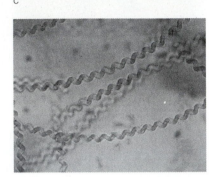

e

Figure 2.4
Oscillatorian filaments.
(*a*) *Oscillatoria*. (*b*) *Lyngbya*
with a firm sheath extending
beyond the terminal cell.
(*c*) Formation of a
hormogonium by the
separation of a small
segment of a trichome
(*Lyngbya*). (*d*) *Arthrospira*
with a regular spiral.
(*e*) *Spirulina* with a spiraling
filament in which the cross
walls between adjacent cells
are not visible.

Figure 2.5
Anabaena. (*a*) Akinetes.
(*b*) Filaments with
heterocysts.

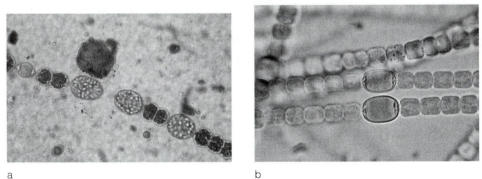

a

b

many years. At germination, the protoplast is released by rupture of the wall or through a pore before growth commences.

Heterocysts are specialized cells for **nitrogen fixation** (figs. 2.5*b*, 2.6). In this process, nitrogen (N₂) from the air is converted to ammonium. When dissolved nitrogen compounds are low in the surrounding water, some vegetative cells differentiate into heterocysts. During differentiation, additional layers are added to the wall, and pores through the wall connect heterocysts to adjacent vegetative cells. Heterocysts are usually larger than vegetative cells, have a light yellow-green color from loss of their phycobiliproteins, and lack storage granules, such as carboxysomes, starch granules, and cyanophycin granules. In heterocysts, the respiratory pathways and part of the photosynthetic system provide energy in the form of ATP and reduced compounds for nitrogen fixation. Sugars are transported into heterocysts from adjacent cells for oxidation in respiration. Photosystem II of photosynthesis and the Calvin cycle do not function in heterocysts, but photosystem I generates ATP and reduced ferredoxin. The enzyme nitrogenase catalyzes the conversion of nitrogen to ammonium using energy from ATP and electrons from

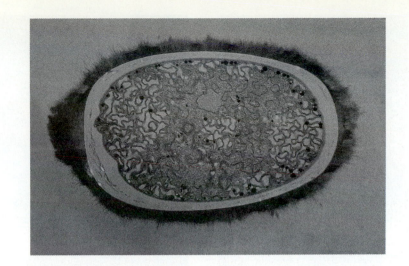

Figure 2.6
Heterocyst of *Anabaena*.
Note the thick wall and
dispersed thylakoids as
compared to vegetative cells.
(Courtesy G. B. Chapman.)

Figure 2.7
Summary of nitrogen fixation
in a heterocyst.

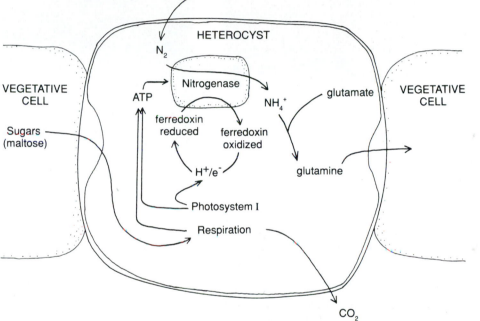

ferredoxin (fig. 2.7). The ammonium produced is used to form glutamine, which is transferred to other cells or released into the surrounding water. Nitrogenase only catalyzes the conversion of nitrogen to ammonium in the absence of oxygen. Thus, heterocyst formation separates reactions that are incompatible with the photosynthetic activity of normal vegetative cells. Heterocysts are weak points, where breaks in a filament are more likely.

    Anabaena is a typical heterocystous cyanobacterium (see fig. 2.5). Constrictions between adjacent cells of *Anabaena*'s unbranched filaments sometimes give the appearance of a string of beads. The formation of heterocysts is inversely related to the availability of dissolved inorganic nitrogen compounds in the surrounding water. Ordinarily, akinetes only develop when conditions become unfavorable for continued growth.

a

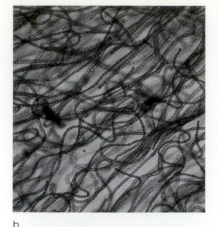

b

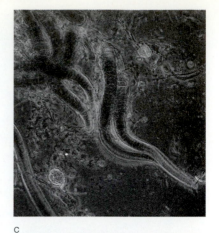

c

d

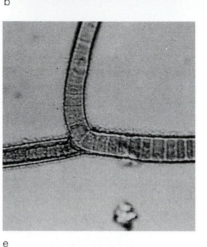

e

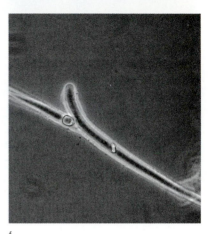

f

Figure 2.8
Heterocystous cyanobacteria.
(*a, b*) *Nostoc* colonies with a
firm sheath enclosing
trichomes. (*c, d*) Tapering
filaments of *Calothrix* and
Gloeotrichia. (*e*) *Scytonema*
with false branching.
(*f*) *Tolypothrix* with false
branches adjacent to
heterocysts. (*g*) *Stigonema*
showing true branching.
(*e* courtesy A. C. Hall.)

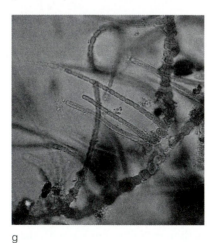

g

Heterocystous cyanobacteria vary in the following features: (1) filaments in bundles surrounded by a mucilaginous sheath, (2) tapering filaments with heterocysts at their base, (3) filaments with false branching, and (4) branching filaments. In true branching, cells divide laterally to form a branch. False branching occurs in some cyanobacteria with a firm sheath. After a break in the series of cells, one or both parts of a filament continue to grow and bend to the side (see fig. 2.8*e*). Because the sheath holds the parts together, the filament appears to branch.

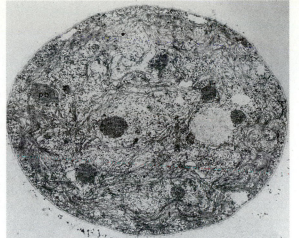

Figure 2.8 shows representative cyanobacteria. *Nostoc* is similar to *Anabaena,* but has a firm mucilaginous sheath (fig. 2.8*a, b*). *Aphanizomenon* forms bundles of parallel filaments surrounded by a watery mucilage (see fig. 2.12). Some cyanobacteria have tapering filaments ending in colorless hairs (fig. 2.8*c, d*). These hairs form in response to phosphate deficiency (Livingstone and Whitton 1983) by increased vacuolation and loss of pigments in the terminal cells. *Calothrix* has a heterocyst at one end and tapers at the other end (this is more obvious when filaments are short). *Scytonema* is a common genus with false branching (fig. 2.8*e*). In the genus *Tolypothrix,* a false branch develops adjacent to a heterocyst (fig. 2.8*f*). *Stigonema* shows true branching and a multiseriate condition in its older axes (fig. 2.8*g*).

Prochlorophytes

In 1975, Ralph Lewin reported the discovery of a unicellular prokaryotic alga lacking phycobilins and containing chlorophyll *b.* The alga, later named *Prochloron,* resembled other cyanobacteria in its cell structure and biochemistry, but lacked phycobilisomes on its thylakoids (fig. 2.9). Since then, two other prochlorophytes have been found.

Two genera of prochlorophytes are nonfilamentous, and the other genus is a nonheterocystous filament. Both *Prochloron* and *Prochlorococcus* have spherical cells without obvious sheaths and grow in marine environments. *Prochloron* grows symbiotically on tunicates (sea squirts), while *Prochlorococcus* is widespread in the marine plankton. *Prochlorothrix* is filamentous with gas vesicles in its cells and grows in freshwater.

Ecology

Cyanobacteria are common in a wide range of marine and freshwater environments, where they may grow floating in the water as members of the plankton or attached to the bottom as part of benthic communities. Some cyanobacteria are moderately successful in terrestrial environments, while others live inside other organisms as symbionts. Cyanobacteria make two important contributions in biologic communities. First, they are photosynthetic producers of organic material and oxygen. Second, many cyanobacteria fix nitrogen (N_2). Nitrogen in this form is unavailable to most living organisms because breaking the triple bond between the nitrogen atoms is difficult. Cyanobacteria convert N_2 to ammonium (NH_4^+), which is then incorporated into amino acids and a variety of other nitrogen-containing compounds.

Since the conversion of nitrogen gas to ammonium by nitrogen fixation is energetically "expensive," it occurs only when other, more usable forms of nitrogen, such as nitrate and ammonium, are not available to meet cellular needs. Under these conditions, many filamentous cyanobacteria form heterocysts, as described earlier, to separate the reactions of nitrogen fixation from exposure to oxygen. However, in the absence of oxygen, some nonheterocystous cyanobacteria show nitrogenase activity in ordinary vegetative cells either at night or in oxygen-depleted regions within dense filament bundles.

In waters with very low concentrations of nitrate, ammonium, and other forms of dissolved nitrogen, such as tropical oceans and some lakes, nitrogen fixation by cyanobacteria is an important source of usable nitrogen. Many symbiotic cyanobacteria fix nitrogen and thus benefit their hosts as nitrogen sources as well as photosynthetic producers. The ability of many cyanobacteria to fix nitrogen may give them an advantage over other algae in waters with low nitrogen concentrations.

The success of the cyanobacteria is due, in part, to their ability as a group to tolerate a wide range of environmental conditions, perhaps because of the prokaryotic nature of their cells, although individual species have more limited ranges. Cyanobacteria are successful at both high and low temperatures, may grow in very hypersaline environments, and tolerate extended periods of desiccation. Only pH limits many species to neutral or basic conditions. Cyanobacteria can grow at extreme temperatures, including in the hot water from geothermal discharges. The unicellular cyanobacterium *Synechococcus* can tolerate temperatures up to 74° C. In Yellowstone National Park, colorful algal mats occur in many thermal areas (plate 1*b*). The distribution of cyanobacteria in these heated waters is limited sometimes by low pH from high sulfide concentrations (producing sulfuric acid). At the other extreme, cyanobacteria survive in Antarctic lakes permanently covered by several meters of ice. Mats composed of filaments of *Phormidium* and *Lyngbya* are able to grow in the water beneath the ice, where temperatures are close to freezing and light levels penetrating the ice during the short Antarctic summer are low (Parker et al. 1981; Wharton, Parker, and Simmons 1983).

Cyanobacteria also tolerate high salt concentrations resulting from evaporation of water from tidepools and lakes. The filamentous cyanobacterium *Spirulina* is common in lakes with a high soda content and high pH (but is not confined to these conditions) (see fig. 2.4*e*). Its abundance in some of the lakes of the African Rift Valley makes it a major food for large flamingo populations. Native peoples near Lake Chad (Africa) and Lake Texcoco (Mexico) have used *Spirulina* for food, and interest in growing *Spirulina* commercially as food for humans and domestic animals is increasing because of *Spirulina's* high protein content (up to 70% of its dry weight), vitamins, and trace metals. As a crop, *Spirulina* has the advantage of growing well in saline ponds in arid environments that do not support other crops. However, production and harvesting costs, and its lack of acceptability limit use of *Spirulina* as a food at the present time.

Some cyanobacteria show an alternative form of photosynthesis under anoxic (anaerobic) conditions, similar to photosynthetic sulfur bacteria. Instead of water, hydrogen sulfide serves as the electron donor, and thus, the overall reaction produces no oxygen:

$$2H_2S + CO_2 \rightarrow [CH_2O] + 2S + H_2O$$

Figure 2.10
Mats of *Oscillatoria* that have detached from the bottom and are floating on the water surface.

Benthic cyanobacteria are widespread in coastal marine and freshwater environments. On reefs, free-living and symbiotic cyanobacteria are important fixers of nitrogen. Cyanobacteria are common in salt marshes, seagrass meadows, mudflats, and tidepools, and may blacken rocks wetted by ocean spray. In freshwaters, cyanobacteria grow on lake and stream bottoms. Benthic cyanobacteria often form mats of interwoven filaments. Sediments may stick to the mucilage around cells, and mats may contain a rich community of bacteria, protozoa, and small animals. Oxygen accumulation may buoy a mat, and sometimes, part of a mat detaches from the bottom and continues to grow floating on the water surface (fig. 2.10). Many mat-forming cyanobacteria deposit calcium carbonate in their sheath, creating a hard structure. Thick mats forming dome-shaped or columnar structures are called stromatolites. Fossilized stromatolites, originally formed in shallow ocean waters, have been dated at over 3 billion years.

Cyanobacteria also grow floating in the water as members of planktonic communities. Planktonic cyanobacteria are more common in offshore waters of temperate and tropical oceans than in coastal waters or at high latitudes. They occur primarily in two forms—as either small solitary cells or as large filament bundles. The small cyanobacteria are often part of the picoplankton, a category for organisms less than 2 micrometers in size. Their large surface-to-volume ratio may account for their success in obtaining nutrients in central oceanic regions with very low nutrient concentrations. Picoplankton such as *Synechococcus, Synechocystis,* and the prochlorophyte *Prochlorococcus* are widespread. Chlorophyll is widely used as a measure of phytoplankton abundance in the oceans, and often there is a distinct subsurface chlorophyll maximum in tropical and subtropical oceans.

Trichodesmium (sometimes classified as a species of *Oscillatoria*) is also widespread in the plankton of tropical and subtropical oceans (Carpenter, Capone, and Reuter 1992; Janson et al. 1995). Its filaments aggregate in large bundles, which float because of gas vesicles in the cells (fig. 2.11). Bundles may consist of trichomes arranged parallel to each other or in a radiating pattern. Large bundles of *Trichodesmium* resemble sawdust floating on the water, as Charles Darwin observed during the voyage of the HMS *Beagle*. Since phycoerythrin rather than phycocyanin is the principal light-harvesting pigment, dense growths of *Trichodesmium* may discolor the water red and are what gave the Red Sea its name. Even though it lacks heterocysts, *Trichodesmium* is able to fix nitrogen in its vegetative cells. However, photosynthetic activity and nitrogen fixation are spatially separated in different parts of a bundle (Bergman and Carpenter 1991; Paerl 1994), preventing deactivation by oxygen.

Figure 2.11
Trichodesmium from the Sargasso Sea in the central North Atlantic. (*a*) Bundles of trichomes. (*b*) Cell with numerous gas vesicles. (*b* from Gantt, Ohki, and Fujita 1984, courtesy the authors and *Protoplasma* Springer-Verlag.)

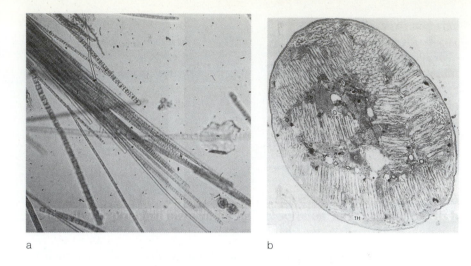

a   b

Figure 2.12
Filaments of *Aphanizomenon* in a bundle.

    While planktonic cyanobacteria are widespread in oceanic waters with low nutrient concentrations, the opposite is usually true in freshwaters, where cyanobacterial blooms are normally associated with nutrient enrichment. Nuisance blooms often occur in eutrophic or nutrient-rich lakes during the summer, producing a slimy scum (plate 1*a*). Many bloom-producing species have gas vesicles that cause their colonies or filaments to accumulate on the water surface. Such scums detract from the recreational appeal of a body of water, and later the blooms may wash ashore and rot, or sink to the lake bottom and decompose, causing severe oxygen depletion. Common species producing nuisance blooms belong to *Anabaena* (see fig. 2.5), *Aphanizomenon* (fig. 2.12), and *Microcystis* (see fig. 2.3*c*), which are sometimes shortened to "Annie, Fanny, and Mike." Recently, blooms of the filamentous cyanobacteria *Nodularia, Aphanizomenon,* and *Anabaena* have

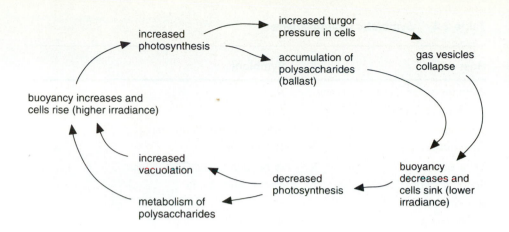

Figure 2.13
Buoyancy regulation in
cyanobacteria by formation of
gas vesicles and
accumulation of heavy
molecules in cells as ballast.

occurred in the Baltic Sea, probably because of increased nutrient enrichment from human discharge.

Many planktonic cyanobacteria contain **gas vesicles,** sometimes filling much of the cell volume (see fig. 2.11*b*). (Collectively, the gas vesicles in a cell are its gas vacuole.) Each cylindrical vesicle is surrounded by a membrane of proteins and contains air modified by metabolic activities. Gas vesicles may produce a positive buoyancy that causes filaments to accumulate on the water surface, but in other cases, they regulate filament position in the water column. Photosynthetic activity affects buoyancy in two ways (fig. 2.13). First, it leads to an increase in low-weight organic molecules, which increase cell turgor pressure, causing collapse of gas vesicles (Reynolds and Walsby 1975). Second, photosynthesis leads to synthesis of heavy molecules such as starch, which add ballast to cells. Thus, cells generally sink when their photosynthetic activity is high. A wide range of environmental conditions may influence buoyancy, however, including light levels and availability of inorganic carbon and other nutrients (Oliver 1994; Klemer et al. 1996). Gas vesicles also may function as light shields, reducing cells' sunlight exposure when they float near the surface.

Some planktonic cyanobacteria produce toxins (Gorham and Carmichael 1988; Carmichael 1994). In freshwaters, they include bloom-producing species of *Anabaena, Aphanizomenon, Microcystis, Oscillatoria,* and *Nodularia.* The toxins, released when animals ingest the algal cells, include both neurotoxins (which affect neurons) and hepatotoxins (which affect liver cells) (see table 7.2). The toxins can be fatal to mammals, birds, and fishes, and but no human fatalities are known. Anatoxin, which *Anabaena* strains primarily produce, overstimulate muscles, leading to fatigue and paralysis. Microcystin and nodularin are hepatotoxins (Watanabe, Kaya, and Takamura 1992). In marine waters, *Lyngbya* releases chemicals (lyngbyatoxin, aplysia) that cause dermatitis or "swimmer's itch."

Some cyanobacteria grow successfully in terrestrial environments, where they tolerate temperature and moisture variations, and exposure to high levels of light. Many survive freezing and thawing, and extended periods of desiccation when moisture is insufficient for active growth. After rain, cyanobacteria in the soil may quickly produce a conspicuous growth on the soil's surface and in puddles. A thick sheath of firm mucilage protects trichomes of *Nostoc,* a common soil alga (review of *Nostoc* by Dodds, Gudder, and Mollenhauer 1995). Because of their photosynthetic and nitrogen-fixing abilities, the cyanobacteria may act as biofertilizers to

Table 2.4 *Symbiotic Cyanobacteria*

Host	Symbiont
Pennate diatoms (*Rhizosolenia*)	*Richelia*
Protozoa (glaucophytes)	Cyanelles
Sponges	*Aphanocapsa, Phormidium*
Tunicates	*Prochloron*
Polar bear hairs	*Aphanocapsa* or *Gloeocapsa*
Fungi (lichens)	*Nostoc* and others
Bryophytes	*Nostoc*
Azolla (fern)	*Anabaena*[1]
Roots of cycads (gymnosperm)	*Nostoc*
Glands of *Gunnera* (angiosperm)	*Nostoc*

Source; After Rai 1990.
[1]Canini et al. (1992) consider the symbiont to be *Nostoc*.

enrich the soil and may affect soil texture by secreting mucilage that helps aggregate soil particles. *Gloeocapsa,* with a thick, mucilaginous sheath, sometimes forms a conspicuous purple discoloration on rocks (plate 5*a*). It also lives in cracks in rocks under desert conditions.

Cyanobacteria enter into symbiotic associations with a wide range of other organisms, providing organic material, oxygen, and nitrogenous compounds to their hosts (table 2.4). In tropical oceans where nutrients are low, cells of the planktonic diatom *Rhizosolenia* may contain the nitrogen-fixing cyanobacterium *Richelia* (Janson, Rai, and Bergmann 1995). Among benthic invertebrates, sponges are common hosts. Cycads have symbionts in their roots, while *Gunnera,* a tropical angiosperm, has *Nostoc* in glands at the base of its leaves (Osborne et al. 1991). Lichens are associations between fungi and algae (see p. **233**). Approximately 10% contain a cyanobacterium, usually a nitrogen-fixing genus such as *Calothrix* or *Nostoc*. The hairs of polar bears are an unusual habitat for unicellular cyanobacteria, sometimes giving the fur of zoo animals a distinct color (Lewin, Farnsworth, and Yamanaka 1981).

One of the better-studied associations is between the water fern *Azolla* and *Anabaena azollae. Azolla* grows unattached, floating on the surface of freshwater ponds and marshes. Special cavities in its leaves contain filaments of *Anabaena* with a higher ratio of heterocysts to vegetative cells than in free-living *Anabaena*. The fern, however, uses most of the ammonium heterocysts produce. *Azolla* stimulates nitrogen fixation by inhibiting glutamine synthetase, thus preventing the alga's use of the ammonium. In Asia, *Azolla* is grown widely in rice paddies as a natural fertilizer.

Glaucophytes, unicellular eukaryotes found in freshwater environments, contain one to several endosymbiotic cyanobacteria called **cyanelles** in vesicles within their cells (fig. 2.14) (Löffelhardt and Bohnert 1994). Cyanelles act as chloroplasts, providing their hosts with organic material, primarily glucose. Cyanelles are sur-

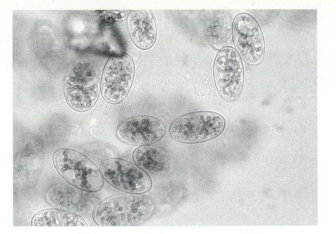

Figure 2.14
Glaucocystis, a eukaryotic cell with an endosymbiotic cyanobacterium.

rounded by a peptidoglycan wall and contain other structures found in free-living cyanobacteria, such as carboxysomes and thylakoids. However, their DNA content is relatively small (similar to the genome size of chloroplasts) compared to other cyanobacteria, and cyanelles cannot survive outside their hosts. Cyanelles of glaucophytes may be an intermediate stage in chloroplast evolution, discussed in the next section.

Origin of Chloroplasts

Two general theories explain the origin of the chloroplasts of eukaryotic algae. In one theory, proliferation of internal membranes developed into all cell organelles, including chloroplasts. These membranes may have originated as invaginations of the cell membrane and progressively developed into the organelles of present-day eukaryotic cells. In an alternative theory, heterotrophic cells, which already had nuclear envelopes and other organelles, acquired chloroplasts by incorporating cyanobacteria as endosymbionts (mitochondria were derived from other endosymbionts). If a host took up a cyanobacterium by endocytosis, the vesicular membrane of the host and the cell membrane of the symbiont represent the two membranes of the chloroplast envelope.

Strong evidence for this endosymbiotic theory comes from the DNA and ribosomes in chloroplasts. Chloroplast DNA resembles prokaryotic chromosomes in being circular and lacking histones. Ribosomes come in two sizes, large and small, designated 80S and 70S, respectively, based on their sedimentation characteristics in ultracentrifuges. Chloroplast ribosomes are similar to the small 70S ribosomes of prokaryotic cells and contrast with the larger 80S ribosomes in the cytosol of eukaryotic cells. Other evidence supporting the endosymbiotic theory is the similarity of nucleotide sequences in DNA and ribosomal RNA of the cyanobacteria and chloroplasts (Kuhsel, Strickland, and Palmer 1990; Douglas and Turner 1991). However, chloroplast DNA codes for only a small fraction of the proteins composing a chloroplast. The majority are encoded in nuclear DNA, synthesized in the cytoplasm, and then incorporated into the chloroplast. Proponents of the endosymbiotic theory argue that genes have progressively transferred from chloroplast DNA to nuclear DNA during evolution of different algal lines

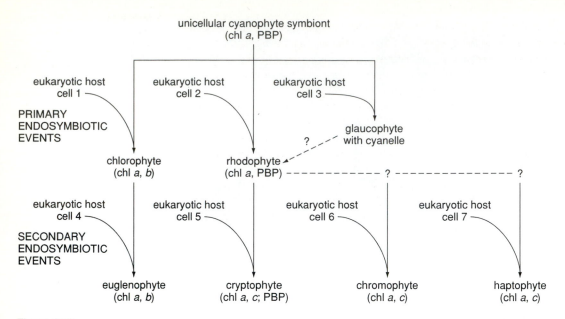

Figure 2.15
Origins of algal chloroplasts by primary and secondary endosymbiotic events. In primary events, cyanobacterial endosymbionts became chloroplasts of chlorophytes (green algae), rhodophytes (red algae), and glaucophytes. Secondary endosymbiotic events involving eukaryotic endosymbionts produced chloroplasts of euglenophytes, cryptophytes, chromophytes, and haptophytes. Glaucophytes may have been an intermediate step leading to rhodophytes.
Chl = chlorophyll;
PBP = phycobiliproteins.

(Baldauf and Palmer 1990). For example, ribulose bisphosphate carboxylase, which catalyzes the carbon-dioxide–fixing step in the Calvin cycle, consists of two types of subunits. Both subunits are coded in the chloroplasts of most algae (Fujiwara et al. 1993). However, in some green algae, the gene for the small subunit is in the nucleus, suggesting a transfer from the chloroplast.

Analyses of ribosomal RNA sequences strongly suggest that chloroplasts probably arose from a single ancestor among cyanobacteria (Bhattacharya and Medlin 1995). The initial event was the incorporation of a cyanobacterium into a host cell. Different eukaryotic hosts may have acquired endosymbionts to become the ancestors of red algae and green algae (fig. 2.15) (Bhattacharya and Medlin 1995). An envelope composed of two associated membranes surrounds the chloroplasts of both of these groups.

For red algae, the initial step may have been the incorporation of a unicellular cyanobacterium into a nonflagellated eukaryotic cell to produce an initial glaucophyte-like stage. Subsequently, the symbiont was modified into the modern rhodophyte chloroplast surrounded by two membranes (vesicle membrane of host and cell membrane of symbiont), containing thylakoids with granular phycobilisomes on their outer surfaces, and having chlorophyll *a* and phycobilins as principal photosynthetic pigments. When viewed with electron microscopes, rhodophyte chloroplasts strikingly resemble unicellular cyanobacteria (compare fig. 2.1*b* with fig. 6.1*d* or 6.27*a*).

Molecular comparisons indicate that green algal chloroplasts are derived from a cyanobacterial symbiont rather than a prochlorophyte (Turner et al. 1989; Douglas and Turner 1991; Palenik and Haselkorn 1992; Urbach, Robertson, and Chisholm 1992). The chlorophyte chloroplast surrounded by two membranes lacks phycobilisomes and phycobiliproteins, and has chlorophylls *a* and *b* as its principal photosynthetic pigments.

The chloroplasts of other eukaryotic algae may have arisen from photosynthetic *eukaryotic* endosymbionts (fig. 2.15) (Gibbs 1990). In these cases, the three or four membranes that surround chloroplasts are interpreted as representing the chloroplast envelope, the symbiont's cell membrane, and a vesicular membrane of the host cell. Cryptomonads show the clearest picture of such a secondary endosymbiotic event by having other remnants of the endosymbiont's cells besides the chloroplast. Membranes (chloroplast endoplasmic reticulum) separate the chloroplast and these remnants from the rest of the cell. The presence of phycobiliproteins and molecular comparisons (see fig. 1.11) indicate that the symbiont was a red alga (McFadden, Gilson, and Hill 1994; Douglas 1994).

Chromophytes and haptophytes also have chloroplast endoplasmic reticulum, but without other remnants of the symbiont. Pigmentation and molecular comparisons do not clearly indicate the source of their chloroplasts. Chloroplast envelopes with three membranes also support the origin of the chloroplasts of dinoflagellates and euglenophytes from eukaryotic symbionts.

Summary

1. The cellular structure of prokaryotic algae or cyanobacteria resembles that of eubacteria, but their photosynthetic system is similar to eukaryotic algae. On the basis of their photosynthetic pigments, the cyanobacteria are divided into two groups. Members of the largest group contain chlorophyll *a* and phycobiliproteins organized into phycobilisomes. The few known representatives of the second group, the prochlorophytes, have chlorophylls *a* and *b*. However, the prochlorophytes are not a distinct evolutionary group within the cyanobacteria.

2. Peptidoglycan walls and a mucilaginous sheath surround cyanobacterial cells. Cells contain flattened vesicles, or thylakoids, with phycobiliproteins organized into phycobilisomes on their surfaces, except in prochlorophytes. Cells also contain a variety of storage granules, ribosomes, and circular chromosomes. The morphology of cyanobacteria ranges from solitary cells and small aggregations to large colonies and filaments. Some filaments have akinetes and heterocysts as specialized cells. Akinetes are perennating stages, while heterocysts separate the reactions of nitrogen fixation from the photosynthetic activity of normal vegetative cells.

3. Cyanobacteria are the oldest group of algae, found in fossils over 3.5 billion years old. Their success is due, in part, to their tolerance of a wide range of environmental conditions. Planktonic cyanobacteria are common in low-nutrient waters of tropical and temperate oceans. In eutrophic lakes, planktonic cyanobacteria with gas vesicles in their cells may form surface scums. Other cyanobacteria form mats of interwoven filaments on submerged surfaces and soil. Some cyanobacteria enter into symbiotic associations with other organisms. Cyanobacteria are important for their photosynthetic production and nitrogen fixation.

4. The chloroplasts of eukaryotic algae are probably derived either directly or indirectly from cyanobacteria. The two closely associated membranes of the chloroplast envelopes of green and red algae indicate that their chloroplasts originated directly from symbiotic cyanobacteria. The additional membranes surrounding chloroplasts of other algal groups suggest an origin from a secondary symbiotic event in which the endosymbiont was another eukaryotic alga.

Further Reading

Balows, A., H. G. Trüper, M. Dworkin, W. Harder, and K. H. Schleifer, eds. 1992. *The Prokaryotes*, Vols. 1–4. Springer-Verlag.

Bryant, D. A., ed. 1994. *The Molecular Biology of Cyanobacteria*. Kluwer Academic Publishers.

Carmichael, W. W. 1994. The toxins of cyanobacteria. *Scientific American* 270(1): 78–86.

Carpenter, E. J., D. G. Capone, and J. G. Rueter. 1992. *Marine Pelagic Cyanobacteria:* Trichodesmium *and Other Diazotrophs*. Kluwer Academic Publishers.

Fay, P., and C. V. Baalen, eds. 1987. *The Cyanobacteria*. Elsevier.

Lewin, R. A., ed. 1993. *Origins of Plastids*. Chapman and Hall.

Lewin, R. A., and L. Cheng, eds. 1989. *Prochloron*. Chapman and Hall.

3

Green Algae (Division Chlorophyta)

The green algae, belonging to the division Chlorophyta, are distinguished by their photosynthetic pigments, carbohydrate reserve, chloroplast structure, and flagella (table 3.1). The green algae's principal photosynthetic pigments are chlorophylls *a* and *b*. The carotenoid lutein is also present, but its function is uncertain. The chloroplast has an envelope composed of two membranes, indicating an origin from an endosymbiotic cyanobacterium. Thylakoids are stacked in groups of two to six, but their arrangement is less regular than that of plant grana. Starch as a reserve is in the chloroplast. Flagellated stages have smooth flagella.

The divison Chlorophyta has four classes, as summarized in table 3.2. The term *chlorophyte* will refer to members of the division Chlorophyta, and the term *chlorophycean algae* will refer to members of the class Chlorophyceae. The classes are distinguished primarily on ultrastructural characteristics related to microtubular arrangement during cell division, basal body orientation, and arrangement of microtubular roots in flagellated cells. The prasinophytes are considered primitive and vary considerably in basic characteristics. In general, comparisons of nucleotide sequences from ribosomal RNA support separation of four classes (Zechman et al. 1990; Kantz et al. 1990; Steinkötter et al. 1994). Some phycologists recognize a fifth class—Pleurastrophyceae—but is included here with the Chlorophyceae.

The flagellar apparatus consists of the flagella, basal bodies, and associated cytoskeleton. Green algae usually have two or four flagella without appendages other than delicate hairs. Basal body orientation helps distinguish the Charo-

Table 3.1 *Characteristics of Division Chlorophyta*

Major photosynthetic pigments	Chlorophylls *a, b*
Carbohydrate reserve	Starch
Chloroplast structure	
Thylakoid associations	Stacks of two to six
Envelope	Two membranes
Chloroplast ER	Absent
Cell covering	Scales, wall (or lacking)
Flagella	Smooth

Figure 3.1
Basal body orientation
relative to hands of a clock.
(*a*) Clockwise (1 o'clock–
7 o'clock). (*b*) Opposite
(12 o'clock–6 o'clock).
(*c*) Counterclockwise
(11 o'clock–5 o'clock).
(*d*) Parallel.

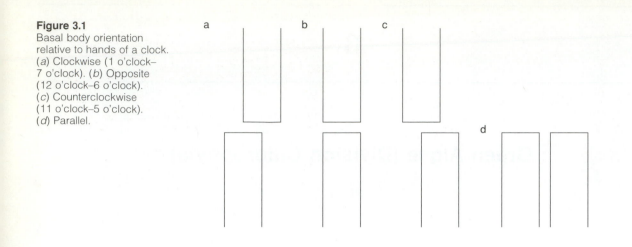

Table 3.2   *Classes of the Division Chlorophyta*

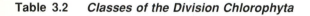

	Prasinophyceae	Chlorophyceae[1]	Ulvophyceae	Charophyceae
Cell covering	Scales or naked	Wall	Wall	Wall
Mitotic spindle	Usually closed	Closed	Closed	Open
Microtubule organization during cell division	Phragmoplast or phycoplast	Phycoplast	(absent)	Phragmoplast
Basal body orientation (clock orientation)	Variable	Clockwise (1:00/7:00 orientation[2])	Counterclockwise (11:00/5:00 orientation)	Parallel (unilateral)
Microtubular roots in flagellated cells	Variable	Cruciate roots	Cruciate roots	Broad band and small root
Principal habitat	Marine and freshwater	Freshwater	Marine	Freshwater
Life cycle	Haplont	Haplont (dormant zygote)	Diplohaplont	Haplont (dormant zygote)

Sources: Kantz et al. 1990; Steinkötter et al. 1994; Hoek, Mann, and Jahns 1995.

[1] Includes Pleurastrophyceae with counterclockwise orientation of basal bodies.

[2] Some with 12:00/6:00 or 11:00/5:00 (counterclockwise) orientation

phyceae, Chlorophyceae, and Ulvophyceae. As a cell is viewed from the front, basal body orientation is described relative to the hands of a clock (fig. 3.1). A clockwise orientation is characteristic of most chlorophycean algae, the ulvophytes show a counterclockwise orientation, and basal bodies are parallel in the charophytes. Bundles or roots of microtubules radiate from the basal bodies around the periphery of flagellated cells. The chlorophyceans and ulvophytes have four microtubular roots in a cruciate pattern (figs. 3.2*a*, 3.9). The number of microtubules in the roots is X-2-X-2, where X is a variable number. In contrast, the charophytes have two microtubular roots. One is a broad band, and the other is a small root with

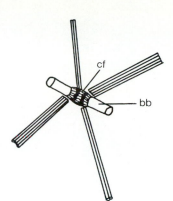

a

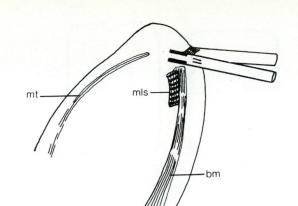

b

Figure 3.2
Arrangements of
microtubules in flagellated
cells. (*a*) Microtubules in a
cruciate arrangement as
seen from anterior end of
cell. bb=basal body;
cf=connecting fiber.
(*b*) Broad band of
microtubules (bm) and a
second small microtubular
root (mt). A multilayered
structure (mls) is associated
with the broad band.

a

b

Figure 3.3
Arrangements of
microtubules during cell
division. (*a*) Phycoplast with
microtubules parallel to the
plane of division, the nuclear
envelope intact, and
daughter nuclei (nu) near
each other during cell
division. (*b*) Phragmoplast
with microtubules
perpendicular to the plane of
cell division, the nuclear
envelope dispersed, and
daughter nuclei (nu)
separated during cell
division. In both (*a*) and (*b*),
division occurs by furrowing.

only a few microtubules (fig. 3.2*b*). Connecting fibers of centrin link basal bodies, and striated roots are also part of the cytoskeleton in the Prasinophyceae, Chlorophyceae, and Ulvophyceae.

The charophytes, chlorophyceans, and ulvophytes also show differences during cell division. In the chlorophyceans and ulvophytes, mitosis occurs without the complete breakdown of the nuclear envelope (closed spindle). In the chlorophyceans, the daughter nuclei from mitosis remain close together, and the spindle microtubules reorganize into a phycoplast in which the microtubules are oriented parallel to the plane of cell division (fig. 3.3*a*). The ulvophytes have a persistent telophase spindle, but the microtubules do not organize into a phycoplast. In contrast, the nuclear envelope breaks down, and the spindle is open during mitosis in the charophytes. At the completion of mitosis, daughter nuclei move to opposite ends of the parent cell, and the spindle persists as a phragmoplast (fig. 3.3*b*). Cell division occurs through a phycoplast or phragmoplast by either ingrowth of a furrow or formation of a cell plate.

The distinction between a furrow and a cell plate is important. **Furrowing,** the common means of cell division among algae, is division by ingrowth of the cell membrane usually in a band around the cell's equator (fig. 3.4*a*). In contrast, a few algae and most land plants (bryophytes and vascular plants) form **cell plates** that begin in the center of a cell and progress outward (figs. 3.4*b*, 3.5). Vesicles from Golgi bodies form a cell plate along a cell's equatorial plane. As vesicles containing materials for the wall are added, the cell plate develops outward and eventually joins the parental cell membrane.

Figure 3.4
Cell division. (*a*) Furrowing involves ingrowth of the cell membrane from the edge of a cell. (*b*) A cell plate is initiated at the center of a cell and develops outward by addition of Golgi vesicles.

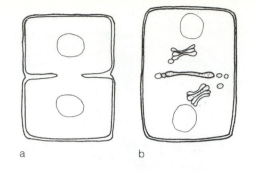

a   b

Figure 3.5
Formation of a cell plate during cell division. (From Floyd, Stewart, and Mattox 1971, courtesy *Journal of Phycology*.)

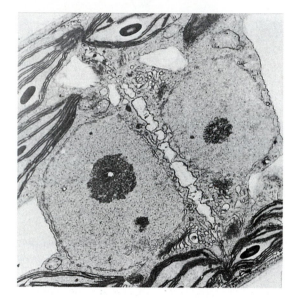

Green algal life cycles show three basic patterns, depending on which stage undergoes meiosis. In the primitive condition, vegetative cells are haploid, and zygotes are the only diploid stage in a haplontic life cycle (fig. 3.6*a*) Zygotes are often dormant stages formed in response to stressful conditions. Meiosis occurs when zygotes germinate. A diplohaplontic life cycle has an **alternation of generations** (fig. 3.6*b*). Instead of undergoing meiosis, the zygote divides mitotically to form a diploid thallus. Meiosis occurs during formation of reproductive cells, and the resulting spores give rise to a haploid thallus. Thus, a haploid vegetative phase alternates with a diploid vegetative phase. If both phases appear similar and are distinguishable only by chromosome number or the type of reproductive cells formed, the generations are **isomorphic.** In an alternation of **heteromorphic generations,** haploid and diploid phases have distinctly different appearances. Finally, a diplontic life cycle found in a few green algae originated from an alternation of generations by a shift in meiosis to the time of gamete formation (fig. 3.6*c*). The zygote resulting from gametic fusion develops into a new diploid thallus. Gametes are the only haploid cells.

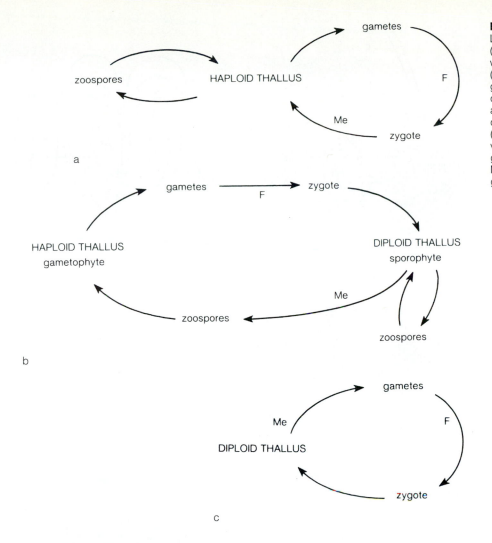

Figure 3.6
Life cycles of green algae.
(*a*) Haploid vegetative phase
with zygotic meiosis
(haplont). (*b*) Alternation of
generations with haploid and
diploid vegetative phases as
a result of sporic meiosis
during zoospore formation
(diplohaplont). (*c*) Diploid
vegetative phase with
gametic meiosis (diplont).
Me=meiosis; F=fusion of
gametes.

Most prasinophytes, chlorophyceans, and charophytes have haploid vegetative phases, the zygote is the only diploid stage, and zygotic meiosis occurs. The ulvophytes have either an alternation of haploid and diploid phases with sporic meiosis, or a diploid vegetative phase and gametic meiosis.

The green algae have a wide ecologic range, occurring in the oceans, freshwaters, and moist terrestrial habitats. The chlorophyceans and charophytes are primarily found in freshwaters, most ulvophytes are marine, and the prasinophytes are found in both freshwater and seawater.

Class Prasinophyceae

Most prasinophytes are unicellular flagellates. Delicate organic scales, composed largely of carbohydrates, usually cover prasinophytes' cells. Scales are also present on their flagella. In other structural features, the prasinophytes vary considerably. Analysis of nucleotide sequences of ribosomal RNA suggests that the prasinophytes may consist of two distinct groups (Steinkötter et al. 1994). Sometimes, primitive green algae are placed in an alternative class—the Micromonadophyceae—with a more restricted grouping of species (Mattox and Stewart 1984).

Figure 3.7
Prasinophytes.
(*a*) *Micromonas.*
(*b*) *Pyramimonas.*
ch=chloroplast; fl=flagellum;
fr=fibrous root;
mb=microbody;
mi=mitochondrion;
mt=microtubules;
nu=nucleus; py=pyrenoid
with starch grains on surface.
(*a* after Marchant et al. 1989;
Sieburth, Johnson, and
Hargraves 1988. *b* after
Pienaar and Aken 1985.)

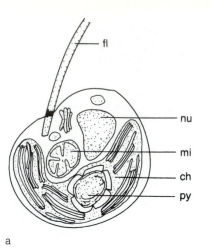

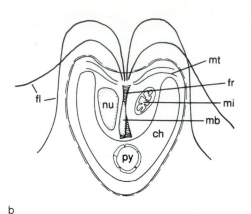

a   b

Figure 3.8
Prasinophytes. (*a*) Electron
micrograph of *Pyramimonas.*
(*b*) Light micrograph of
Tetraselmis. (*a* courtesy
Norris and Pienaar.)

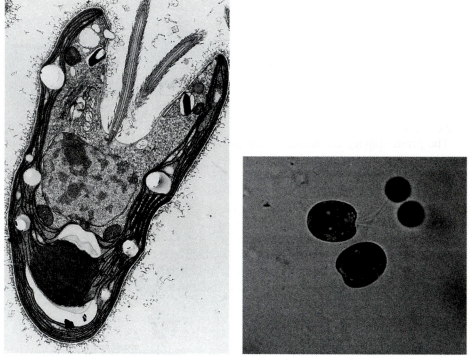

a   b

Micromonas, a very small (less than 2 micrometers) marine flagellate, exemplifies the primitive condition (fig. 3.7*a*). Its naked cells have a single, laterally inserted flagellum, which arises from a basal body with a root of two microtubules. Each cell has a single nucleus, chloroplast, and mitochondrion. Prior to cell division, the organelles and flagellar apparatus divide. Cell division involves a furrow and phragmoplast. Chlorophyll *c* has been reported in *Micromonas* (Wilhelm et al. 1986). The related genus *Mantoniella* has scales and a rudimentary second flagellum.

Pyramimonas and *Tetraselmis* (=*Platymonas*), both unicellular flagellates, exhibit more advanced conditions. *Pyramimonas* has an ovoid cell covered by several layers of scales (figs. 3.7*b*, 3.8*a*). Four flagella arise from a depression at the cell's anterior end. Besides connecting fibers between the basal bodies, the cytoskeleton consists of four bundles of microtubules radiating from the basal bodies around the anterior depression and fibrous roots extending from the basal bodies toward the nucleus. In the cell's posterior is a chloroplast, containing pyrenoid and an eyespot. Golgi bodies are present in the vicinity of the nucleus. Scales are formed in Golgi vesicles and then deposited on the cell's outer surface.

Tetraselmis is similar in general appearance to *Pyramimonas,* with four flagella arising from an anterior depression (fig. 3.8*b*). However, instead of several layers of distinct scales, the scales fuse to form a continuous covering called a theca. The internal structure of *Tetraselmis* is generally similar to that of *Pyramimonas.*

Both *Pyramimonas* and *Tetraselmis* undergo mitosis and then division by furrowing to produce daughter cells, but some cell division events differ significantly. In *Pyramimonas,* the nuclear envelope disperses during mitosis, and the mitotic spindle is open. The spindle persists as a phragmoplast, as the cell divides longitudinally by ingrowth of a furrow. In contrast, the nuclear envelope of *Tetraselmis* remains intact, producing a closed spindle. Daughter nuclei remain relatively close together at the conclusion of mitosis, and the spindle reorganizes into a phycoplast prior to division by furrowing. The phycoplast may have arisen in cells, such as *Tetraselmis,* where a rigid theca prevents rapid cell expansion as daughter nuclei separate (Mattox and Stewart 1977). Division in *Pyramimonas* is similar to that in charophytes, while division in *Tetraselmis* is similar to that in chlorophyceans. Sexual reproduction among the prasinophytes is uncommon and poorly known (see Suda, Watanabe, and Inouye 1989).

The prasinophytes are found in oceanic and freshwater plankton. Their small size means that they are often overlooked until they become abundant.

Class Chlorophyceae

The class Chlorophyceae contains the green algae with a cruciate arrangement of microtubules in their flagellated cells and a phycoplast during cell division (figs. 3.9, 3.10). Most have a clockwise orientation of their basal bodies. Chlorophycean greens range in morphology from single-celled flagellates and colonies of flagellated cells to nonflagellated cells and colonies, and filaments. Normally, vegetative cells are haploid, and the zygote produced by gametic fusion is the only diploid stage and enters a period of dormancy before germinating. Meiosis occurs at germination. Chlorophycean greens are found primarily in freshwater and terrestrial habitats.

The primitive morphologic type in the chlorophyceans is a solitary, flagellated cell derived from a prasinophyte ancestor. Such unicellular algae grow by cellular enlargement, and cell divisions are associated with reproduction. In advanced filamentous species, vegetative divisions also accompany growth to increase the number of cells in the thallus. In a few algae, mitosis occurs without cell division to produce multinucleate cells called coenocytes.

Three major lines of evolution probably arose from ancestral flagellated cells (fig. 3.11). In one line, cells remain flagellated but associate into colonies. A firm framework of mucilage often holds the cells together. In a second line, cells lose their flagella in the vegetative condition (only reproductive cells are flagellated). These nonmotile cells may be solitary or associated in aggregations or colonies. In the third line, cells are linearly arranged to form filaments. Filaments vary from unbranched types with little cell specialization to branched forms with cell

Figure 3.9
Flagellar apparatus of a
chlorophycean alga.
ABB=accessory basal body;
BB=functional basal body;
DF=distal connecting fiber;
EDM=electron dense
material; PF=proximal
connecting fiber; R=rootlet
with two or four microtubules.
(From Greuel and Floyd
1985, courtesy *Journal of
Phycology*.)

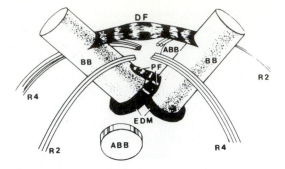

Figure 3.10
Cell division in *Stigeoclonium*
showing phycoplast with
microtubules in cross section.
(From Floyd, Stewart, and
Mattox 1972, courtesy
Journal of Phycology.)

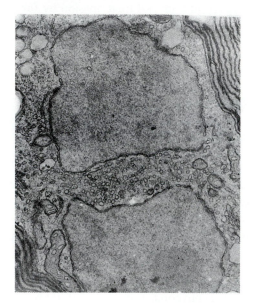

Figure 3.11
Evolution in the class
Chlorophyceae. Ancestral
flagellated cells gave rise to
colonies of flagellated cells,
palmelloid aggregations,
nonflagellated cells and
colonies, and filaments.

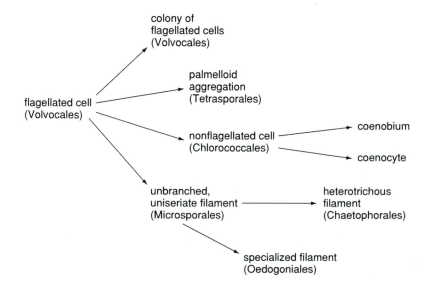

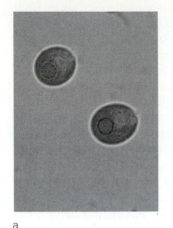

a

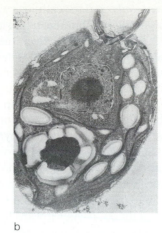

b

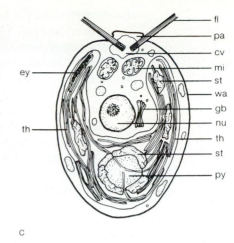

c

Figure 3.12
Chlamydomonas. (*a*) Light micrograph (flagella not visible. (*b*) Electron micrograph. (*c*) Cell structures. cv=contractile vacuole; ey=eyespot; fl=flagellum; gb=Golgi body; mi=mitochondrion; nu=nucleus; pa=papilla; py=pyrenoid; st=starch grain; th=thylakoid; wa=wall. (*b* from Ringo 1967, reproduced from the *Journal of Cell Biology,* 1967, 33:545, by copyright permission of the Rockefeller University Press.)

Table 3.3   *Orders of the Class Chlorophyceae*

Order Volvocales—flagellated cells and colonies

Order Tetrasporales—palmelloid aggregations and colonies; nonmotile flagellates; cells with contractile vacuoles, basal bodies, and eyespots

Order Chlorococcales—nonflagellated cells, aggregations, and colonies; cells without contractile vacuoles; divisions only associated with formation of reproductive stages (exceptions)

Order Microsporales—unbranched filaments without intercellular connections

Order Chaetophorales—heterotrichous filaments with intercellular connections

Order Oedogoniales—filaments; cell division involving formation of a "ring"; stephanokontous zoospores and sperm

specialization. Table 3.3 summarizes the classification within the Chlorophyceae based on the structure of the vegetative thallus.

Chlorophycean Diversity

Flagellated Cells (Order Volvocales)

Chlamydomonas is a single-celled flagellate showing the primitive morphologic conditions for the class Chlorophyceae. *Chlamydomonas* grows in a wide range of freshwater environments, including moist soil and snow. Because of ease of growth in culture and a short generation time, *Chlamydomonas* is a valuable experimental system for genetic studies, especially those about flagella mutants, haploid genetics, and chloroplast inheritance (see Harris 1989). *Chlamydomonas* is a large genus with over 400 species.

Two anterior flagella propel cells of *Chlamydomonas* through water (fig. 3.12). A delicate wall of glycoproteins rather than cellulose surrounds each ovoid cell. This wall may represent an intermediate condition between the theca of prasinophytes, such as *Tetraselmis,* and the cellulose wall of most green algae. Each flagellum arises from a basal body within the cell. The basal bodies have a clockwise orientation (see fig. 3.1*a*) when viewed from the front of the cell and a "V" arrangement relative to each other when seen from the side. Connecting fibers join the basal bodies, and four bundles of microtubules radiate from the basal bodies around the cell periphery in the typical cruciate pattern of chlorophyceans.

A single, large chloroplast shaped like a thick cup fills much of the cell. Within the chloroplast, thylakoids are stacked in groups of two to six. A conspicuous structure in the chloroplast posterior is a spherical pyrenoid, which stores ribulose bisphosphate carboxylase. Starch grains accumulate on the surface of the pyrenoid, and other starch grains may be free within the chloroplast. Another distinct feature in the chloroplast anterior is the eyespot, composed of several layers of red carotene granules against the chloroplast envelope (Satoh et al. 1995). *Chlamydomonas* shows a positive phototaxis (swims toward light). The eyespot helps sense the direction of light but is not the actual photoreceptor (Melkonian and Robenek 1984). Between the lipid bilayers of the cell membrane immediately next to the eyespot are particles of a compound resembling rhodopsin. Rhodopsin is probably the photoreceptor, changing membrane permeability in response to light stimulus, while the eyespot is a shading structure.

The chloroplast surrounds the nucleus, which is not usually visible through a light microscope without special staining. The nucleus has a haploid set of chromosomes and a large nucleolus. Golgi bodies are usually present near the nucleus, and elongate mitochondria with flattened cristae are scattered throughout the cell. In freshwater species, a pair of contractile vacuoles near the flagella's insertion into the cell function in osmoregulation, secretion, uptake by endocytosis, and membrane recycling (Domozych and Nimmons 1992).

Since *Chlamydomonas* exists as a solitary cell, every cell division is associated with reproduction. Under favorable conditions, *Chlamydomonas* undergoes asexual reproduction, producing two to sixteen daughter cells by a series of mitotic and cell divisions (fig. 3.13*a*). Prior to the first division, flagella are discarded. Before each division, the DNA replicates, and the nucleus undergoes mitosis. Many of the organelles, including the chloroplast, also divide. The spindle microtubules reorganize into a phycoplast before the cell divides by furrowing. Mitosis and cell division may be repeated several times. Each of the resulting daughter cells has a haploid nucleus genetically identical to the parent cell. While still contained within the parental wall, each new cell forms its own wall. Then, the enzyme lysin (or "hatching enzyme") breaks down the parental wall, releasing the daughter cells (fig. 3.13*b*). Daughter cells develop flagella and enlarge. If conditions remain favorable for growth, the daughter cells will later divide asexually themselves.

Under moist conditions but in the absence of a fluid medium, *Chlamydomonas* forms a **palmelloid stage** (fig. 3.13*c*). Cells growing on agar are in this condition. A palmelloid stage is initiated when daughter cells are released. Instead of developing flagella and swimming off, the daughter cells remain together in a mucilaginous sheath derived from the partial breakdown of the parental wall. Palmelloid cells retain features of flagellated cells, such as basal bodies and eyespots. Cells may secrete more mucilage and continue to divide, adding more cells to the aggregation. When flooded, the mucilage dissolves, and the cells develop flagella and become motile.

Sexual reproduction is a relatively uncommon event in natural populations. Normally, stress induces sexual reproduction when conditions are no longer favor-

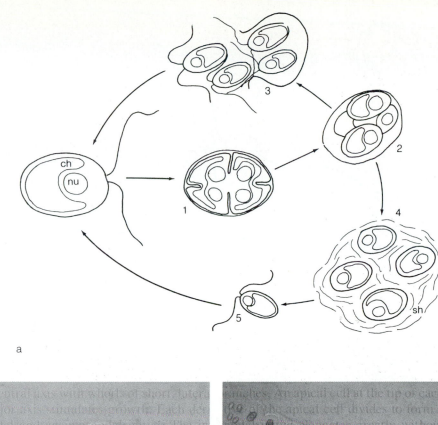

Figure 3.13
Asexual reproduction in *Chlamydomonas*.
(*a*) (1) Division of parent cell;
(2) formation of wall around daughter cells (four daughter cells are shown, but the number may vary);
(3) release of daughter cells by breakdown of the parent wall; (4) formation of a palmelloid stage in the absence of water; (5) cells developing flagella when flooded. ch=chloroplast; nu=nucleus; sh=mucilaginous sheath.
(*b*) Daughter cells formed by asexual reproduction.
(*c*) Palmelloid stage with daughter cells remaining in mucilaginous sheath.

a

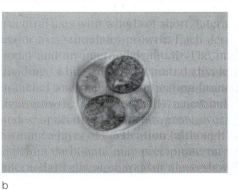

b   c

able for continued growth. In many strains of *Chlamydomonas,* nitrogen depletion in the medium stimulates gamete formation. A process similar to asexual reproduction forms gametes, which resemble daughter cells but behave differently (fig. 3.14*a*). They do not swim as actively as daughter cells and appear to have sticky flagella. In some strains of *Chlamydomonas,* only cells with complementary mating types can mate. A group of linked genes that are inherited together and thus act much like a single gene with different alleles determine mating type. Mating types are designated mt+ and mt–. The mating type genes control gamete recognition and fusion, organelle inheritance, and zygote development (Goodenough et al. 1995). Other strains of *Chlamydomonas* do not have distinct mating types, and cells from the same clone can fuse. Most species of *Chlamydomonas* show isogamy or anisogamy, but in some, oogamy occurs with a nonflagellated egg and flagellated sperm.

Figure 3.14
Sexual reproduction by
Chlamydomonas.
(*a*) (1) Gamete release;
(2) pairing of compatible
gametes; (3) fusion;
(4) planozygote;
(5) thick-walled dormant
zygote; (6) release of haploid
daughter cells after a period
of dormancy. Me=meiosis.
(*b*) Pairing of compatible
gametes. (*c*) Fusion of
gametes. (*d*) Planozygote
resulting from gametic fusion.

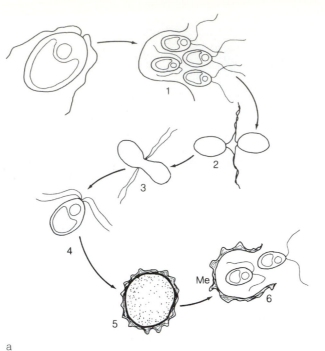

a

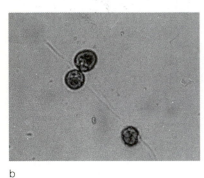

b

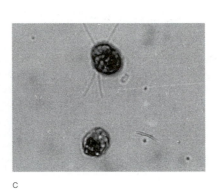

c

d

Initially, gametes accumulate in clumps, possibly due to the release of a chemical attractant. Within a clump, compatible gametes pair (fig. 3.14*b*). Initial contact between cell pairs is at their flagellar tips, which adhere. Glycoproteins called agglutinins on flagellar surfaces control recognition between compatible gametes. Upon contact, the flagella cease beating and join along their lengths, drawing the anterior ends of the cells closer together. Enzymes dissolve the wall around each gamete prior to fusion. A fertilization tube forms between the anterior ends of paired cells and provides initial cytoplasmic contact. Gradually, the cells merge, and their nuclei fuse (fig. 3.14*c*).

Besides controlling the recognition process, the agglutinins are also responsible for activating a sequence of cellular events associated with gametic fusion (Goodenough et al. 1995): (1) a rise in the concentration of cyclic adenosine monophosphate in the cells, (2) activation of "mating structures" consisting of small organelles associated with the cell membrane, (3) development of a microvil-

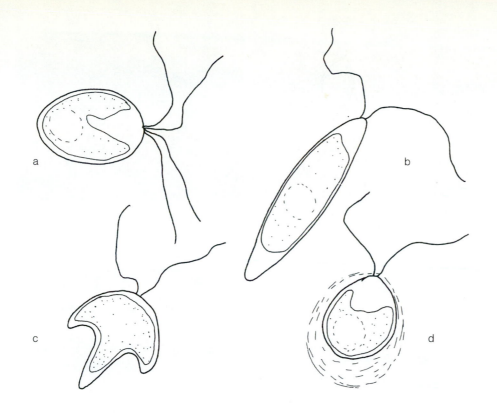

Figure 3.15
Unicellular flagellates.
(*a*) *Carteria* with four flagella.
(*b*) *Chlorogonium* with an
elongate cell.
(*c*) *Brachiomonas*.
(*d*) *Phacotus* with lorica
surrounding its cell.

lus with actin from each cell to form the fertilization tube, (4) fusion of gametes to form a binucleate cell, and (5) fusion of nuclei to form the diploid zygote nucleus.

Only the mt+ gamete contributes a chloroplast to the zygote, while the chloroplast of the mt– gamete is destroyed. Thus, unlike nuclear inheritance in which both parents contribute to the zygote's nucleus in a typical Mendelian fashion, chloroplast inheritance is uniparental (or maternal), with only one parental genome persisting in the zygote's chloroplast.

Following gametic fusion, the zygote with four flagella is called a **planozygote** (fig. 3.14*d*). Soon, the flagella are discarded, the contents of the zygote become dense, and a thick wall forms. The zygote enters a dormant state, which allows survival when conditions are unfavorable for vegetative growth.

When external conditions are favorable, a zygote germinates. This process involves breaking dormancy and releasing the protoplast by breakdown of the zygote wall. Meiosis occurs, and haploid vegetative cells form. Half of these cells are mt+, and half are mt–.

Several points about the life cycle of primitive algae such as *Chlamydomonas* are important. First, vegetative cells are haploid, and the zygote is the only diploid stage in the life history. Second, sexual reproduction is a response to deteriorating conditions and results in the formation of a dormant zygote. Third, all divisions are reproductive (vegetative growth is only by cell enlargement), with daughter cells escaping from the parent cell wall. In contrast, in vegetative divisions of multicellular algae, daughter cells usually retain and share the parent cell wall.

Chlamydomonas is only one of a number of flagellated green algae that differ in cell shape and in the presence or absence of particular organelles. Figure 3.15 and plate 2*a* show other examples. A **lorica** instead of a wall surrounds some

Figure 3.16
Colonial green algae.
(*a*) Colony held together by
direct cell contact
(Pascherina). (*b*) Colony held
together by mucilage
(*Gonium*). (*c*) Sixteen-celled
colony of *Gonium*.
(*d*) *Pandorina*. (*e*) *Eudorina*.
(*f*) *Volvox*, with daughter
colonies inside parent colony.
(*g*) Cells in a peripheral layer
in a colony of *Volvox*.

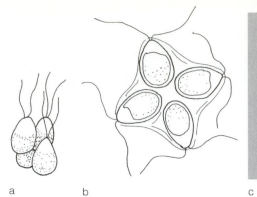

a   b

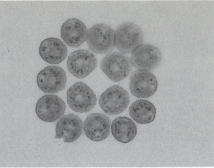

c

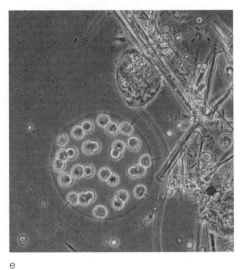

d   e

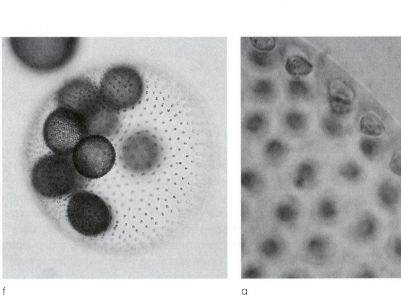

f   g

Table 3.4 *Volvocine Series*

	Number of Cells in Colony	Polarity to Colony	Special Reproductive Cells	Sexual Reproduction
Gonium	4–32	None	No	Isogamy
Pandorina	4–32	Yes	No	Isogamy
Eudorina	16–64	Yes	No	Anisogamy
Pleodorina	32–128	Yes	Yes	Anisogamy or oogamy
Volvox	500–50,000	Yes	Yes	Oogamy

unicellular green flagellates, such as *Phacotus* (fig. 3.15*d*). A lorica differs in form and composition from a wall. It is often larger than the cell and appears sculptured. Flagella may extend through pores in the lorica. The genus *Polytoma* has arisen from *Chlamydomonas* by loss of the photosynthetic system (Rumpf et al. 1996).

Colonies of Flagellated Cells (Order Volvocales)

Colonies result when daughter cells from asexual reproduction remain together. Direct contact (fig. 3.16*a*), or more commonly, a framework of mucilage (fig. 3.16*b*) may hold cells together. Colonies are flat or spherical, and are composed of cells similar to *Chlamydomonas*.

Some of the more common colonial green algae belong to the volvocine series, in which genera can be arranged to show a progression from small colonies without cell specialization to large colonies with clear cell specialization (table 3.4). These green algae share a common developmental feature. During asexual reproduction, a series of successive cell divisions determines the number of cells in the new colony, and then the colony inverts to orient its cells with their anterior ends outward. The initial colony prior to inversion is called a **plakea.** In platelike colonies, the inversion is only a slight change in colony's curvature to bring the anterior ends of the cells to the convex side, but in spherical colonies, the plakea must turn itself inside out. After inversion, cells no longer divide. A colony in which the number of cells is fixed during initial formation is called a **coenobium.** Other colonial green algae do not have plakeal stages, but their cells are oriented correctly from the start.

One of the simplest members of the volvocine series is a species of *Gonium* with four cells arranged in a flat, mucilaginous sheath (fig. 3.16*b*). The cells resemble *Chlamydomonas* and are functionally alike. Even in larger colonies of *Gonium,* with sixteen or thirty-two cells in a flat plate (fig. 3.16*c*), or in spherical colonies of *Pandorina* (fig. 3.16*d*), cells specialize very little. Colonies of *Eudorina* are larger, often with sixty-four cells, and have cells arranged around the periphery of a spherical framework of mucilage (fig. 3.16*e*). The colony has polarity, and anisogametes form in sexual reproduction.

Volvox is the largest and most advanced member of the volvocine series (fig. 3.16*f*). Its colonies with hundreds of cells may be visible without magnification. The biflagellated cells are arranged in a peripheral layer around the

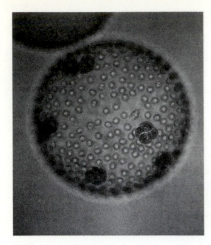
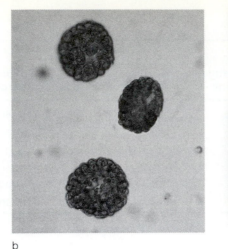
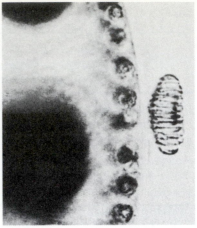

a   b   c

Figure 3.17
Reproductive stages of
Volvox. (*a*) Gonidia divided
several times to initiate new
colonies. (*b*) Plakea.
(*c*) Sperm packet. (*c* from
Hutt and Kochert 1971,
courtesy *Journal of
Phycology.*)

spherical colony, while watery mucilage fills the interior (fig. 3.16*g*). As a colony
swims, it maintains definite anterior and posterior ends. In a colony's posterior are
a few large cells (often sixteen) without flagella. Only these gonidia are capable of
asexual reproduction (fig. 3.17*a*). Each **gonidium** undergoes a series of cell divisions to produce a plakea. The plakea is a hollow sphere with the anterior ends of
its cells oriented toward the inside of the sphere (fig. 3.17*b*). A pore develops in the
plakea, and the colony turns itself inside out through the pore, with the intercellular
connections acting as hinges. Changes in cell shape cause the movement producing
the inversion. Once oriented properly, the cells secrete mucilage and separate from
each other (intercellular connections may disappear), as flagella form. No further
divisions occur after inversion. Newly formed daughter colonies remain in the
interior of the parent colony until it breaks apart and its vegetative cells die (see
fig. 3.16*f*).

A chemical inducer controls the formation of sexual colonies in *Volvox*
(fig. 3.18). In the presence of the inducer, new colonies formed asexually are
sexual, with either eggs or sperm instead of gonidia. The inducer is a glycoprotein
derived from a few spontaneously occurring sexual colonies. In a dense population
of colonies, the chemical inducer reaches effective levels.

In male colonies, sperm initials divide to form small packets of sperm. When
released, a packet swims to a female colony, where it attaches and comes apart
(fig. 3.17*c*). The eggs are large cells in the posterior of the female colony. Sperm
enter the female colony and fertilize the eggs. The resulting zygotes form thick
walls and enter into a dormant state. At germination, each zygote undergoes meiosis. Only one product of meiosis produces a new colony.

Palmelloid Aggregations (Order Tetrasporales)

The volvocalean algae form temporary palmelloid stages. In the tetrasporalean
algae, the palmelloid condition is the normal state. Cells are nonflagellated, and an
extensive mucilaginous sheath surrounds them. In some genera, the sheath has a
definite shape, while in others, its form is indefinite. Tetrasporalean cells are similar
to *Chlamydomonas,* possessing basal bodies and eyespots typical of flagellated

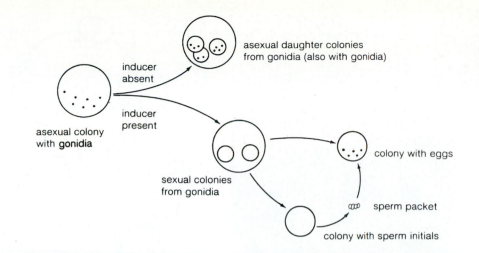

Figure 3.18
A chemical inducer stimulates sexual reproduction in *Volvox*. The inducer causes the formation of sexual colonies (with eggs or sperm initials) instead of asexual colonies (with gonidia).

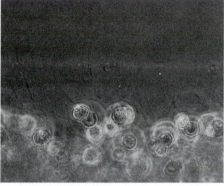

a

b

Figure 3.19
Tetraspora (*a*) Colony in calm water. (*b*) Cells with pseudoflagella.

cells. The order Tetrasporales traditionally has been recognized, even though its characteristics overlap with the Volvocales and Chlorococcales.

Two representatives showing the structural range of this order are *Gloeocystis* and *Tetraspora*. *Gloeocystis* forms small aggregations of cells surrounded by a layered sheath (see fig. 1.12*c*). In contrast, colonies of *Tetraspora* are much larger, forming tubes or sheets up to several centimeters long in calm water (fig. 3.19*a*). *Tetraspora* cells are arranged in groups of four within an extensive mucilaginous sheath. Each cell bears two long extensions, called **pseudoflagella** (or pseudocilia), which resemble flagella but are rigid and lack the 9+2 microtubular structure of flagella (fig. 3.19*b*). Their function is not known. Cells divide by a process similar to asexual reproduction in *Chlamydomonas*. Four daughter cells usually form at a time.

Nonflagellated Cells, Aggregations, and Colonies (Order Chlorococcales)

The chlorococcalean algae are a large group commonly found in the freshwater plankton and terrestrial habitats. They occur as solitary cells, aggregations of cells, and well-defined colonies. Vegetative cells lack flagella and structures associated with flagellated cells, such as eyespots, contractile vacuoles, and basal bodies.

Figure 3.20
Chlorococcum. (*a*) Spherical
vegetative cells and elongate
zoospores. (*b*) Life cycle.
(1) Parent cell undergoes a
series of cell divisions; in
asexual reproduction, either
(2) aplanospores
(autospores) or (3) zoospores
are produced; (4) zoospores
lose their flagella and round
up; (5) sexual reproduction
occurs between isogametes;
(6) zygote is produced;
(7) meiosis (Me) occurs at
the time of germination, and
either zoospores or
aplanospores are produced.

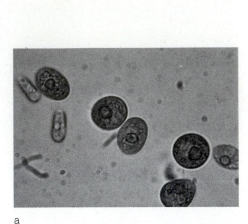

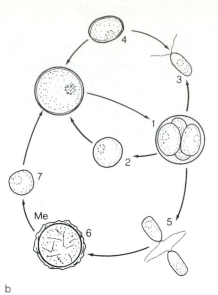

a

b

Arrangements of basal bodies among species presently classified in the Chlorococcales vary. Some show the typical clockwise orientation of other chlorophycean algae, but others have an opposite or counterclockwise orientation (see fig. 3.1). Comparison of nucleotide sequences of genes for ribosomal RNA also suggests that the Chlorococcales is a polyphyletic group (Wilcox et al. 1992). Sometimes, species with a counterclockwise orientation are placed in a separate class—Pleurastrophyceae.

Chlorococcum is a relatively unspecialized member of this order. Its spherical cells may be solitary or loosely aggregated (fig. 3.20*a*). Each cell has a large cup-shaped chloroplast with a distinct pyrenoid and a haploid nucleus. Asexual reproduction involves formation of either flagellated or nonflagellated spores by a series of mitotic and cell divisions (fig. 3.20*b*). Breakdown of the parental wall releases the spores. Normally, in a fluid medium, biflagellated zoospores, with an elongate shape, are produced. They swim for awhile before losing their flagella and becoming spherical. **Aplanospores** are nonflagellated spores usually produced in terrestrial habitats. Sexual reproduction is similar to *Chlamydomonas.*

Starting with *Chlorococcum* as a relatively unspecialized genus, we can identify the following trends among chlorococcalean algae: (1) association of cells to form colonies, (2) development of multinucleate cells, and (3) loss of flagellated stages and sexuality in some planktonic species.

Association of Cells to Form Colonies

Associations of cells among chlorococcalean algae range from loose aggregations to well-defined colonies with a definite number and arrangement of cells. In *Oocystis* and *Dictyosphaerium,* daughter cells (aplanospores) from asexual reproduction remain together. A vesicle derived from the wall of the parent cell encloses cells of *Oocystis* (figs. 3.21*a*, 3.22*a*). In *Dictyosphaerium,* cells are connected by strands derived from the walls of the parent cells and are surrounded by mucilage (figs. 3.21*b*, 3.22*b*).

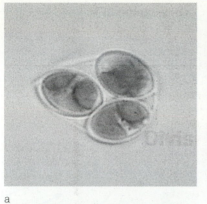

a

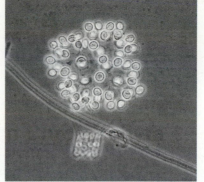

b

Figure 3.21
Association of cells into aggregations and colonies. (*a*) *Oocystis* with cells in a vesicle. (*b*) *Dictyosphaerium* with cells joined by wall strands and surrounded by mucilage. (*c, d*) *Scenedesmus* with laterally joined cells. (*e*) Daughter colony of *Scenedesmus* being released. (*f*) *Pediastrum*. (*g*) Daughter colony of *Pediastrum* being released in a vesicle (*arrow*).

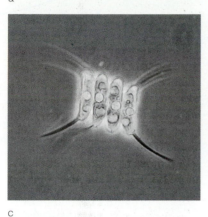

c

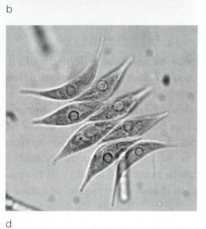

d

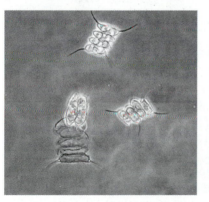

e

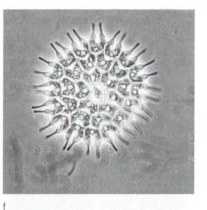

f

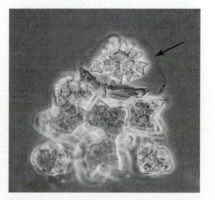

g

Figure 3.22
Reproduction in the
Chlorococcales. (*a*) *Oocystis.*
(*b*) *Dictyosphaerium.*
(*c*) *Scenedesmus.*
(*d*) *Pediastrum.* Polyedes are
nonflagellated cells. Sexual
reproduction as in
Chlorococcum.

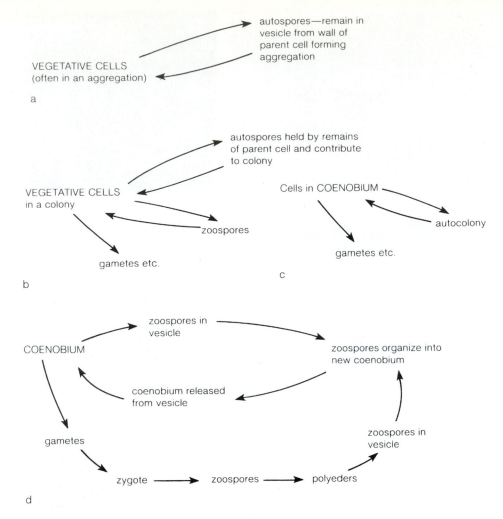

In other colonies, a series of divisions at the time a colony forms inside its parent cell determines the number of cells, and no further divisions occur after the colony is released. Such a colony is a **coenobium.** Cells in a coenobium usually have a distinctive arrangement. Coenobia of *Scenedesmus* often have four cells joined laterally, but unicells and colonies with two, eight, and sixteen cells also occur, depending on environmental conditions (figs. 3.21*c–e*, 3.22*c*). Species differ in cell shape and in whether spines are present. *Pediastrum* forms a larger coenobium with its cells arranged in a circular plate (plate 2b; fig. 3.21*f*). Reproduction by any cell in a colony leads to formation and release of zoospores inside a vesicle (fig. 3.21*g*). The zoospores organize into a new coenobium before the vesicle opens (fig. 3.22*d*).

Formation of Multinucleate Cells

In some chlorococcalean algae, cell division does not follow mitosis, which results in multinucleate cells called **coenocytes.** As they enlarge, they may continue to increase their number of nuclei. Most coenocytes have a central vacuole, and the surrounding cytoplasm is a relatively thin layer immediately inside the wall. *Protosiphon* is an example of a simple coenocyte (fig. 3.23*a*). Its cells normally have a

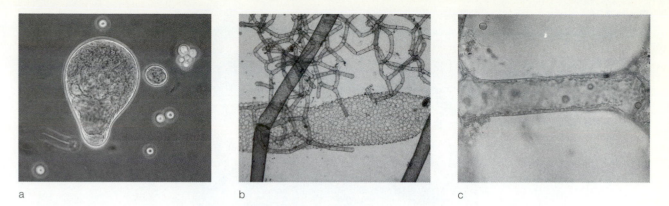

a   b   c

Figure 3.23
Multinucleate greens.
(*a*) *Protosiphon.*
(*b*) *Hydrodictyon* or "Water
Net," showing variation in
colony sizes. (*c*) Netlike
chloroplast and central
vacuole in a cell of
Hydrodictyon.

rhizoidal extension for attachment. A central vacuole fills most of each cell. In the cytoplasmic layer between the vacuole and wall are numerous nuclei and a reticulate (netlike) chloroplast. *Hydrodictyon* ("Water Net") forms a tubular coenobium of elongate cells arranged to form a net (fig. 3.23*b*, *c*). Each of the cells is a coenocyte.

Loss of Flagellated Stages

Many common planktonic chlorococcaleans do not form flagellated stages, including gametes for sexual reproduction. *Oocystis* is a representative of a family that reproduces only by aplanospores. The aplanospores are often miniature copies of the parent cell and are called **autospores** (fig. 3.24). Since resistant zygotes do not form, a small population of cells presumably persists through unfavorable conditions and initiates population growth when conditions are favorable.

In natural environments, *Scenedesmus* is only known to reproduce asexually by forming autocolonies (coenobia resembling the parent colony). Laboratory conditions, however, have induced flagellated gametes to form in some strains of *Scenedesmus,* indicating that the capacity for sexual reproduction has not been lost entirely (Cain and Trainor 1976).

Filamentous Green Algae (Orders Microsporales and Chaetophorales)

Filamentous algae occur in several orders of the Chlorophyceae. In filaments, growth occurs by cell enlargement and division. In vegetative divisions, mitosis is followed by reorganization of the mitotic spindle into a phycoplast. Either a cell plate or a furrow develops at the parent cell equator to form a new cross wall between daughter cells. Typically in vegetative divisions, the daughter cells retain the parent wall. In contrast, the parent wall is normally discarded after a reproductive division.

The simplest filaments are unbranched members of the Microsporales with uninucleate cells, a single large chloroplast, and no intercellular connections. More advanced heterotrichous filaments are placed in the Chaetophorales. The order Oedogoniales contains several filamentous algae with specialized features.

Filaments may have originated by retention of spores in a linear series within the parent cell wall (Pickett-Heaps 1975). *Microspora* shows a primitive condition. Instead of a continuous wall, each cell shares with neighboring cells overlapping H-shaped wall pieces (fig. 3.25). During cell division, a new H-piece is

Figure 3.24
Daughter cells (autospores) about to be released by breakdown of the parent cell wall in *Chlorella*, a relative of *Oocystis*. (Courtesy Barchi and Chapman.)

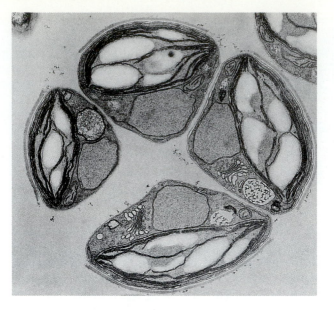

Figure 3.25
Microspora. (*a*) Part of the unbranched filament with an H-shaped wall piece visible at end. (*b*) Interpolation of new wall piece during cell divison.

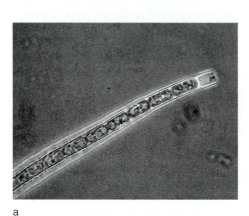

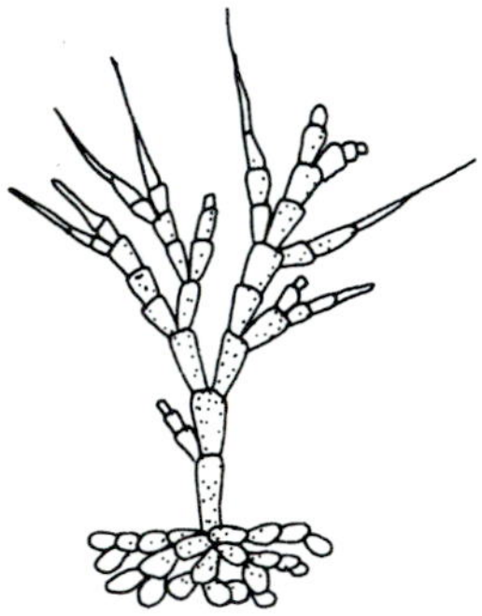

a

b

Figure 3.26
Heterotrichous filament showing differentiation into basal branches growing over substrate and openly branching erect system.

interpolated between existing H-pieces of the parent cell. The H-pieces may represent an intermediate condition between asexual reproduction and the formation of cross walls, as seen in the Chaetophorales.

Microspora grows attached by a holdfast or unattached as a mass of floating filaments. Its cells contain a single nucleus, central vacuole, and parietal chloroplast without pyrenoids. It reproduces by fragmentation, when part of the filament breaks off, or by zoospore formation.

The chaetophoralean green algae have **heterotrichy**—differentiation into a system of prostrate branches and a system of erect branches (fig. 3.26). The prostrate branches grow over the substrate and are often compact. The erect filaments are free from the substrate and have a more openly branching pattern. The erect

Figure 3.27
Stigeoclonium. (*a*) Branches
of erect system with terminal
hairs. (*b*) Basal branches.

a   b

filaments are the principal sites of photosynthesis and often of reproduction. In many species, formation of hairs at the end of branches is a response to nutrient limitation (Gibson and Whitton 1987a, 1987b). The exact function of hairs is not known, but they may increase the surface area for nutrient uptake and secrete enzymes.

The common freshwater alga *Stigeoclonium* has a relatively unspecialized heterotrichous condition (fig. 3.27). Cells in the basal branches are normally shorter than cells in the erect system. *Stigeoclonium* shows considerable morphologic variation, depending on environmental conditions. When nutrient levels are low in the surrounding water, the erect system is more branched, and branches terminate in long, tapering hairs of nonpigmented cells. Individual cells of *Stigeoclonium* contain a single nucleus, a central vacuole, and a band-shaped chloroplast with pyrenoids in the surrounding cytoplasmic layer. Intercellular connections (plasmodesmata) extend through the cross walls between adjacent cells. Filaments enlarge by vegetative growth involving mitosis and cell plate formation with a phycoplast (see fig. 3.10). Reproduction is usually by release of quadriflagellated zoospores. Sexual reproduction is poorly known (see Simons and van Beem 1987).

Other heterotrichous genera show greater specialization. In *Draparnaldia,* the erect system differentiates into primary and secondary branches (fig. 3.28*a*, *b*). Cells in primary branches have reduced chloroplasts, while chloroplasts in the more elongate cells of secondary branches extend most of the cell length. In *Fritschiella,* the erect system of uniseriate filaments arises from basal regions, which may become parenchymatous (fig. 3.28*c*). Some heterotrichous algae show reduced morphology. *Apatococcus* commonly grows on tree bark (fig. 3.28*d*). Its thallus is usually a compact cluster of compressed spherical cells and only rarely shows an obvious filamentous structure.

Specialized Filaments (Order Oedogoniales)

The three genera in the Oedogoniales exhibit a number of specialized features. They have a unique form of new cell formation, flagellated stages with a ring of flagella **(stephanokontous condition)**, and an advanced oogamous form of sexual reproduction.

Oedogonium is a common green alga in relatively calm freshwater habitats. Its thallus is an unbranched filament (fig. 3.29*a*). A modified holdfast cell usually attaches the filament, but filaments may become free-floating in a dense mat (see

Figure 3.28
Heterotrichous filaments.
(*a, b*) *Draparnaldia.*
(*c*) *Fritschiella*
(parenchymatous regions at
base not visible).
(*d*) *Apatococcus.*

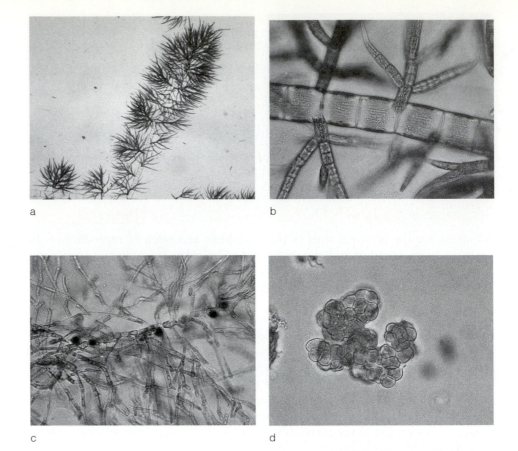

a

b

c

d

fig. 8.2). Sometimes, cells have distinct striations or caps resulting from vegetative divisions (fig. 3.29*b, c).* Each cell is uninucleate with a large, central vacuole (fig. 3.29*d).* Adjacent to the cell wall is a netlike (reticulate) chloroplast (in contrast to the band-shaped chloroplast of the chaetophoralean algae).

During vegetative division, *Oedogonium* shows one of its characteristic features (fig. 3.29*e*). Following mitosis, the wall at a cell's upper end thickens internally with a distinct ring of noncellulosic polysaccharides (fig. 3.29*f*). The parent wall adjacent to this ring splits, and the ring expands as the cell lengthens due to increased turgor pressure (fig. 3.29*g*). The expanded ring becomes the outer wall layer of the upper daughter cell, and cellulose is deposited on the inside of the expanded ring. Meanwhile, a cell plate develops into a cross wall to separate daughter cells. A flared edge often remains where the original wall split, producing a distinct "cap" on the cell. Cells involved with several divisions may have a series of caps that appear as striations at their distal ends (fig. 3.29*b*).

Oedogonium reproduces both asexually and sexually. Asexual reproduction produces one zoospore per cell. The zoospore is unusual in having an anterior ring of flagella. In sexual reproduction, specialized reproductive cells form. Antheridial cells are much shorter than vegetative cells and often occur in a series (fig. 3.30*a*). One or two sperm form in each antheridial cell. Sperm have a ring of flagella and are chemically attracted to oogonia. Oogonia are enlarged cells with a single egg (fig. 3.30*b*). A sperm enters through a pore in the wall of an oogonium to fertilize

a

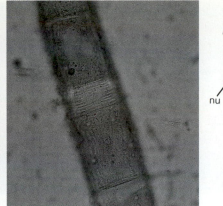

b

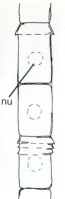

c

Figure 3.29
Oedogonium.
(*a*) Unbranched filament.
(*b, c*) Caps on cells.
nu=nucleus. (*d*) Vegetative
cell with netlike chloroplast
and central vacuole. The
nucleus and pyrenoids are
visible. (*e*) Cell division.
(1) Migration of nucleus (nu)
toward distal end of cell;
(2) ring (r) and cell plate;
(3) expanding ring (xr)
(*f*) (1) Ring; (2) cell
beginning to expand.
(*g*) Expanding cell.

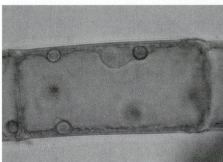

d

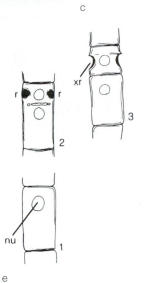

e

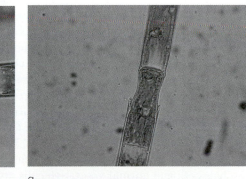

f

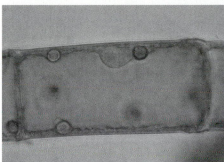

g

the egg (fig. 3.30*c*). The resulting zygote becomes dormant. Meiosis occurs at germination.

Sexual reproduction, as described, involves the formation of reproductive structures on normal filaments (**macrandry**) (fig. 3.31*a*). Some species have an additional dwarf-male filament (**nannandry**) (fig. 3.31*b*). Androsporangia form in

Figure 3.30
Sexual stages of
Oedogonium. (*a*) Antheridial
cells (*arrows*). (*b*) Oogonium
with egg. (*c*) Sperm next to
oogonium. (*d*) Dwarf male
filaments (*arrows*) in a
nannandrous species.
(*c* from Hoffman 1973,
courtesy *Journal of
Phycology. d* courtesy
P. Biebel.)

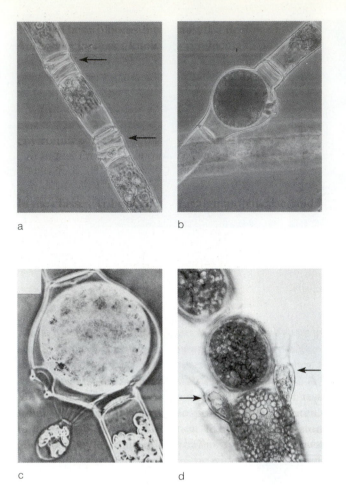

a   b

c   d

place of antheridia on normal filaments. They resemble antheridia but produce
flagellated cells called **androspores.** Androspores are attracted to oogonia, where
they attach and form reduced filaments, consisting of a holdfast and several antheridial cells from which sperm are derived (fig. 3.30*d*). A series of chemicals is
involved in attracting androspores, controlling development of male filaments and
oogonia, and attracting sperm to eggs (Rawitscher-Kunkel and Machlis 1962).

Bulbochaete, a branching relative of *Oedogonium,* has a distinctive hair cell or
seta at the end of each branch (fig. 3.32). Each unpigmented hair cell has a swollen
base and a long, tapering extension. When a branch is initiated, the first cell formed
is a hair cell, and normal vegetative cells are added beneath it. The large number
of Golgi bodies and extensive endoplasmic reticulum in hair cells suggest a secre-
tory function.

Chlorophycean Ecology and Commercial Uses

The chlorophycean green algae grow in a wide range of freshwater habitats as
members of either planktonic or benthic communities. Occasionally, they are also

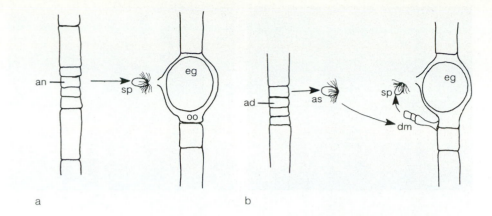

Figure 3.31
(*a*) Macrandrous and
(*b*) nannandrous
reproduction in *Oedogonium*.
an=antheridium;
ad=androsporangium;
as=androspore;
dm=dwarf-male filament;
eg=egg; oo=oogonium;
sp=sperm.

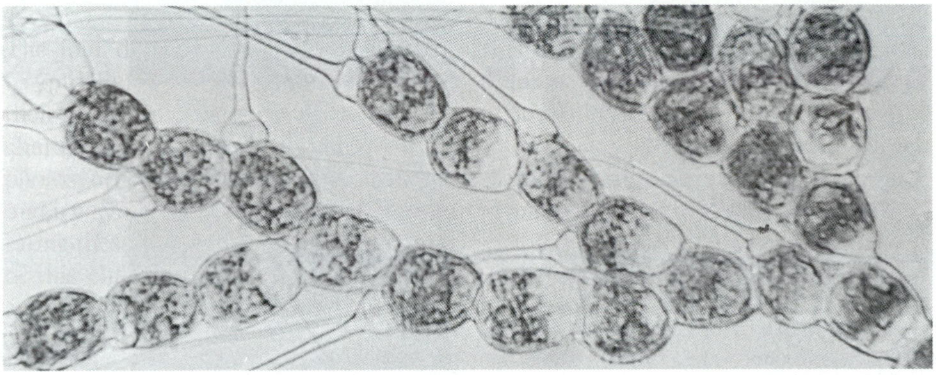

Figure 3.32
Bulbochaete, a branching
filament with terminal setal
(hair) cells. (Courtesy
P. Biebel.)

common in estuarine waters. Other species are adapted to live in symbiotic associations or in moist terrestrial habitats.

The volvocalean greens and many chlorococcalean greens are planktonic. They are often abundant in eutrophic lakes and ponds during summer, but flagellated species sometimes bloom under ice covers or in meltwater from snow. Both groups must stay near the surface, where light is sufficient for photosynthesis, but show contrasting adaptations. The volvocalean algae are flagellated and attracted to light (positive phototaxis) and thus swim upward toward the surface. The chlorococcalean algae are nonflagellated and show form modifications that increase their surface areas and create drag with the water, such as small size, elongate shapes (fig. 3.33*a*), spines (fig. 3.33*b*), and cell arrangements in platelike (fig. 3.33*c*; plate 2*b*) or stellate colonies (fig. 3.33*d*).

Filamentous chlorophyceans grow on submerged surfaces in both standing and flowing waters. Normally, a holdfast attaches them at their base, and flexible filaments extend into the water. Under calm conditions, filaments may detach or release fragments that continue to grow as floating masses of entangled filaments, often buoyed by oxygen bubbles trapped among the filaments (see fig. 8.2).

In contrast to planktonic algae, terrestrial algae reduce their surface areas to minimize evaporation while exposed to air. Since a spherical shape has the smallest surface area for a given volume, many terrestrial greens, such as *Chlorococcum*

Figure 3.33
Planktonic genera in
Chlorococcales.
(*a*) *Ankistrodesmus.*
(*b*) *Micractinium.*
(*c*) *Crucigenia.*
(*d*) *Actinastrum.*

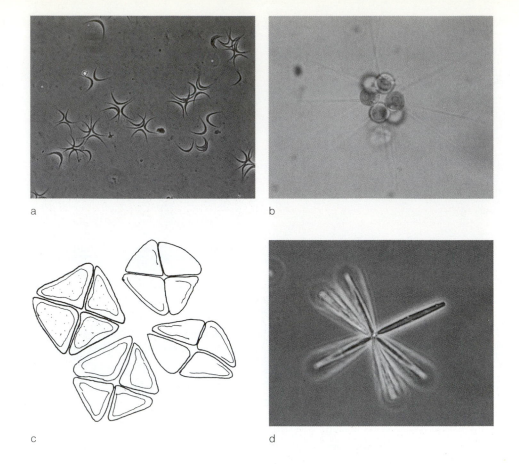

a

b

c

d

(see fig. 3.20*a*), are "little round green things" (or "LRGTs"), a designation that reflects the difficulty in distinguishing species and genera with spherical shapes.

Some green algae enter into symbiotic associations with other organisms. Green algae are common components of lichens, especially *Trebouxia.* Some species of *Hydra, Paramecium,* and other freshwater invertebrates contain *Chlorella* in their cells. Such green symbionts are called zoochlorellae. A few algae are parasites. *Prototheca,* a colorless relative of *Chlorella* that normally grows on soil or trees, may also infect the skin of animals and is reported to cause lesions on human skin.

Most algae are active for only part of the year. Although some of the larger algae may persist even when not actively growing, microscopic algae normally disappear when conditions are unfavorable for vegetative growth. During periods of inactivity, they may survive as thick-walled, dormant cells. In many members of the Chlorophyceae, sexual reproduction is a response to stressful conditions, such as nutrient deficiency, and leads to the formation of a dormant zygote that survives until external conditions are again favorable for vegetative growth. In zygotes and other resistant cells, photosynthesis and respiration are reduced, and chlorophyll levels decrease.

A few microscopic green algae have commercial value. One is *Dunaliella,* with biflagellated cells lacking a surrounding wall (review by Avron and Ben-Amotz

Table 3.5   Orders of the Class Ulvophyceae

Order Ulotrichales—filaments with uninucleate cells

Order Ulvales—parenchymatous thalli with uninucleate cells

Order Cladophorales—multicellular algae with multinucleate cells; filaments or saccate thalli

Order Caulerpales—single coenocytic cell composing thallus; walls of cellulose, mannans, or xylans; siphonaxanthin present as an accessory pigment

Order Dasycladales—thallus a single cell with radial symmetry (whorls of branches); gametes formed in a cyst; walls of mannans

Order Trentepohliales—terrestrial filaments with differentated reproductive structures

---

1992). *Dunaliella* commonly grows in hypersaline lakes, where it tolerates high salt concentrations, high levels of light, and acidic pH sometimes below 4. When *Dunaliella* is abundant, the high carotenoid content of its cells produces a pink discoloration. High irradiance stimulates the production of β-carotene, which accumulates as droplets in the chloroplast, probably to protect against high irradiance. *Dunaliella* is a commercial source of β-carotene used for food coloring and in pharmaceuticals. *Dunaliella* may also produce large quantities of glycerol for osmoregulation.

Other green algae with possible potential use include *Haematococcus* (plate 2*a*) as a source of the carotenoid astaxanthin. *Chlorella* is commercially grown in Asia as a health food with proclaimed "cure-all" properties, but its sale in the United States is limited. Rapidly growing algae, such as *Chlorella* and *Scenedesmus,* have potential food value, especially as protein supplements. However, production costs, taste, and appearance limit their use. These and other algae may have potential for use in life-support systems during long space expeditions to produce oxygen, take up carbon dioxide, and process wastes.

Class Ulvophyceae

Most ulvophytes are macroalgae—large enough to be easily visible to the unaided eye during at least one phase of their life cycle. Some are multicellular filaments or parenchymatous sheets, while others have a thallus that is a single, large coenocyte. Ulvophytes normally grow attached to submerged substrates in shallow marine environments; a few occur in freshwaters. Their flagellated stages are characterized by basal bodies in a counterclockwise orientation. (see fig. 3.1*c*) and cruciately arranged microtubular roots. A persistent mitotic spindle occurs during mitosis, but neither a phycoplast or phragmoplast is present during cell division. Life cycles are either diplohaplontic (sporic meiosis, see fig. 3.6*b*) or diplontic (gametic meiosis, see fig. 3.6*c*). In cycles with an alternation of generations, both gametophytes and sporophytes may have similar forms (isomorphic) or distinctly different forms (heteromorphic). Zygotes normally germinate immediately without becoming dormant. Some ulvophytes have siphonaxanthin as an accessory photosynthetic pigment. Table 3.5 lists the orders of the Ulvophyceae.

Figure 3.34
Ulothrix. (*a*) Unbranched
filaments with band-shaped
chloroplast in each cell.
(*b*) Life cycle with (1) asexual
reproduction by
quadriflagellated zoospores;
(2) sexual reproduction by
biflagellated gametes;
(3) gametic fusion;
(4) planozygote; (5) diploid
Codiolum stage;
(6) zoospores released after
meiosis (Me). (*c*) Zoospores.
(*d*) Codiolum stage.

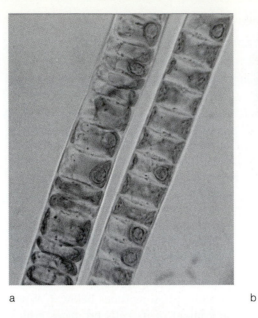

a

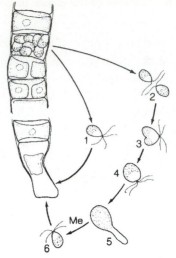

b

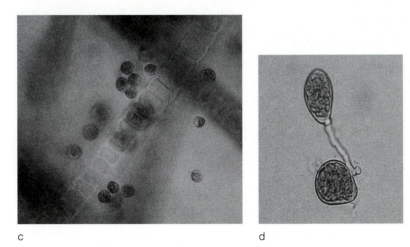

c

d

Ulvophyte Diversity

Filamentous Green Algae (Order Ulotrichales)

Ulothrix has a simple filamentous thallus. It grows in both freshwater and marine environments, often during the winter. Cells in an unbranched filament of *Ulothrix* are arranged in a single series (uniseriate) and are alike in structure and function except for a modified holdfast cell at the base for attachment to the substrate (fig. 3.34*a*). Each cell has a haploid nucleus, one or more vacuoles in its center, and a band-shaped chloroplast with pyrenoids extending around the periphery. Mitosis followed by cell division adds new cells to a filament.

Ulothrix shows three methods of reproduction: vegetative propagation, asexual reproduction, and sexual reproduction (fig. 3.34*b*). Vegetative propagation

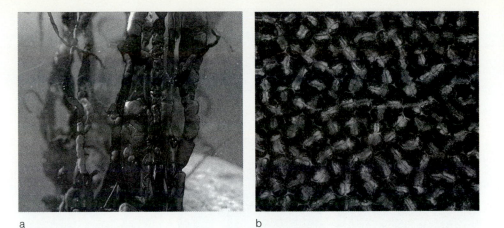

Figure 3.35
Ulvalean algae.
(*a*) *Enteromorpha* with a
tubular thallus. (*b*) Cells in
surface view of a blade of
Ulva.

a   b

occurs by fragmentation, when part of a filament breaks off and grows into a separate thallus. In asexual reproduction, one or more cells repeatedly divide to form zoospores (fig. 3.34*c*) or, less commonly, aplanospores. Quadriflagellated zoospores give rise directly to new filaments. Sexual reproduction involves release of biflagellated gametes, and their fusion to form a zygote, which normally remains green and metabolically active as it enlarges and develops a rhizoidal extension for attachment. This unicellular diploid phase is a **codiolum stage** (originally described as the genus *Codiolum*) (fig. 3.34*d*). Later, meiosis produces haploid zoospores, which develop into new filaments. The codiolum stage in the life cycle may represent an initial step toward an alternation of generations, as other ulvophytes with multicellular diploid phases show.

Parenchymatous Green Algae (Order Ulvales)

The ulvalean algae form simple parenchymatous thalli with their cells arranged in one or two layers to form flat sheets or tubes. Compared to advanced brown algae and vascular plants, construction is relatively simple. The ulvalean algae do not differentiate into tissue layers or show much specialization among cells, all of which contain photosynthetic pigments and are capable of reproduction. Nor do they form distinct reproductive structures or have intercellular connections (plasmodesmata). Hence, in these respects, they are little more than complex colonies of cells.

Enteromorpha and *Ulva* ("Sea Lettuce") are two widespread members of the order Ulvales. The thallus of *Enteromorpha* is a hollow tube of cells arranged in a single layer (fig. 3.35*a*). Oxygen bubbles from photosynthesis may accumulate in the tube's interior. Each cell has an isodiametric shape and contains a large chloroplast. Cells near the base produce rhizoidal outgrowths that form the holdfast (fig. 3.36*a*). *Ulva* is similar to *Enteromorpha* except its thallus is "ironed out" into a blade with two layers of cells (fig. 3.35*b*; plate 2*c*). In culture, *Ulva* has been induced to form tubular thalli resembling *Enteromorpha* (Bonneau 1977; Provasoli and Pintner 1980). *Monostroma* and *Ulvaria* resemble *Ulva* but have a single layer of cells composing their blades.

Ulvalean algae have an alternation of generations (see fig. 3.6*b*). In *Ulva, Ulvaria,* and *Enteromorpha,* haploid and diploid phases are multicellular and

Figure 3.36
Ulvalean algae. (*a*) Holdfast
formed by rhizoids from cells
near base (*Ulva*).
(*b*) Zoospore formation
(*Enteromorpha*). (*c*) Germling
(*Ulva*).

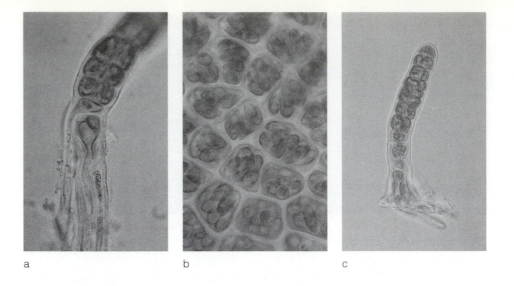

a   b   c

isomorphic. In *Monostroma,* a diploid codiolum phase alternates with a parenchy-matous haploid phase, as in *Ulothrix*. In a typical life cycle, cells in the diploid thallus undergo meiosis, leading to the formation of haploid zoospores (fig. 3.36*b*). The zoospores give rise to the haploid thallus, which later forms flagellated gametes. The zygote formed from gametic fusion develops, without becoming dormant, into the diploid phase. During development of either multicellular haploid or diploid thalli, there is an initial filamentous phase resembling *Ulothrix* (fig. 3.36*c*). Normal development in some cases depends on unidentified chemials from bacteria, red algae, or brown algae (Tatewaki, Provasoli, and Pintner 1983; Nakanishi et al. 1996).

Siphonous Green Algae (Orders Cladophorales, Caulerpales, Dasycladales)

The coenocytic condition in the orders Cladophorales, Caulerpales, and Dasy-cladales results from mitosis without accompanying cell divisions. Most siphonous algae are macroalgae, having ovoid, club-shaped, tubular, or filamentous forms. Life cycle and ultrastructural features suggest that the siphonous algae are derived from ulotrichalean ancestors (Sears and Brawley 1982). Siphonous algae are divided into three orders. The order Cladophorales contains multicellular algae, in which the cellular units are multinucleate. In the Caulerpales, the thallus is a single, multinucleate cell. Only during the formation of reproductive structures or in response to injury do membranes partition the thallus. Algae in the order Dasy-cladales have unicellular thalli and a number of specialized features.

Multicellular Genera

Cladophora is widespread in the oceans, where most of its species occur; a few species grow in freshwater. *Cladophora* is a branching filament attached by a holdfast composed of rhizoidal branches from basal cells (fig. 3.37*a, b*). A thick wall that lacks a mucilaginous sheath and may become overgrown with epiphytes surrounds each elongate cell (see fig. 8.1). A large vacuole fills most of the cell (fig. 3.37*c*). The cytoplasmic layer surrounding the vacuole contains a reticulate

a

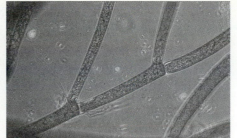

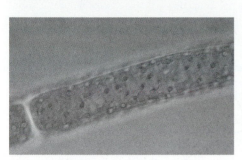

c

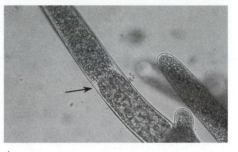

b

d

Figure 3.37
Cladophora. (*a*) Branching filament. (*b*) A branch typically arises from the upper end of its supporting cell, and a cross wall occurs at the base of the branch. (*c*) Reticulate chloroplast in cytoplasmic layer between wall and central vacuole. (*d*) Apical cell dividing by furrowing (*arrow*).

(netlike) chloroplast and numerous nuclei, which divide synchronously. Cell division, which does not accompany mitosis, occurs by ingrowth of a new cross wall (fig. 3.37*d*).

Fragmentation is probably an important means of reproduction in *Cladophora* and its relatives. Some species commonly grow entangled with larger algae rather than attached by a holdfast. New thalli are initiated from fragments that break off and grow if they become caught on suitable hosts (see *Rhizoclonium* in fig. 3.39*a*) or form floating mats.

Most species of *Cladophora* have an alternation of isomorphic generations (fig. 3.38*a*). Terminal cells function as gametangia or sporangia. Sporangial cells of sporophytes (diploid phase) release zoospores. The zoospores are haploid and give rise to gametophytes. Isogametes from gametophytes fuse to form zygotes, which develop (without dormancy) into sporophytes. Some species, such as the common freshwater species *Cladophora glomerata,* lack a haploid thallus. Filaments are diploid and reproduce by forming diploid zoospores or haploid gametes (fig. 3.38*b*).

Other cladophoralean algae may have arisen from an ancestor resembling *Cladophora* by: (1) reduction in branching as seen in the unbranched condition of *Rhizoclonium* (fig. 3.39*a*) and *Chaetomorpha* (fig. 3.39*b*); (2) branching in one plane, as in the blade of *Anadyomene,* composed of closely appressed, club-shaped cells (fig. 3.39*c, d*); or (3) cell enlargement to form balloonlike coenocytes (fig. 3.39*e*).

Unicellular Genera

Among the Caulerpales, *Derbesia* has a simple morphology and an alternation of generations. The sporophyte is a branching tube that resembles a filament, but

Figure 3.38
Life cycle of *Cladophora*.
(*a*) Alternation of isomorphic generations. (1) Sporophyte forms haploid zoospores; (2) gametophyte forms isogametes; (3) fusion forms a zygote, which develops without a period of dormancy into sporophyte. Me=meiosis in sporangium.
(*b*) Modification of life cycle by elimination of haploid phase. (1) Asexual reproduction by diploid zoospores; (2) sexual reproduction by formation of gametes; (3) diploid thallus develops from zygote. Me=meiosis in gametangium.

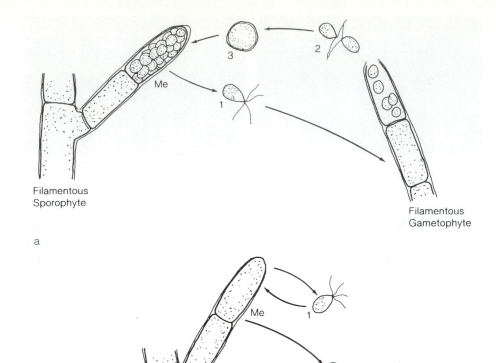

Filamentous
Sporophyte

Me

Filamentous
Gametophyte

a

Me

Diploid
Filament

b

a

b

c

Figure 3.39
Cladophoralean genera.
(*a*) *Rhizoclonium* with unbranched filaments, growing entangled with blades of a brown alga.
(*b*) Unbranched filaments of *Chaetomorpha*.
(*c, d*) *Anadyomene* with inflated cells closely appressed to form a blade.
(*e*) Inflated thallus of *Valonia*.
(*e* courtesy J. F. Storr.)

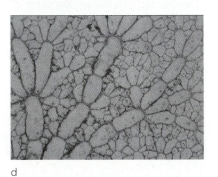

d

e

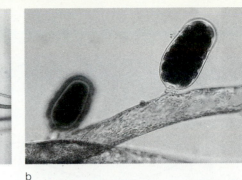

a

b

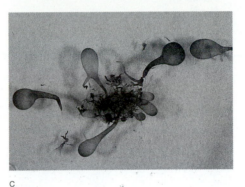

c

Figure 3.40
Derbesia. (*a*) Diploid thallus
of branching tubes without
cross walls. (*b*) Sporangium
separated from rest of thallus
by wall at its base.
(*c*) Haploid *Halicystis* phase.

cross walls do not partition it into cellular units (fig. 3.40*a*). A large, central vacuole is continuous throughout the thallus, and a thin cytoplasmic layer containing nuclei and chloroplasts surrounds the vacuole. Inflated sporangia form at the ends of side branches and are separated from the rest of the thallus by walls at their bases (fig. 3.40*b*). Meiosis in sporangia leads to formation and release of multinucleate zoospores. Zoospores have a ring of flagella at their anterior ends (stephanokontous condition).

The gametophyte, which develops from a zoospore, is an inflated vesicle that sometimes reaches the size of a large marble (fig. 3.40*c*). Originally, it was described as the genus *Halicystis*. Since its relationship to *Derbesia* was established in culture, it has been referred to as the Halicystis phase of *Derbesia*. The Halicystis thallus has a thin cytoplasmic layer containing nuclei and chloroplasts adjacent to the wall and a large, central vacuole. Cytoplasm accumulates on one side of the thallus and is separated from the rest of the thallus by a membrane to form a gametangium. Gametangia forcibly discharge uninucleate gametes. Compatible anisogametes fuse, and the resulting zygote produces the tubular thallus of the sporophytic phase (fig. 3.41). Although gametes fuse, their nuclei remain separate and divide as the tubular sporophyte develops (Eckhardt, Schnetter, and Seibold 1986). Thus, instead of diploidy, a heterokaryotic condition occurs, in which two genetically different types of haploid nuclei are in the thallus. Fusion between nuclei is delayed and occurs in sporangia at the time of reproduction, followed immediately by meiosis. In a gametophyte, all the nuclei are genetically similar, and the condition is homokaryotic.

Figure 3.41
Life cycle of *Derbesia*.
(1) Sporophyte produces
haploid zoospores; haploid
thallus forms gametangial
regions; (2) fusion occurs
between anisogametes to
form (3) nonresting zygote.
N+N=heterokaryotic;
Me=meiosis in sporangium.

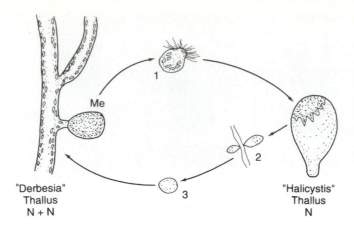

"Derbesia"
Thallus
N + N

"Halicystis"
Thallus
N

Figure 3.42
Bryopsis. (*a*) Featherlike
thallus. (*b*) Detail of branch
tip.

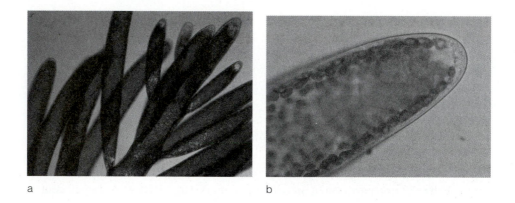

a

b

Bryopsis is similar to the diploid phase of *Derbesia* with tubular branches, but the branching pattern is regular, producing a featherlike thallus several centimeters long (fig. 3.42). In the life cycle of *Bryopsis,* the upright macrothallus produces gametes and alternates with a reduced protonema stage of branching tubes (Tanner 1981; Schnetter et al. 1990). While both *Derbesia* and *Bryopsis* have alternating generations, many other caulerpalean genera are diplontic, with meiosis occurring during gamete formation and the zygote developing directly into the macrothallus (see fig. 3.6*c*).

Caulerpa may grow over a meter in length but is still a single cell. Its thallus consists of a rhizome that runs horizontally over the surface of the substrate (usually sand) and is anchored at intervals by clusters of rhizoids. Also at intervals along the rhizome are upright **assimilators** with a characteristic morphology for different species (fig. 3.43). The assimilators are the principal sites of photosynthesis. Growth in *Caulerpa* occurs at the ends of the rhizome.

While *Caulerpa* has an open growth pattern, *Codium* is a compact mass of threadlike tubes that produce a spongy thallus resembling green fingers, a cushion, or a mass of worms (fig. 3.44). When viewed in section, the thallus has a central medullary region of longitudinally oriented tubes. Side branches form the outer

a   b

Figure 3.43
(*a, b*) Two species of
Caulerpa, showing variation
in assimilators.

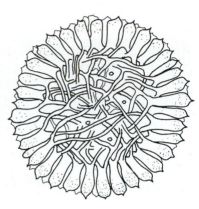

a   b

Figure 3.44
(*a*) *Codium*. (*b*) Cross section
of thallus showing outer layer
of swollen utricles and inner
medulla.

region of the thallus and terminate in enlarged structures called **utricles,** which
appress to form a palisade-like layer at the outer surface of the thallus. Photosynthesis occurs in the utricles.

A number of genera show the same basic construction pattern as *Codium,* with
entangled tubular threads and utricles. Some are spongy and bladelike, such as
Avrainvillea (fig. 3.45*a*), while others deposit calcium carbonate over their utricles
as a defense against herbivores (fig. 3.45*b–e*). *Halimeda* has flattened calcified
segments separated by flexible, uncalcified joints (fig. 3.45*b*). An external layer of
calcium carbonate (with lesser amounts of magnesium carbonate) is deposited in
intercellular spaces between the utricles.

The walls and plastids of the caulerpalean algae vary in composition. The
major structural components of the wall may be polysaccharides of xylose (xylans),
mannose (mannans), or glucose (cellulose). In some genera, the only plastids are
chloroplasts, while in other genera (*Caulerpa, Halimeda, Avrainvillea*), starch is
stored in separate leucoplasts. The term ***heteroplasty*** describes the presence of both
chloroplasts and leucoplasts.

a

b

c

Figure 3.45
Caulerpalean algae.
(*a*) *Avrainvillea*. (*b*) *Halimeda*.
(*c*) *Udotea*. (*d*) *Penicillus*.
(*e*) *Rhipocephalus*.

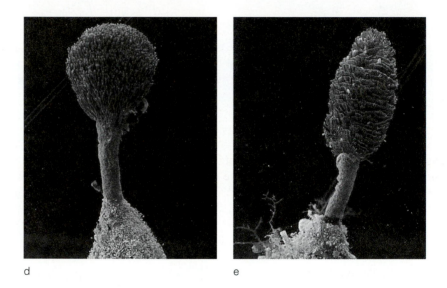

d

e

Dasycladalean Algae

The eleven living genera of the dasycladalean algae are largely confined to shallow tropical and subtropical oceans, and possess a number of distinctive characteristics. The dasycladalean algae are unicellular but may reach several centimeters in length. Their thalli show strong polarity and radial symmetry, with whorls of lateral branches arising from a central upright axis anchored by basal rhizoids. Walls of mannose polymers are covered to varying degrees by a calcareous layer. The diploid vegetative phase may be uninucleate (*Acetabularia, Batophora*) or become multinucleate early in development (*Cymopolia*). Uninucleate species have a large primary nucleus (diploid or polyploid) at the base of the thallus near the rhizoids, which produces secondary haploid nuclei prior to reproduction. In gametangia (or gametophores) associated with lateral branches, cysts form, and each cyst contains one haploid nucleus. Cysts are released into the water and, in turn, release gametes.

a   b   c   d

Figure 3.46
Dasycladalean algae.
(*a*) *Batophora* with whorls of
lateral branches.
(*b*) *Cymopolia* with calcified
segments. (*c*) *Acetabularia*
growing in shallow water.
(*d*) Thallus with a whorl of
gametangial rays. (*b* courtesy
L.R. Hoffman. *c* courtesy R.S.
Blanquet.)

Batophora, with its whorls of lateral branches arising from a central axis, resembles a green brush (fig. 3.46*a*). In *Cymopolia,* a layer of calcium carbonate covers branches, making them appear segmented (fig. 3.46*b*). As dasycladalean algae evolved, the number of lateral branches decreased. In *Acetabularia,* the vegetative thallus is an elongate cell, with a lightly calcified wall of mannose. It differentiates into an upright stalk bearing whorls of hairlike branches at the apex and rhizoidal outgrowths at the base (fig. 3.46*c, d*). A single whorl (sometimes two) of gametangial branches forms for reproduction. The wedge-shaped gametangial branches give the thallus the appearance of a wineglass. The primary nucleus at the base undergoes meiosis followed by mitotic divisions to produce numerous secondary nuclei (haploid), which migrate into the rays of the cap. Here, uninucleate cysts with cellulose walls form (fig. 3.47). After release into the water, sometimes following a period of dormancy, cysts divide to produce biflagellated isogametes. All gametes released from one cyst are the same mating type. Compatible gametes from different cysts fuse, and new thalli develop directly from the zygotes.

Acetabularia has been an important model system for basic research in cellular and molecular biology. In classic studies in the 1930s and 1940s, Joachin Hammerling grafted thallus parts and transferred nuclei among different species and showed the role of "morphogenic substances" from the nucleus in controlling cellular development. With the development of modern molecular techniques, *Acetabularia* has continued to be a "white rat" for cell biology studies.

While few dasycladalean algae survive today, their calcareous thalli have left an extensive fossil record. Almost 200 fossil genera, dating back 500 million years to the early Cambrian, are known.

Ulvophyte Ecology and Commercial Uses

Most ulvophytes grow in shallow, marine habitats, but some species are found in freshwaters. As members of benthic communities, ulvophytes grow attached to submerged rocks or other solid substrates. *Ulva* and *Enteromorpha* are widespread and common. Their success is related to their high photosynthetic capacities, high growth rates, and tolerance of a wide range of environmental conditions. However, neither is overly resistant to herbivores, which may limit their local abundance.

Figure 3.47
Life cycle of *Acetabularia*.
(1) Juvenile phase;
(2) mature thallus with
gametangial rays; (3) primary
nucleus breaks up;
(4) daughter nuclei migrate
into gametangial rays where
uninucleate cysts are formed;
(5) cysts are released; (6)
gametes from different cysts
(7) fuse to form a zygote.

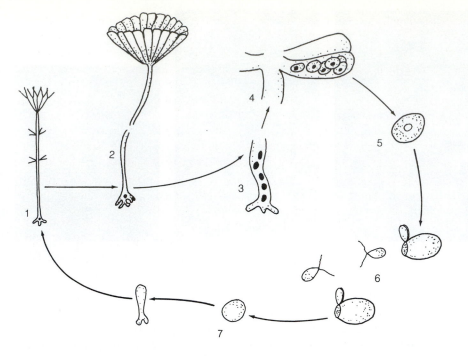

Enteromorpha is often one of the first macroalgae to colonize newly cleared surfaces on rocky shores. It is especially tolerant of variation in salinity, and thus grows well in tidepools, in estuaries, and on the hauls of ships passing from salt water into freshwater. *Ulva* is also a good colonizer and tolerates pollution better than most macroalgae, often thriving where competition with other species is reduced.

The siphonous green algae are most diverse in shallow tropical oceans, but many species extend into colder waters. Some produce rhizoids that allow them to grow well on sandy bottoms. Besides being important as photosynthetic producers, those with calcareous coverings contribute to carbonate sediments when they die. *Halimeda* is especially important in this respect (see fig. 3.45*b*). Its calcareous segments commonly cover the bottom of tropical lagoons and ultimately break down into sand.

Some benthic algae, such as *Enteromorpha,* produce nuisance growths when they foul boats or clog beaches. *Codium fragile* is a problem along parts of the temperate Atlantic coast when it interferes with shellfish beds or its remains accumulate on beaches (see fig. 3.44). This Asian species has spread to Europe and the East Coast of North America, probably transported on ship bottoms. It was first reported on Long Island in 1957 and today has a range extending from New Jersey to Massachusetts, with isolated populations in North Carolina, Virginia, and Maine (Carlton and Scanlon 1985). *Codium*'s ability to reproduce by fragmentation has facilitated its spread. It also forms gametes without meiosis. Recently, *Caulerpa taxifolia* invaded the Mediterranean Sea, where it is becoming a major weed.

Human consumption of green algae is limited, compared to that of some brown and red algae, but *Codium, Caulerpa, Enteromorpha,* and *Ulva* are consumed to some extent, mostly in the Pacific region.

Some ulvophytes, including species of *Ulothrix, Rhizoclonium, Pithophora,* and *Cladophora,* are common in freshwaters. *Ulothrix* often forms conspicuous growths at the water edge in winter and early spring. *Cladophora* grows during spring and fall in both standing water and rapid flows. Dense growths of *Cladophora glomerata* are a nuisance when they clog the water (review of *Cladophora* by Dodds and Gudder 1992). *Pithophora* and *Cladophora* may form floating mats in calm water.

The order Trentepohliales consists of four genera of filamentous algae, in which gametes and zoospores form in specialized structures (Chapman and Good 1983). The relationship of this group to other green algae is uncertain, but sometimes, they are classified in the Ulvophyceae on the basis of basal body orientation and molecular comparisons (Zechman et al. 1990). Trentepohlian algae grow in terrestrial environments, especially in the tropics. They often produce orange-red growths as a result of the accumulation of a carotenoid (hematochrome) in their cells. *Trenepohlia* grows on a variety of substrates (rocks, leaves, bark, etc.) (see fig. 8.7), while *Cephaleuros* produces orange spots on the leaves of citrus fruits, coffee, tea, and other plants. Some trentepohlian algae associate with fungi to form lichens.

Class Charophyceae

The charophytes have flagellated cells with a broad band of microtubules along one side of the cell, an associated multilayered structure near the basal bodies, and a second small microtubular root (fig. 3.48). Basal bodies have a parallel orientation, and the flagella arise from one side slightly below the cell's anterior end. Flagellated cells do not have eyespots. During mitosis, the nuclear envelope breaks down, and the mitotic spindle is open. At the conclusion of mitosis, daughter nuclei form at opposite ends of the cell, and the mitotic spindle persists as a phragmoplast during cell division by furrowing or, less commonly, by cell plate formation. The charophytes range in structure from unicellular forms to complex filaments. Vegetative cells are haploid, and a dormant zygote usually forms. Meiosis occurs when the zygote germinates. The charophytes grow primarily in freshwater, but a few live on moist soil in terrestrial environments. The class is divided into four orders (table 3.6). Their cell structures and molecular comparisions indicate that, of the classes of green algae, the charophytes are most closely related to the bryophtes and vascular plants.

Charophyte Diversity

Filaments in the Orders Klebsormidiales and Coleochaetales

Klebsormidium is an unbranched filament resembling *Ulothrix* but usually lacking a holdfast (fig. 3.49). Its cells have a band-shaped chloroplast and a single nucleus. Cell divisions by furrowing add new cells to a filament. *Klebsormidium* reproduces primarily by fragmentation into short segments of cells. Less commonly, the entire cell protoplast transforms (without division) into a single, biflagellated zoospore.

Coleochaete grows as an epiphyte on larger algae and freshwater plants. Its heterotrichous thallus is usually a compact disk of filaments spreading over the substrate, but sometimes, parenchymatous growth occurs (fig. 3.50). Setal cells with unicellular hairs are present. Cell divisions involve a cell plate and phragmoplast, a condition similar to land plants. In asexual reproduction, *Coleochaete*

Figure 3.48
Electron micrograph of a zoospore of *Coleochaete*, showing broad band of microtubules (lower) and multilayered structure (MLS). (From Graham and McBride 1979, courtesy *American Journal of Botany*.)

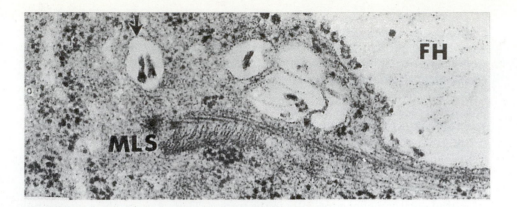

Figure 3.49
Unbranched filament of *Klebsormidium*.

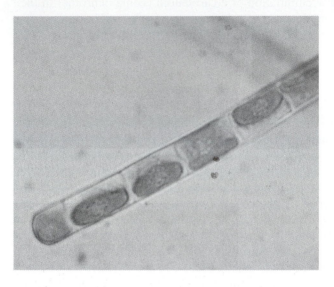

Table 3.6 *Orders of the Class Charophyceae*

Order Klebsormidiales—solitary cells, aggregations, and unbranched filaments

Order Coleochaetales—heterotrichous filaments

Order Charales—branching filaments with apical growth and complex reproductive structures

Order Zygnematales—filaments and unicells, flagellated stages lacking, conjugation of amoeboid gametes

---

forms a single zoospore from the contents of a cell. Sexual reproduction is oogamous. Unpigmented sperm form singly in antheridial cells, and one egg forms in each oogonium. After fertilization, the zygote remains in the oogonium, and a layer of sterile cells develops around it. After a period of dormancy, meiosis occurs at germination, and zoospores are produced.

Coleochaete shows advanced features in: (1) progression from filamentous (pseudoparenchymatous) to parenchymatous construction, (2) division with a cell plate, (3) formation of one zoospore per cell, (4) oogamy, and (5) retention of the zygote inside a protective covering.

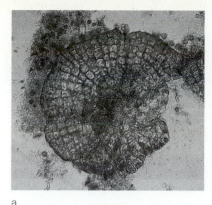

a

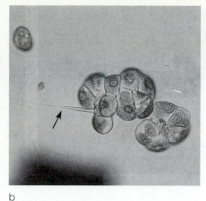

b

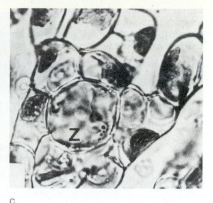

c

Figure 3.50
Coleochaete: (*a*) Discoid thallus. (*b*) Young plant with seta (*arrow*). (*c*) Vegetative cells growing around zygote (*Z*). (*c* from Graham 1984, courtesy *American Journal of Botany.*)

Order Charales

The six genera in the order Charales have a more complex structure and developmental pattern than other green algae. The common representative *Chara* grows in shallow areas of lakes or ponds, where it is often mistaken for a vascular plant because of its size (50 centimeters or more) and form (fig. 3.51). The thallus has a central axis with whorls of short, lateral branches. An apical cell at the tip of each major axis stimulates growth. Each derivative of the apical cell divides to form a nodal and an internodal initial. The internodal initial elongates greatly without dividing, while the nodal initial divides to produce whorls of radiating lateral branches and compact corticating branches. Most lateral branches are limited in their growth, but occasionally, a new indeterminate axis is produced. Also from the nodes, specialized branches grow over the surfaces of adjacent internodal cells, forming a layer of **cortication** (although not all charalean genera have cortication). Calcium carbonate may precipitate on the surface of corticating cells. The large internodal cells of charalean algae show cytoplasmic streaming and have proven useful for studies of membrane transport.

Chara does not form zoospores, but vegetative propagation is an important means of reproduction. Sexual reproduction is oogamous. The reproductive structures borne at the nodes are complex multicellular structures (fig. 3.51*c*). Sperm form in **globules,** and an egg forms in each **nucule.** After the egg is fertilized, the cellular covering of the nucule breaks down, and the zygote overwinters in a dormant state. Meiosis occurs during germination. A protonemal stage develops from one product of meiosis and later gives rise to the macrothallus.

Order Zygnematales

The zygnemataleans are a distinct group of freshwater green algae that lack flagellated stages and reproduce by conjugation. The primitive members of the order are unbranched filaments, while unicellular species, called **desmids,** have arisen by cell separation after division.

Of the thirteen genera of filamentous zygnemataleans (family Zygnemataceae), three—*Spirogyra, Mougeotia,* and *Zygnema*—are widespread and often common in ponds and other freshwater habitats. They may grow attached to the

Figure 3.51
Chara. (*a*) Whorls of lateral
branches. (*b*) Longitudinal
section showing apical cell
(ac) and ring of nodal cells
(nc) alternating with
internodal cells (in.).
(*c*) Reproductive structures at
a node, spherical globule (g),
and elongate nucule (n),
shown in section.

a

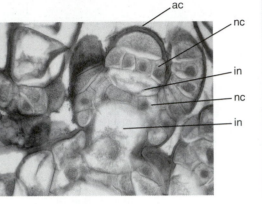

b

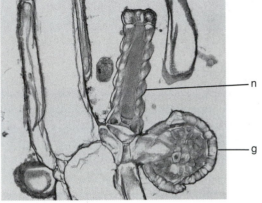

c

bottom by rhizoids from their basal cells, or may become detached and form floating masses of entangled filaments. *Spirogyra* ("Water Silk") is easily recognized by the one or more spiraling chloroplasts in its cells (fig. 3.52*a*). Each chloroplast has a series of pyrenoids along its length. A vacuole occupies the center of a cell, and the nucleus is suspended in a hammock of cytoplasm extending through the vacuole. Extensive mucilage is secreted around filaments. A large number of species are in the genus *Spirogyra*, and complexes of related species, differing in chromosome levels and filament widths, have been identified (McCourt, Hoshaw, and Wang 1986). These complexes may have arisen by polyploidy (Hoshaw, Wells, and McCourt 1987). Chloroplasts distinguish other filamentous genera. *Zygnema* has a pair of stellate chloroplasts on either side of the nucleus in the center of each cell (fig. 3.52*b*). *Mougeotia* has a single, plate-shaped chloroplast in each cell that rotates in response to irradiance (fig. 3.52*c*). The pigment phytochrome in the cell membrane controls chloroplast movement (Haupt 1983; Wagner and Grolig 1992). Phytochrome controls a wide range of development responses in vascular plants but is rare in algae.

In *Spirogyra* and other filaments, growth involves new cell production and subsequent cell enlargement. Cell division involves a phragmoplast and either a

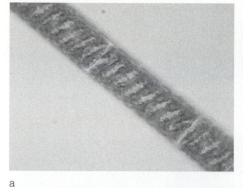

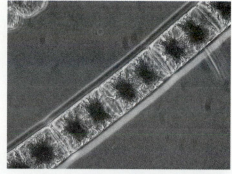

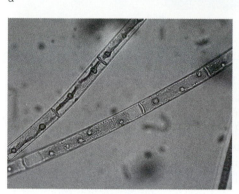

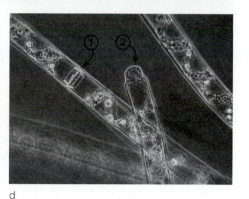

a

b

c

d

Figure 3.52
Filamentous zygnemataleans.
(*a*) *Spirogyra* with spiraling
parietal chloroplasts.
(*b*) *Zygnema* with central
stellate chloroplasts.
(*c*) *Mougeotia* with a platelike
chloroplast seen in surface
and profile. (*d*) *Spirogyra* with
replicate end walls. (1) Folds
in end wall and (2) expanded
end wall separate cells.

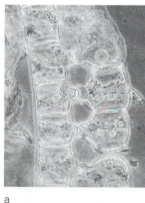

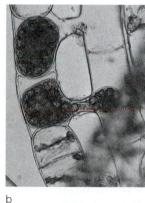

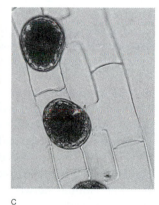

a

b

c

Figure 3.53
Conjugation in filamentous
zygnemataleans.
(*a*) Development of papillae
from adjacent filaments.
(*b*) Conjugation tube between
adjacent cells and fusion of
gametes. (*c*) Zygotes.

furrow or cell plate, or a combination of both. There are no asexual reproductive
stages, but filaments readily fragment as a means of vegetative propagation. In
some species, folds in the wall between adjacent cells (**replicate end wall**) aid cell
separation (fig. 3.52*d*). When the end wall unfolds and expands due to increased
turgor pressure, the cells are pushed apart.

Sexual reproduction is by **conjugation.** In scalariform conjugation, cells in two
parallel filaments form conjugation tubes. In a typical pattern, parallel filaments
secrete extensive mucilage, and opposing cells in the two filaments develop papillae
that grow toward each other and join (fig. 3.53*a*). At the point of contact, the wall

Figure 3.54
Lateral conjugation between
adjacent cells in a filament.

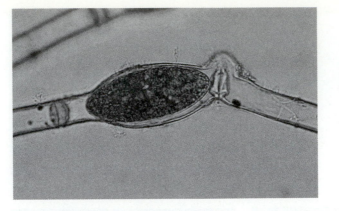

dissolves, creating a conjugation tube. Meanwhile, the protoplast of each cell rounds up to become an amoeboid gamete. Gametes from one filament usually migrate through the conjugation tubes (fig. 3.53b) and fuse with gametes in the other filament to produce zygotes (fig. 3.53c). Zygotes develop a thick wall and become dormant. At germination, meiosis occurs, and one of the resulting haploid cells initiates a new filament. Less common than scalariform conjugation is lateral conjugation, in which adjacent cells in the same filament conjugate (fig. 3.54). Conjugation may have evolved as an adaptation to a terrestrial environment to avoid the need for water to disperse flagellated gametes, and the living zygnematalean algae may represent a secondary return to an aquatic existence (Stebbins and Hill 1980). Environmental conditions stimulating conjugation are poorly known.

Two other groups in the Zygnematales are mostly unicellular and are called desmids. In one group (saccoderm desmids), solitary cells resemble filamentous genera in their chloroplasts and in forming conjugation tubes. Figure 3.55 shows representative genera. Recent molecular comparisons do not support the separation of saccoderm desmids and filamentous zygnemataleans, and indicate that chloroplast shape is more important as a classification criterion than cellular arrangement (McCourt et al. 1995).

More advanced desmids (placoderm desmids) have elaborate cells, often with a distinct constriction at the cell equator. *Cosmarium* is a typical representative. Its cell is divided into two **semicells** by a deep constriction (fig. 3.56a). The nucleus is in the isthmus between the two semicells, and each semicell contains a large chloroplast with pyrenoids. Walls of some species of *Cosmarium* may appear sculptured or bear short spines.

Cosmarium reproduces by dividing to produce two daughter cells, each of which receives one of the original semicells. At division, the isthmus lengthens, the nucleus undergoes mitosis, and a furrow develops between the two daughter nuclei. The new semicell of each daughter cell gradually expands until, normally, symmetry is achieved (fig. 3.56b). Meanwhile, the chloroplast divides, and organelles rearrange to position the nucleus between the two chloroplasts.

Conjugation by *Cosmarium* involves pairing of cells, which then secrete an envelope of mucilage around themselves (fig. 3.56c). The contents of each cell contract, becoming a gamete, and then the cell wall splits at the isthmus. Amoeboid

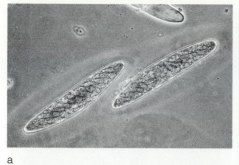

a

b

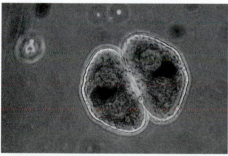

c

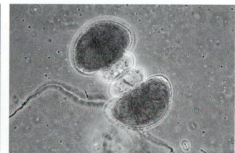

d

Figure 3.55
Saccoderm desmids.
(*a*) *Spirotaenia* with parietal chloroplasts. (*b*) *Roya* with platelike chloroplast.
(*c*) *Cylindrocystis* with stellate chloroplasts. (*d*) *Netrium* with complex stellate chloroplasts.

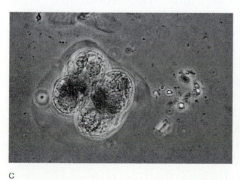

a

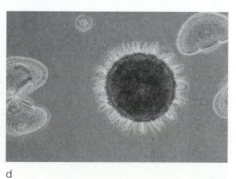

b

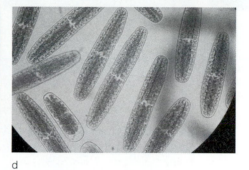

c

d

Figure 3.56
Cosmarium. (*a*) Vegetative cell with semicells joined by isthmus. (*b*) Sexual reproduction involving cell division and formation of new semicells. (*c*) Conjugating pair in mucilaginous envelope. (*d*) Zygote between empty semicells.

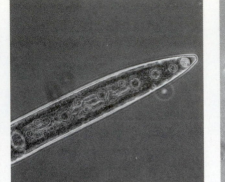
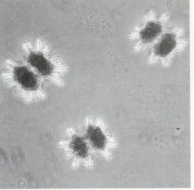

a   b   c

Figure 3.57
Placoderm desmids.
(*a*) *Closterium* without a
median constriction. (*b*) Row
of pyrenoids in each
chloroplast and vacuole with
granules of calcium sulfate
(gypsum) or barium sulfate at
end of each semicell (clear
circle). (*c*) *Staurastrum* with
radial extensions and spines.
(*d*) Flat cell of *Micrasterias*
with secondary constrictions.
(*e*) *Desmidium*, a filamentous
desmid.

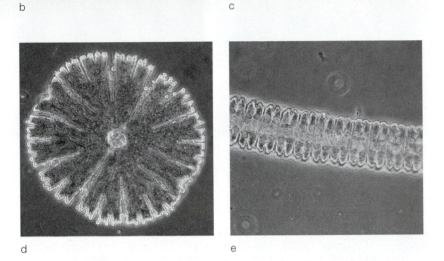

d   e

gametes fuse, producing a zygote. The zygote forms a thick wall with a sculptured outer surface and enters into a dormant state (fig. 3.56*d*). Meiosis occurs during germination.

Some of the thirty-two genera of placoderm desmids have elaborate and beautiful cells. Most have an obvious median constriction and are unicellular, but in a few genera, cells remain joined in chains or colonies. *Closterium* is a common genus without a constriction, but the position of its chloroplasts make its semicells recognizable (fig. 3.57*a, b*). Figure 3.57 shows other placoderm desmids. Many desmids move by secreting mucopolysaccharides (Domozych et al. 1993).

Charophyte Ecology

The charophytes are largely restricted to freshwater and terrestrial habitats, but occasionally, they occur in brackish water. The filamentous zygnemataleans and members of the Charales are sometimes common in shallow water. *Spirogyra* and its relatives are often abundant in warm, stagnant ponds, sometimes becoming detached as floating mats (fig. 3.58). Desmids are members of planktonic communities. The zygnemataleans, especially *Mougeotia,* are often present in acidic waters.

Figure 3.58
Floating mat of
zygnematalean filaments.

Evolution of Land Plants

Similar photosynthetic pigments and a starch reserve indicate that the bryophytes (mosses and liverworts) and vascular plants evolved from green algae. However, the fossil record does not indicate which group of green algae was the immediate ancestor of land plants. In *Origin of Land Plants,* Graham (1993) discusses the relationship between plants and the charophytes. The phragmoplast, ultrastructure of flagellated stages, and similar enzymes suggest that plants originated from the charophytes rather than another class of chlorophytes (de Jesus, Tabatabai, and Chapman 1989; Laliberté and Hellebust 1991). Additional evidence supporting a charophyte ancestor are: (1) similar introns (Manhart and Palmer 1990), (2) location of the *tuf*A gene in the nucleus rather than in the chloroplast (Baldauf, Manhart, and Palmer 1990), and (3) presence of lignin and sporopollenin in the walls (Delwiche, Graham, and Thomson 1989; Graham 1993).

Among living algae, *Coleochaete* shares the following characteristics with land plants that suggest that it is close to the line that led to the land plants: (1) division with a cell plate, (2) oogamy, (3) retention of the egg and zygote on the gametophyte, (4) formation of a covering of cells around the zygote, and (5) formation of more than four spores from each zygote. Some species of *Coleochaete* also show localized growth, parenchymatous construction, and multcellular antheridia. If land plants arose from an algal ancestor similar to *Coleochaete,* with zygotic meiosis and a haploid vegetative phase, the following events probably occurred after land plants diverged from the algae: (1) a change in the time of meiosis to produce an alternation of generations, (2) retention of the zygote and subsequent development of the sporophyte on the gametophyte, and (3) establishment of nutritional and developmental relationships between the young sporophyte and gametophyte (Graham 1984).

Summary

1. The chlorophytes or green algae have chlorophylls *a* and *b* as their major photosynthetic pigments, starch as the carbohydrate reserve, and chloroplast envelopes with two associated membranes. The division has four classes. Primitive flagellates in the class Prasinophyceae usually have scales rather than walls on their cell surfaces, and their cell structures vary considerably.

2. The chlorophycean greens are characterized by four roots of microtubules in a cruciate arrangement in their flagellated stages, a clockwise orientation of their basal bodies, a phycoplast during cell division, and haploid vegetative cells with zygotic meiosis. Chlorophyceans are found in freshwater and terrestrial environments. The unicellular flagellate *Chlamydomonas* is representative of the primitive condition in the class. Unicellular and colonial forms, both flagellated and nonflagellated, are common as members of the freshwater plankton and in terrestrial environments. Filamentous chlorophyceans are parts of benthic communities.

3. The ulvophytes grow as multicellular filaments, simple parenchymatous blades and tubes, or as coenocytic tubes and vesicles. They commonly are found attached in shallow marine environments, but some species grow in freshwater. The class is characterized by flagellated cells with counterclockwise-oriented basal bodies and microtubular roots in a cruciate pattern. Life cycles are either an alternation of haploid and diploid phases with meiosis during spore formation, or diploid thalli with meiosis at the time of gametogenesis.

4. The charophytes are characterized by a phragmoplast during cell division, and flagellated stages contain a broad band of microtubules and basal bodies in a parallel orientation. The charophytes are filamentous or unicellular in form, and are found in freshwater and terrestrial habitats. The bryophytes and vascular plants probably evolved from a charophyte ancestor.

Further Reading

Berger, S., and M. J. Kaever. 1992. *Dascycladales*. Georg Thieme.

Gerrath, J. F. 1993. The biology of desmids: A decade of progress. In F. E. Round and D. J. Chapman, eds., *Progress in Phycological Research*, vol. 9, 79–192. Biopress.

Goodenough, U. W., E.V. Armbrust, A. M. Campbell, and P. J. Ferris. 1995. Molecular genetics of sexuality in *Chlamydomonas*. *Annual Review of Plant Physiology and Molecular Biology* 46: 21–44.

Graham, L. E. 1993. *Origin of Land Plants*. Wiley.

Harris, E. H. 1989. *The* Chlamydomonas *Sourcebook*. Academic Press.

Hoek, C. van den, D. G. Mann, and H. M. Jahns. 1995. *Algae*. Cambridge University Press.

Sym, S. D., and R. N. Pienaar. 1993. The Class Prasinophyceae. In F. E. Round and D. J. Chapman, eds., *Progress in Phycological Research*, vol. 9, 281–376. Biopress.

4

Division Chromophyta

Phycologists group the ten classes of algae with chlorophyll c in different ways (table 4.1). In this book, the division Chromophyta includes algae with chlorophylls a and c, and heterokontous flagellated stages (one smooth flagellum and one flagellum with mastigonemes). However, what is included in the division Chromophyta (or Chrysophyta) in the published literature varies greatly. As defined here, the chromophytes include the brown algae (Phaeophyceae) but do not include the haptophytes, dinoflagellates, or cryptomonads. Molecular comparisons of sequences of nucleotide bases in ribosomal RNA support this characterization of the chromophytes and indicate a close relationship with the oomycetes, which also

Table 4.1 *Alternative Groupings for Classes of Algae with Chlorophyll* c

Chromophyta As All Algae with Chlorophyll *c*	Chromophyta As Algae with Chlorophyll *c* and Heterokontous Flagellation (used in this book)	Chrysophyta As Unicellular and Colonial Algae
Chrysophyceae	Chrysophyceae	Chrysophyceae
Synurophyceae	Synurophyceae	Synurophyceae
Tribophyceae	Tribophyceae	Tribophyceae
Eustigmatophyceae	Eustigmatophyceae	Eustigmatophyceae
Raphidophyceae	Raphidophyceae	Raphidophyceae
Bacillariophyceae	Bacillariophyceae	Bacillariophyceae
Phaeophyceae	Phaeophyceae	Prymnesiophyceae
Prymnesiophyceae		
Dinophyceae		**Phaeophyta As Complex, Multicellular Algae**
Cryptophyceae	**Haptophyta**	
	Prymnesiophyceae	Phaeophyceae
	Dinophyta	**Dinophyta**
	Dinophyceae	Dinophyceae
	Cryptophyta	**Cryptophyta**
	Cryptophyceae	Cryptophyceae

Table 4.2   *Characteristics of Division Chromophyta*

Major photosynthetic pigments	Chlorophyll a, c_1, c_2, ±fucoxanthin (see table 4.3)
Carbohydrate reserve	β-1,3-linked glucans (chrysolaminarin, laminarin)
Chloroplast structure	
Thylakoid associations	Three
Envelope	Two membranes
Chloroplast ER	Present
Cell covering	Variable (see table 4.3)
Flagella	Heterokontous (one smooth flagellum, one flagellum with mastigonemes)

Table 4.3   *Classes of Division Chromophyta*

Class (common name)	Fucoxanthin	Cell Covering
Chrysophyceae (golden brown algae)	Present	Scales, lorica
Synurophyceae	Present	Siliceous scales
Tribophyceae (=Xanthophyceae) (yellow-green algae)	Absent	Pectic or cellulose wall
Eustigmatophyceae	Absent	Wall (composition unknown)
Raphidophyceae (=Chloromonadophyceae)	Absent	Periplast
Bacillariophyceae (=Diatomophyceae) (diatoms)	Present	Siliceous frustule
Phaeophyceae (=Fucophyceae) (brown algae)	Present	Cellulose wall with alginic acid

have heterokontous flagellation (Bhattacharya et al. 1992; Leipe et al. 1994; Medlin, Lange, and Baumann 1994). Besides their pigmentation and flagellation, the chromophytes have a β-1,3-linked polymer as their principal reserve and a characteristic chloroplast structure (table 4.2). The chromophytes are divided into seven classes, distinguished by their cell coverings, their principal pigments, and the complex multicellular construction in the Phaeophyceae (table 4.3).

The chromophytes have chlorophylls a and, normally, both c_1 and c_2 and sometimes c_3. In the chrysophytes, synurophytes, diatoms, and brown algae, the carotenoid fucoxanthin is an additional important photosynthetic pigment, giving cells a brown color. Members of other classes have a green color due to chlorophyll. β-carotene is also present in the chromophytes.

The principal carbohydrate reserve is a β-1,3-linked polymer of glucose called chrysolaminarin (=leucosin) or laminarin (in Phaeophyceae) (see fig. 1.5b). The reserve is present in cytoplasmic vacuoles rather than in the chloroplast or associated with pyrenoids. Lipid reserves are also important in some chromophytes.

Flagellated stages are **heterokontous.** Heterokontous cells have two flagella—one smooth flagellum and one flagellum with two rows of stiff tubular hairs or mastigonemes. The flagellum bearing mastigonemes usually is longer and directed forward, while the smooth flagellum is directed backward. The posterior flagellum has a flagellar swelling that contains flavin and functions as a photoreceptor (Kawai 1988). An eyespot associated with the flagellar swelling may reflect light or act as a shading structure. The cytoskeleton usually consists of connecting fibers between the basal bodies, four complex microtubular roots, and fibrous roots.

The chromophytes have a distinctive chloroplast structure (see fig. 1.9c, f). Thylakoids associate in groups of three, which may include a set of thylakoids extending around the chloroplast periphery (girdle lamella). A double-membrane envelope and a layer of endoplasmic reticulum (CER) that is continuous with the nuclear envelope surround the chloroplast. The periplastidal space between the chloroplast envelope and CER is narrow and lacks any distinct structures. Ribosomes may be present on the CER's outer surface. By analogy with the cryptomonads (see chapter 5), the four membranes around the chloroplast (envelope and CER) are evidence for a eukaryotic origin, with the CER derived from the vesicular membrane of the host and the cell membrane of the symbiont (Gibbs 1990). However, evidence to identify which eukaryotic alga was the source of the chloroplast is scant.

The chromophytes vary in their cell coverings (table 4.3). Some lack any external covering, while others have organic scales, siliceous scales, loricae, or walls. Scales may be composed largely of carbohydrates or become impregnated with silica. Scales form in Golgi vesicles and are then deposited on the cell surface. They are gradually shed and replaced. Other species form walls and loricae of cellulose or pectic material (polymers of galacturonic acid). The diatoms have a distinctive wall of silica, called a frustule. The brown algae have walls with cellulose fibrils surrounded by the mucopolysaccharide alginic acid.

Classes Chrysophyceae and Synurophyceae

Most members of the classes Chrysophyceae and Synurophyceae are flagellated single cells or colonies of flagellated cells, but other morphologic types are known. Fucoxanthin usually gives cells a brown color (plate 3a). Cell coverings are typically scales or loricae. Andersen (1987) separated *Synura* and *Mallomonas* from the Chrysophyceae and placed them in a new class, Synurophyceae, on the basis of their siliceous scales and ultrastructure, including paired flagellar swellings, lack of an eyespot, and features of the flagellar apparatus. Besides chlorophyll a, the synurophytes have only chlorophyll c_1, while the chrysophytes have both chlorophyll c_1 and c_2. Both groups form distinctive dormant stages called statocysts with siliceous coverings.

Representative Genera

The unicellular flagellate *Ochromonas* shows the basic structure of a chrysophyte (fig. 4.1). Its ovoid cell is naked, and two flagella arise from its anterior end. The longer flagellum bears two rows of mastigonemes, and the shorter flagellum is smooth. Near the base of the smooth flagellum is a swelling containing flavin as a

Figure 4.1
Ochromonas. (*a*) Typical chrysophyte cell with heterokontous flagellation, chloroplast (ch), chloroplast endoplasmic reticulum (CER) continuous with nuclear envelope, nucleus (nu), and chrysolaminarin vacuole (lv). (*b*) Electron micrograph. (*b* from Slankis and Gibbs 1972, courtesy *Journal of Phycology.*)

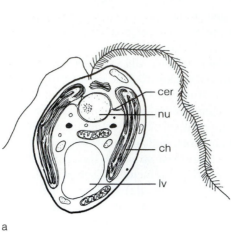

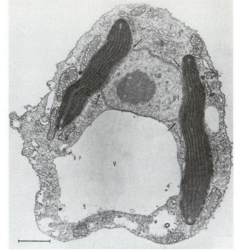

a   b

Figure 4.2
Mallomonas. (*a*) Scales with long bristles as seen through a light microscope (flagellum not visible). (*b*) Scales as seen with a scanning electron microscope. (*b* from Beech, Wetherbee, and Pickett-Heaps, 1990, courtesy *Journal of Phycology.*)

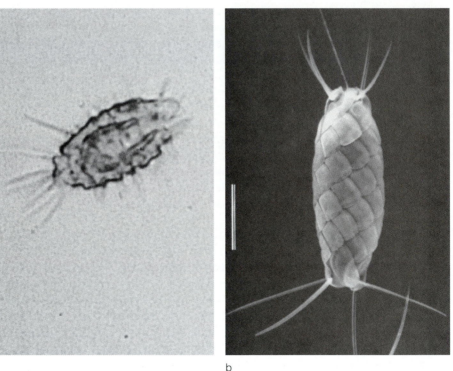

a   b

photoreceptor. Within the cell, basal bodies are oriented perpendicular to each other, and the cytoskeleton consists of connecting fibers, a fibrous root extending from the basal bodies to the nucleus, and four microtubular roots radiating around the cell periphery. The one or two chloroplasts present are surrounded by CER, and contain pyrenoids and an eyespot adjacent to the flagellar swelling. The eyespot and flagellar swelling interact in responding to the direction of light. A Golgi body

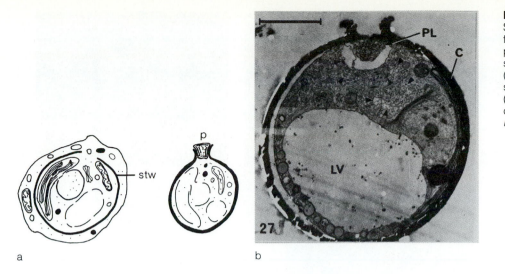

Figure 4.3
Statocyst. (*a*) Siliceous cyst
formed within a cell with a
plug of carbohydrate.
stw=statocyst wall; p=plug.
(*b*) Electron micrograph
showing cyst of *Ochromonas*.
(*b* from Hibberd 1977,
courtesy *Journal of
Phycology*.)

a

b

is near the nucleus. In the cell posterior is a large vacuole containing chrysolaminarin. Mitochondria have tubular cristae, as is characteristic of the chromophytes.
Ochromonas reproduces by dividing longitudinally. In general, sexual reproduction
is rare among the chrysophyte algae (see Sandgren 1981).

Mallomonas is a unicellular synurophyte (review by Siver 1991). It has a single
functional flagellum (and a second rudimentary flagellum) and is covered by
conspicuous siliceous scales with long, delicate extensions or bristles (fig. 4.2). The
flagella arise from parallel basal bodies. Each cell has one or two chloroplasts, a
Golgi body, and chrysolaminarin vacuole, as in *Ochromonas*. Scales form in silica
deposition vesicles derived from the Golgi body and are then positioned on the cell
surface in overlapping rows to form a protective case around the cell. Details of
scale structure are important in taxonomy. As in *Ochromonas, Mallomonas* reproduces by cell division, and sexual reproduction is rare.

Both the chrysophytes and synurophytes form a distinctive resistant stage
called a **statocyst** (statospore) (fig. 4.3). Cyst formation either by differentiation of
a vegetative cell directly or of a zygote following sexual reproduction is usually in
response to a sudden environmental change (Siver 1991). The statocyst forms
within a cell in a silica deposition vesicle and may contain only part of the cell
cytoplasm. The spore wall is composed of silica, except for a pore filled by a
polysaccharide plug. The contents of a statocyst enter into a dormant condition. At
germination, the protoplast is released through the pore when the plug breaks
down.

Examples of other unicellular chrysophytes include *Epipyxis* and *Chrysopyxis*.
Epipyxis is flagellated but grows attached to other algae rather than actively swimming about (fig. 4.4*a*). A cup-shaped lorica attached at the base to the substrate
surrounds *Epipyxis*, and the flagella extend from a wide opening at the anterior end.
A flask-shaped lorica surrounds the epiphyte *Chrysopyxis* (fig. 4.4*b*). From a
narrow opening in the lorica, *Chrysopyrix* extends a delicate pseudopod into the
surrounding medium.

Examples of common colonial forms include *Uroglena, Dinobryon,* and *Synura*. In *Uroglena* colonies, the cells are on the periphery of a mucilaginous sheath

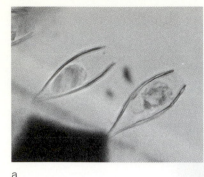
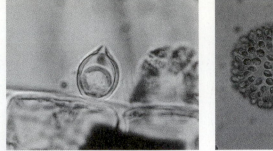
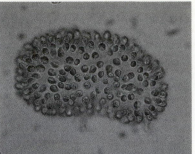

a

b

c

Figure 4.4
Representative chrysophytes.
(*a*) *Epipyxis*, a flagellate with
a lorica attached to a
filamentous green alga.
(*b*) *Chrysopyxis* in a lorica
with pseudopodia extending
through opening (not visible).
(*c*) *Uroglena*. (*d, e*)
Dinobryon.

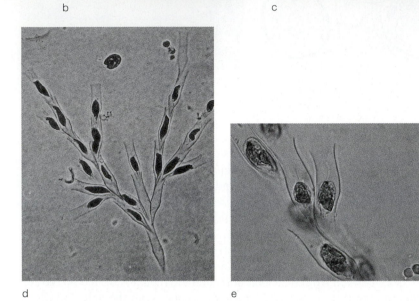

d

e

(fig. 4.4*c*). *Dinobryon* forms dendroid colonies (fig. 4.4*d, e*). A cup-shaped lorica surrounds each cell and attaches to the lip of the lorica beneath it. The synurophyte *Synura* has delicate scales on its cells and is not surrounded by a mucilaginous sheath, but its cells are in direct contact at the center of the colony (fig. 4.5; plate 3*a*).

Ecology of Chrysophytes and Synurophytes

The chrysophytes and synurophytes are more common and diverse in freshwaters, but some species are marine. Flagellated cells and colonies are often abundant in the plankton of oligotrophic (low-nutrient) lakes during late winter and early spring when water temperatures are low. However, representatives of both groups may grow under a wider range of conditions. Many species prefer acidic pH (Siver 1991; Cumming, Smol, and Birks 1992). Common genera are *Dinobryon, Synura, Mallomonas, Uroglena,* and *Chrysosphaerella*. Blooms of some species may give drinking water an unpleasant taste and odor, sometimes described as "fishy."

Many chrysophytes show considerable nutritional flexibility as mixotrophs. In addition to obtaining energy through photosynthesis, they can utilize dissolved organic material as energy sources or ingest particulate organic material, such as bacteria, by phagocytosis (Rothhaupt 1996).

a

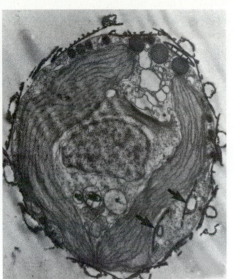

b

Figure 4.5
Synura. (*a*) Scales on cell surface as seen with a scanning electron microscope. (*b*) Section of a cell with external scales, developing scales (*arrows*), chloroplast, and nucleus visible. (*a, b* from Leadbeater 1990, courtesy the author and the British Phycological Society.)

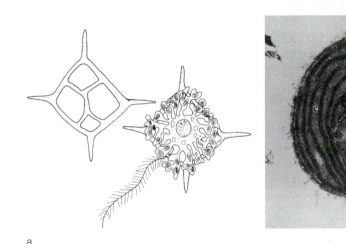

a

Figure 4.6
Marine chrysophytes. (*a*) The silicoflagellate *Dictyocha* showing the siliceous skeleton, chloroplasts in peripheral cytoplasm, and nucleus in central core of cell. (*b*) *Aureococcus* with nucleus (n), chloroplast (*c*), and pyrenoid (p) visible. Bar = 1 micrometer. (*b* from Sieburth, Johnson, and Hargraves 1988, courtesy *Journal of Phycology*.)

Two unicellular chrysophytes, *Dictyocha* and *Aureococcus*, are sometimes common in the marine plankton. Silicoflagellates such as *Dictyocha* have a single flagellum and a characteristic external framework of silica. The protoplast is wrapped around this framework, giving the appearance of an internal skeleton (fig. 4.6*a*). *Dictyocha* is widespread in the oceans (Henriksen et al. 1993). Its blooms have caused fish-kills in the Baltic Sea, possibly as a result of oxygen depletion. The siliceous skeletons may also damage fish gills.

Aureococcus, a small, unicellular chrysophyte, is responsible for "brown tides" in coastal waters (fig. 4.6*b*). Since 1985, brown tides have been reported in bays along the northeast Atlantic coast and in the Gulf of Mexico (Sieburth, Johnson, and Hargraves 1988; Cosper, Bricelj, and Carpenter 1989; Milligan and Cosper

1994). These blooms have caused a decline in seagrasses (*Zostera*) and shellfish. *Aureococcus* is not known to produce a toxin, but when abundant, it reduces light penetration into the water, which harms seagrasses, and it outcompetes other phytoplankton for nutrients, which reduces the availability of these phytoplankton as food for bivalves.

Fossilized remains of chrysophytes and synurophytes include scales and statocysts. These microfossils in lake sediments have been used to determine past environmental conditions (Siver and Hamer 1992).

Classes Tribophyceae, Eustigmatophyceae, and Raphidophyceae

In the classes Tribophyceae, Eustigmatophyceae, and Raphidophyceae, chlorophyll *a* is usually the principal pigment for light absorption, and as a result, cells have a green color. With only a few exceptions, members of these three classes are of secondary importance in freshwater or marine environments, compared to other algal groups.

Class Tribophyceae

The tribophytes, or yellow-green algae, usually are pale green, but sometimes, a starch test is necessary to distinguish them from the chlorophytes. A typical cell has two chloroplasts with pyrenoids, Golgi bodies associated with the nucleus, and cytoplasmic vacuoles with chrysolaminarin. A wall, sometimes consisting of two pieces, usually surrounds cells. Asexual reproduction leads more commonly to formation of aplanospores than zoospores. Sexual reproduction is rare in most tribophytes. Common tribophytes are nonflagellated cells, filaments, and coenocytes, but other thallus types are known. The similar forms in the tribophytes and chlorophytes provide a good example of parallel evolution in unrelated groups.

Nonflagellated, unicellular tribophytes include *Botrydiopsis,* with a spherical cell containing numerous small chloroplasts (fig. 4.7*a*), and *Ophiocytium,* with a cylindrical cell ending in stalklike extensions at one or both ends (fig. 4.7*b*). The planktonic *Centritractus* also has an elongate cell with spines at its ends (fig. 4.7c).

Filamentous genera include *Tribonema* and *Heterococcus. Tribonema* is an unbranched filament (fig. 4.7*d*). Adjacent cells share overlapping H-shaped wall pieces, as in the green alga *Microspora. Heterococcus* has a heterotrichous thallus (fig. 4.7*e*). The prostrate branches may form a compact pseudoparenchymatous mass, while the erect system is normally reduced.

Vaucheria is a coenocytic representative of the Tribophyceae. Its thallus is tubular with irregular branches (fig. 4.7*f*). A central vacuole extends throughout the thallus, and numerous discoid chloroplasts are present in the surrounding cytoplasmic layer. Growth occurs at the ends of branches. Cross walls form to separate reproductive structures. *Vaucheria* shows a number of specialized features in reproduction. In asexual reproduction, the contents of a sporangium fail to divide into uninucleate zoospores but are released as a single compound zoospore with flagellar pairs on its surface and a nucleus associated with each flagellar pair. Sexual reproduction is oogamous. Distinct gametangia form on special branches, with antheridia associated with oogonia (fig. 4.7*g*).

Most tribophytes grow in freshwater but are rarely abundant. Occasionally, *Tribonema* or *Vaucheria* grows extensively in ditches or on damp soil. *Vaucheria* is also common in salt marshes.

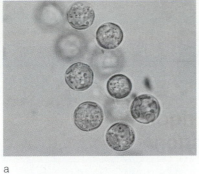

a

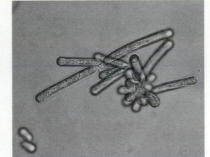

b

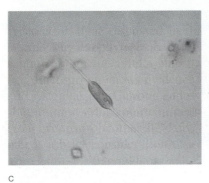

c

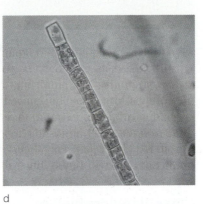

d

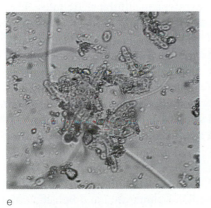

e

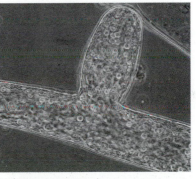

f

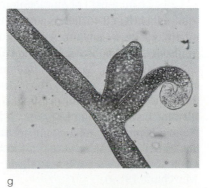

g

Figure 4.7
Representative tribophytes.
(*a*) *Botrydiopsis.*
(*b*) *Ophiocytium.*
(*c*) *Centritractus.*
(*d*) *Tribonema.*
(*e*) *Heterococcus.*
(*f*) *Vaucheria.* (*g*) Oogonium
adjacent to an antheridial
branch on *Vaucheria.*

Figure 4.8
Raphidophytes.
(a) *Vacuolaria.*
(b) *Gonyostomum.*

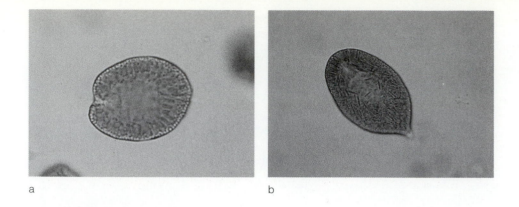

a   b

Class Eustigmatophyceae

The eustigmatophytes have chlorophyll *a* and violaxanthin as their principal pigments (Owens, Gallagher, and Alberte 1987), and distinctive ultrastructural features, especially in their flagellated stages. The zoospores are elongate and often uniflagellated. Their single chloroplast lacks a pyrenoid. An eyespot is in the cytoplasm near the cell's anterior end, rather than as part of the chloroplast, as in most algae.

Seven genera belong to this class. They are all coccoid forms living either in fresh or marine waters or on soil, but none are common.

Class Raphidophyceae

The nine genera in the Raphidophyceae are unicellular flagellates. Their cells have two anterior flagella, with one directed forward and the other trailing. The raphidophytes lack cell walls or scales. Their cells contain many discoid chloroplasts. Chlorophyll usually dominates, giving cells a green color. Chloroplasts normally lack pyrenoids and eyespots.

The raphidophytes occur in freshwater and seawater, sometimes showing a preference for acidic conditions. *Vacuolaria* and *Gonyostomum* are common freshwater representatives (fig. 4.8). Blooms of *Heterosigma, Chattonella,* and *Fibrocapsa* in seawater sometimes produce fish-kills, possibly due to a neurotoxin (Khan, Arakawa, and Onoue 1996).

Class Bacillariophyceae— The Diatoms

The diatoms are important components of planktonic and benthic communities in a wide range of marine and freshwater habitats. They grow as solitary cells, chains of cells, or members of small colonies. A distinctive siliceous wall called a **frustule** surrounds their nonflagellated vegetative cells.

The frustule has two valves and girdle bands. The **epivalve** is the older, usually slightly larger valve; the other valve is called the **hypovalve** (figs. 4.9*b*, 4.10). **Girdle bands** are loops of silica inserted between the valves. Their number varies, depending on the species, and when numerous, may completely separate any value overlap. The terms *epitheca* and *hypotheca* refer to the valves plus their associated girdle bands. The diatoms may be derived from an ancestral cell in which siliceous scales on opposite sides of the cell enlarged as the valves, while other scales became the girdle bands (Crawford and Round 1989). Alternatively, the diatoms may have come from zygotes or statocysts of a chrysophyte or synurophyte (Mann and Marchant 1989).

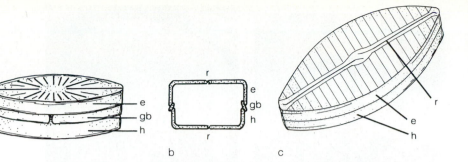

Figure 4.9
Diatom frustule. (*a*) Centric
diatom. (*b*) Transverse
section of pennate diatom.
(*c*) Pennate diatom.
e=epivalve; gb=girdle band;
h=hypovalve; r=raphe.

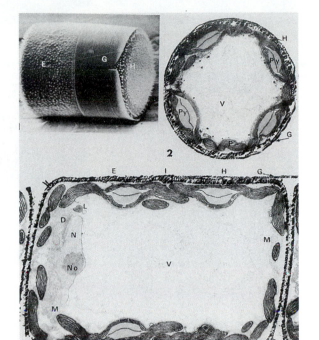

Figure 4.10
Melosira varians.
(1) Scanning electron
micrograph showing epivalve
(E), hypovalve (H), and girdle
bands (G). (2) Transverse
section through a cell
showing central vacuole (V)
and chloroplasts.
Py=pyrenoid.
(3) Longitudinal section
showing nucleus (N),
chloroplasts, vacuole (V), and
frustule. (1–3 from Crawford
1973, courtesy *Journal of
Phycology*.)

Diatom taxonomy is based largely on valve structure and markings, and a great
deal of terminology is used to describe subtle differences. Ordinarily, the protoplast
must be burned away with acid and the frustules mounted to see clearly the markings on valve surfaces. Distinguishing the different views of a cell is important when
working with diatoms. A cell seen from the side is in girdle view. Presentation of
the valve face is a valve view. As seen in valve view, the wall has distinct markings
from pores, thickenings, and extensions. The diatoms are divided into two groups
based on the symmetry of these markings. Centric diatoms (order Centrales) have
radially arranged markings (figs. 4.9*a*, 4.11*a*, *b*), while pennate diatoms (order
Pennales) have markings bilaterally arranged about the valve's long axis (figs. 4.9*c*,
4.12*a–c*). Many pennate diatoms have a **raphe** appearing as a distinct line running
the length of the valve. The raphe is a slit in the valve that functions in motility, and
other valve markings are arranged symmetrically about the raphe (fig. 4.9*c*). Depending on whether a raphe is present or not, pennate diatoms are subdivided into
raphid and araphid types. Some pennate diatoms lack a raphe on one or both valves,
but have a clear area in the center of the valve called a **pseudoraphe** (fig. 4.12*c*).

Figure 4.11
Centric diatoms.
(*a*) *Coscinodiscus* showing
markings on valve surface.
(*b*) Scanning electron
micrograph of *Cyclotella*.
(*c*) *Melosira* forming a chain
of cells. (*d*) *Coscinodiscus*
showing central vacuole
surrounded by cytoplasmic
layer with chloroplasts.
(*b* from Hoops and Floyd
1979, courtesy *Phycologia*.)

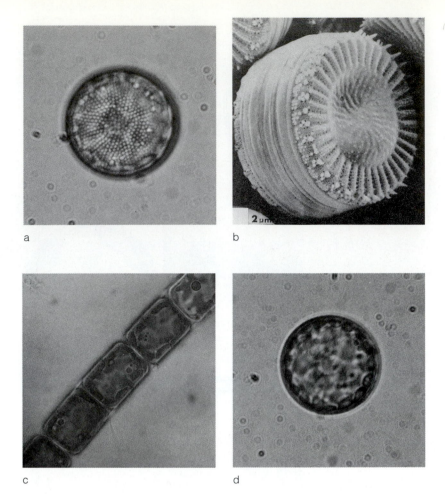

a

b

c

d

Usually, the markings on both the epivalve and hypovalve are the same, but valve patterns differ on a few pennate diatoms. Many centric diatoms have spines and other extensions arising from the valve faces or margins.

As mentioned earlier, the diatoms exist as solitary cells, chains of cells, or members of small colonies. Chains of cells may resemble filaments, but adjacent cells do not share common end walls (fig. 4.11*c*). Extensions of the frustule or mucilage may hold chains and colonies together. Other diatoms secrete mucilaginous stalks for attachment or form mucilaginous tubes.

Centric diatoms are considered more primitive than pennates. Centric diatoms have a large central vacuole and numerous discoid chloroplasts in the cytoplasmic layer between the cell membrane and vacuolar membrane (figs. 4.10, 4.11*d*). A diploid nucleus is located against one valve. Although chrysolaminarin granules may be present, lipids are often more important as reserves.

Pennate diatoms have fewer chloroplasts, which are often elongate (fig. 4.12*d*, *e*). The nucleus is located against one valve, and central vacuoles are often present. Pennate diatoms with a raphe secrete mucilage to produce a gliding motility over a substrate. Vesicles with mucopolysaccharides secrete strands of mucilage through the raphe's slit (Edgar and Pickett-Heaps 1984a). Each strand attaches to the cell membrane at one end, extends through the raphe, and adheres to the substrate at the other end. As the point of attachment to the cell membrane is displaced along the cell surface, the cell moves in the opposite direction

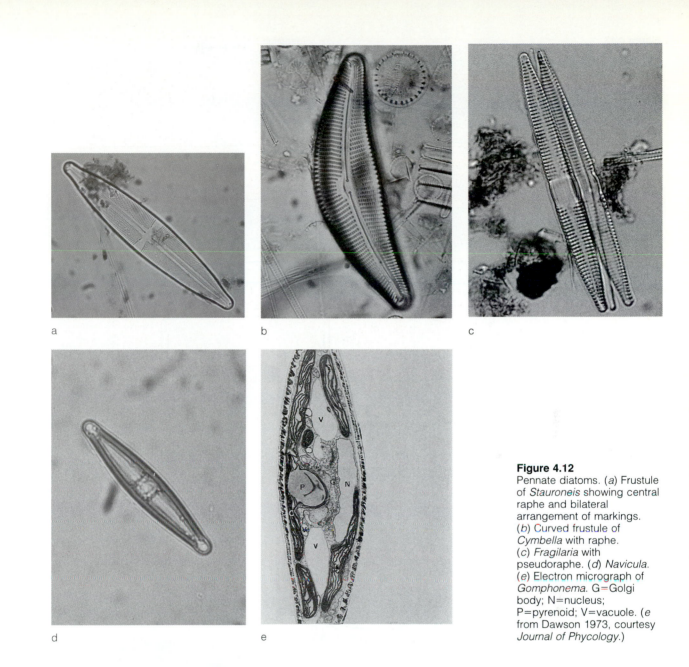

a

b

c

d

e

Figure 4.12
Pennate diatoms. (*a*) Frustule
of *Stauroneis* showing central
raphe and bilateral
arrangement of markings.
(*b*) Curved frustule of
Cymbella with raphe.
(*c*) *Fragilaria* with
pseudoraphe. (*d*) *Navicula*.
(*e*) Electron micrograph of
Gomphonema. G=Golgi
body; N=nucleus;
P=pyrenoid; V=vacuole. (*e*
from Dawson 1973, courtesy
Journal of Phycology.)

(fig. 4.13). Some centric diatoms and araphid pennates show limited movement by secretion of mucilage through pores in their valves (Pickett-Heaps, Hill, and Blaze 1991).

Reproduction in the Diatoms

Diatoms reproduce asexually by cell division to produce two daughter cells. Following mitosis, a cell expands, pushing apart the valves, and divides by furrowing (fig. 4.14*a*). Each daughter cell receives one of the valves of the parent cell and forms a new valve within an elongate **silica deposition vesicle** (fig. 4.14*b*). New girdle bands also form in vesicles. The daughter cell with the original epivalve is the same size as the parent, but the hypovalve of the parent cell becomes the epivalve of the other daughter cell. This daughter cell is usually smaller than

Figure 4.13
Motility in pennate diatoms with a raphe. (*a*) Mucilage is secreted in strands through the raphe and attaches to substrate. (*b, c*) Points of attachment to cell membrane are displaced. Numbers indicate sequential points of attachment of strands to substrate.

direction of movement →

strands of mucilage

1   2   3   4

a

1   2   3   4   5

b

2   3   4   5   6

c

Figure 4.14
Asexual reproduction in diatoms. (*a*) Cell division showing daughter nuclei (nu) after mitosis, cell division by furrowing (fu), and formation of new valves in silica deposition vesicles (sv). Note that the daughter cell receiving the hypovalve of the parent cell is smaller. (*b*) *Navicula*, showing formation of new valves. (*b* from Edgar and Pickett-Heaps 1984b, courtesy *Journal of Phycology*.)

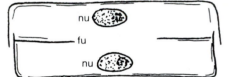
nu
fu
nu

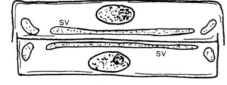
sv
sv

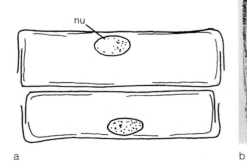
nu

a

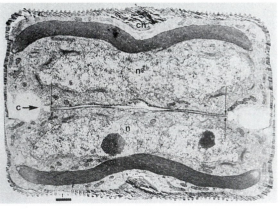

b

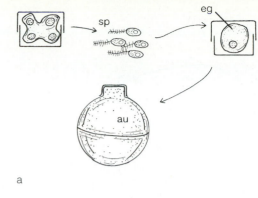

a

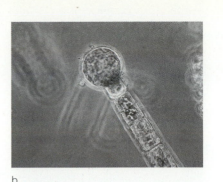

b

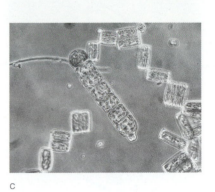

c

Figure 4.15
Sexual reproduction in
centric diatoms.
(*a*) Formation of gametes
(sp=sperm, eg=egg) after
meiosis in gametangial cells,
and auxospore (au) resulting
from fusion of gametes and
enlargement of zygote.
Melosira: (*b*) Spherical
auxospore at end of a
filament. (*c*) Divisions of
auxospore producing large
cells, which remain in a
filament.

the parent. Thus, the average cell size in a population of diatoms may progressively decrease. Sexual reproduction restores the maximum size.

The decrease in cell size as diatoms divide is a means of timing the interval between sexual reproduction (Lewis 1984). Since the rate of cell division in a diatom population is related to environmental conditions, some diatoms show seasonal patterns in cell size and sexual reproduction (Mizuno and Okuda 1985). Sexual reproduction occurs in cells that are less than half the maximum size for the species. Vegetative cells are diploid, and gametes are the only haploid stages. Sexual reproduction differs in centric and pennate diatoms.

In centric diatoms, gametangial cells (cells stimulated to form gametes) undergo meiosis to form eggs and sperm (fig. 4.15*a*). Antheridial cells produce sperm; sometimes, more than four form in a cell if additional divisions occur. Each sperm has a single flagellum with mastigonemes. Oogonial cells usually produce one egg, which may be released into the water or retained in the parental frustule. Fertilization of an egg by a sperm results in a zygote, which then enlarges to several times its original size by uptake of water and forms valves. This enlarged zygote, called an **auxospore,** is often spherical and has a different valve morphology than valves of vegetative cells (fig. 4.15*b*). As the auxospore divides, normal valves form (fig. 4.15*c*). The cells derived from an auxospore are the maximum size for the species.

In pennate diatoms, gametes are not flagellated but conjugate in a process resembling that of desmids. Gametangial cells pair in mucilage, and each cell undergoes meiosis (fig. 4.16*a,b*). Usually, two of the haploid nuclei (sometimes, only one) in each cell are incorporated into amoeboid gametes. Paired cells exchange gametes, resulting in the formation of two zygotes (fig. 4.16*c*). As in

Figure 4.16
Sexual reproduction in
pennate diatoms. (*a*) Pairing
of gametangial cells in
mucilage (sh). Gametes (g)
result from meiosis, and
auxospores (au) develop
from zygotes. fu=empty
frustules of parent cells.
Cocconeis: (*b*) Paired cells
with one gamete in each cell.
(*c*) Fusion. (*d*) Mature
auxospore. (*e*) Large
vegetative cell. (*b–e* from
Mizuno 1987, courtesy
Journal of Phycology.)

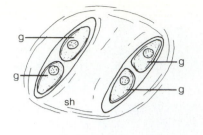
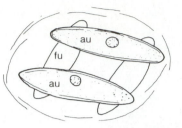

a

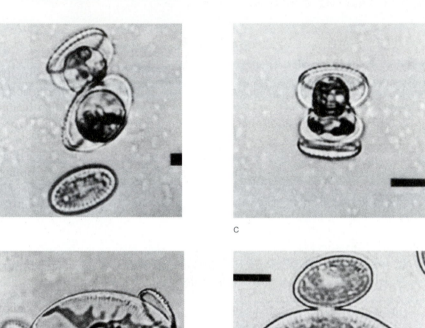

b

c

d

e

centric diatoms, each zygote enlarges to become an elongate auxospore, enclosed by valves and girdle bands, and the maximum cell size for the species is restored (fig. 4.16*d, e*).

In some diatoms, vegetative enlargement occurs by release of the protoplast from its surrounding frustule, enlargement, and formation of a new frustule.

Some centric diatoms form dormant cysts for survival when conditions are unfavorable for active growth. Cyst formation is independent of cell size and unrelated to sexual reproduction. Environmental conditions inducing this event vary among species, but nitrogen depletion and decreasing temperatures are often important (French and Hargraves 1985; McQuoid and Hobson 1995). Cysts usually form internally and have heavily silicified valves with a different morphology than vegetative cells (see fig. 4.17*i*).

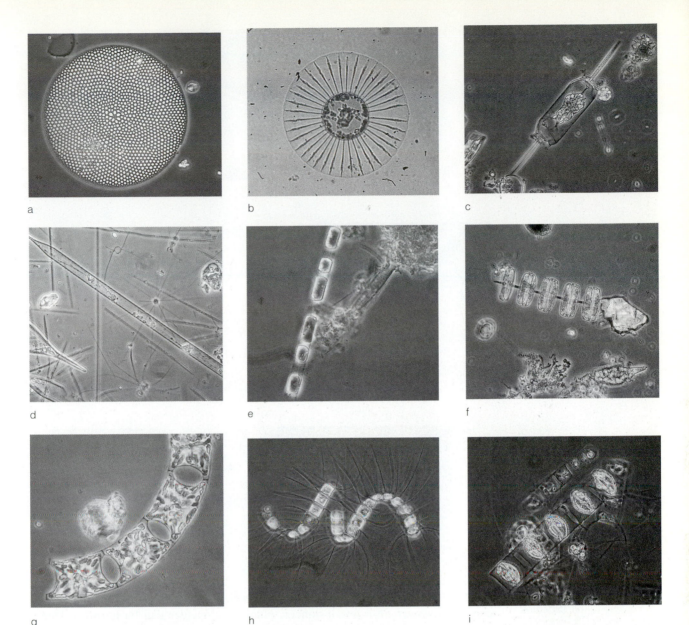

a

b

c

d

e

f

g

h

i

Ecology and Diversity of the Diatoms

The diatoms are usually major components of planktonic and benthic communities in the oceans and freshwaters. It is sometimes convenient to divide the diatoms into three types based on ecologic strategy: (1) euplanktonic diatoms, (2) benthic (periphytic) diatoms, and (3) meroplanktonic (tycoplanktonic) diatoms (Kuhn et al. 1981).

Euplanktonic diatoms are permanent members of the plankton and include nearly all centric diatoms and many pennate diatoms. Euplanktonic diatoms often produce spring and fall blooms in temperate lakes and oceans, and summer blooms at higher latitudes (described in chapter 7). In coastal waters and lakes, they may survive between growing seasons as cysts that settle to the bottom and become dormant (Smetacek 1985). Figures 4.17 and 4.18 show examples of euplanktonic diatoms.

Figure 4.17
Marine euplanktonic diatoms. (*a*) Valve view of *Coscinodiscus*. (*b*) *Planktoniella* with winglike extensions around central area. (*c*) *Ditylum* with spines. (*d*) Elongate cell of *Rhizosolenia*. (*e*) Chain of *Skeletonema* with cells joined by spines from valve margins. (*f*) *Thalassiosira* with cells joined by mucilaginous threads. (*g*) *Euchampia* in a ribbonlike chain. (*h*) *Chaetoceros*. (*i*) Cysts of *Chaetoceros* (all are centric diatoms). (*d* courtesy R. M. Fabricant.)

Figure 4.18
Freshwater euplanktonic
diatoms. (*a*) *Stephanodiscus*
(centric). (*b*) *Nitzschia*
(pennate). (*c*) *Asterionella*
(pennate).

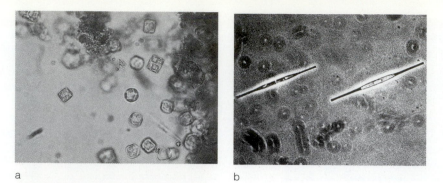

a

b

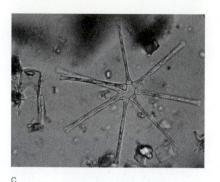

c

a

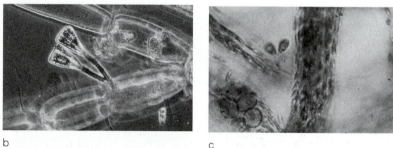

b

c

Figure 4.19
Marine benthic diatoms.
Licmophora (*a*) on *Chondrus*
and (*b*) on *Cladophora*.
(*c*) Tubiferous diatom
Berkeleya.

Most blooms of marine diatoms are beneficial as the first step in aquatic food chains. However, in a few instances, planktonic species can be harmful. Dense growths of diatoms such as *Chaetoceros* may damage fish gills, and *Pseudo-nitzschia* produces a neurotoxin called domoic acid that accumulates in food chains. The first report of a toxic diatom bloom was from Prince Edward Island (Canada) in 1987, when over a hundred people became ill and three people died from eating contaminated mussels. Subsequently, toxic blooms of *Pseudonitzschia* have been reported along both coasts of North America, and in Europe and Japan. The first known occurrence on the U.S. West Coast was in 1991, when seabirds in Monterey Bay, California, died from eating anchovies containing the toxin. The toxin domoic acid accumulates primarily in bivalves (clams, mussels, scallops) and fishes, but also has been found in the commercially important Dungeness crab in the Pacific. Symptoms of "amnesic shellfish poisoning" from eating seafood containing domoic acid include abdominal cramps, diarrhea, nausea, and in severe cases, disorientation, loss of short-term memory, and coma. Domoic acid is an amino acid that overexcites neurons. Birds that have eaten contaminated fish may show abnormal behavior.

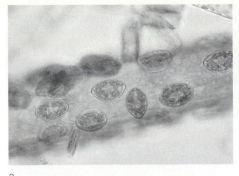

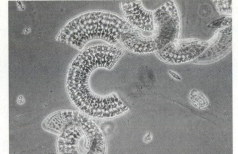

a

b

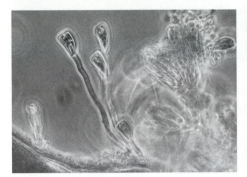

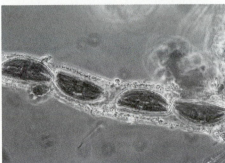

c

d

Figure 4.20
Freshwater benthic diatoms.
(*a*) *Cocconeis* on a
filamentous green alga.
(*b*) *Meridion*.
(*c*) *Rhoicosphenia* with
mucilaginous stalks.
(*d*) *Cymbella* in mucilaginous
tube.

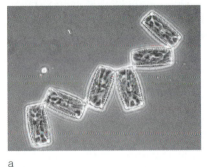

a

b

c

Figure 4.21
Meroplanktonic diatoms.
(*a*) Zigzag chain of *Diatoma*.
(*b*) Ribbonlike colony of
Fragilaria. (*c*) Frustule of
Diatoma showing septa.

Most benthic diatoms are pennates. In both freshwater and marine habitats, diatoms are often the first algae to colonize submerged substrates. Diatoms and bacteria secrete mucilage to form a biofilm that "prepares" the substrate for other organisms to colonize. Diatoms also grow on living substrates, such as other algae and aquatic plants. Dense growths of diatoms may produce golden brown discolorations. Benthic diatoms show a number of different habits: solitary cells growing fixed to the substrate (fig. 4.20*a*), solitary cells showing gliding motility (fig. 4.12*d*), upright clusters of cells in rosettes or fans attached by pads of mucilage at their bases (fig. 4.20*b*), cells growing at the ends of long, mucilaginous stalks (figs. 4.19*a, b,* 4.20*c*), chains of cells (fig. 4.21*a*), and cells in mucilaginous tubes (figs. 4.19*c,* 4.20*d*). Many diatoms on nonliving surfaces move about by means of their raphe system. On beaches, diatoms may migrate down into the sand to avoid disturbance by waves. Epiphytic diatoms often firmly attach to their hosts. The entire valve face may be appressed to the host's surface, or the cells may be attached by a mucilaginous pad or raised on a stalk of mucilage. Mucilage is derived from Golgi vesicles and secreted through pores or a raphe.

Meroplanktonic diatoms are temporary members of the plankton. They are often pennate diatoms that live loosely associated with substrates. During periods of turbulence, they are carried out into the open waters, where they continue to grow suspended in the water (fig. 4.21).

In the Arctic and Antarctic, sea ice covers extensive areas of the ocean. Associated with sea ice are communities of microalgae living in the meltwater on the ice surface, in brine pockets within the ice, and on the bottom of the ice. Pennate diatoms are the most common group. Melting ice releases cells into the water that may seed planktonic growths.

Diatoms may form symbiotic associations with other organisms. Some diatoms contain endosymbiotic cyanobacteria. The planktonic diatom *Rhizosolenia* is able to grow in nutrient-poor seawater because its cells contain the nitrogen-fixing cyanobacterium *Richelia*. Other diatoms are themselves endosymbionts and in this condition may not form frustules. Foraminifera that contain diatoms are common in tropical sand (Lee 1995).

For construction of their walls, the diatoms require silicon in the form of silicic acid $Si(OH)_4$, which cells actively take up from the surrounding water. Silicon often controls diatom seasonality. Diatom blooms may occur when silicon is available, but as this nutrient is depleted, the diatoms decline, and other algae replace them. Because diatom species differ in their requirements, the silica levels may also determine which species are common.

Diatom frustules resist decay and may accumulate in sediments after cells die. Deposits of diatomaceous earth are commercially mined for use in polishes, filters, and paints (to control luster), and as insulators, fillers, and clearing agents. Fossil diatoms are sometimes useful as environmental indicators. Frustules from lake or ocean sediments sometimes provide a record of past ecologic conditions.

Class Phaeophyceae

Unlike other classes of the Chromophyta, the brown algae (class Phaeophyceae) form complex multicellular thalli and for this reason sometimes are placed in a separate division Phaeophyta. However, they have similar pigments, reserves, chloroplast structures, and heterokontous flagellated stages as other members of the Chromophyta. Ultrastructural evidence suggests they are derived from the Chrysophyceae (O'Kelly and Floyd 1985), but nucleotide sequences of ribosomal RNA indicate a closer relationship with the Tribophyceae (Bhattacharya et al. 1992; Leipe et al. 1994). Brown algal walls are composed of cellulose fibrils and the commercially important mucopolysaccharide **alginic acid.** In the brown algae, mannitol is important in osmoregulation (Reed et al. 1985) and as a means of transporting organic material to different parts of the thallus in large species. Polyphenolic compounds protect the brown algae against herbivores and, possibly, parasites (Regan and Glombitza 1986; Steinberg 1985).

No brown algae are unicellular or colonial, but the simplest are branching filaments. More advanced browns have pseudoparenchymatous or parenchymatous thalli and often show differentiation into outer layers of photosynthetic cells composing the **cortex** and inner layers of nonpigmented cells of the **medulla** primarily for storage or transport. Some browns are over 40 meters long.

The brown algae show several growth patterns. In some, growth is diffuse with new cells produced throughout the thallus. In others, new cell production is localized in a **meristem.** Meristematic regions may be at the ends of branches (terminal

Table 4.4 *Orders of Brown Algae (Class Phaeophyceae)*

Order	Macrothallus Construction	Macrothallus Growth
Isomorphic generations		
Ectocarpales	Filamentous, pseudoparenchymatous	Diffuse
Sphacelariales	Filamentous, parenchymatous	Apical
Cutleriales[1]	Parenchymatous	Trichothallic
Tilopteridales[2]	Filamentous	Trichothallic
Dictyotales	Parenchymatous	Apical
Heteromorphic generations		
Chordariales	Filamentous, pseudoparenchymatous	Variable
Desmarestiales[3]	Pseudoparenchymatous	Trichothallic
Dictyosiphonales	Parenchymatous	Variable
Scytosiphonales	Parenchymatous (gametophyte)	Diffuse
Laminariales	Parenchymatous	Meristodermal
No free-living haploid stage		
Fucales	Parenchymatous	Apical
Durvillaeales	Parenchymatous	Diffuse

[1] Heteromorphic generations in *Cutleria*.
[2] See Kuhlenkamp, Müller, and Whittick (1993); Kuhlenkamp and Hooper (1995).
[3] Includes Sporochnales (Müller, Clayton, and Germann 1985).

or apical) or within branches (intercalary). A special type of intercalary meristem at the base of a terminal hair is a **trichothallic meristem** (see fig. 4.33*b*).

Brown algal life cycles are alternation of generations (diplohaplontic). In some orders, the sporophytes (diploid phase) and gametophytes (haploid phase) are isomorphic, with a similar structure and general appearance. Other brown algae have distinctly dissimilar haploid and diploid stages or heteromorphic phases. The sporophyte is usually more complex structurally (parenchymatous or pseudoparenchymatous), while the gametophyte is filamentous. Advanced groups of the brown algae do not have a free-living haploid phase.

The brown algae are divided on the basis of life cycle (table 4.4). Orders are further distinguished by their growth patterns, macrothallus construction, and sexual reproduction. The Dictyotales, Desmarestiales, Laminariales, Fucales, and Durvillaeales have oogamy.

Isomorphic Generations

Filaments and Pseudoparenchymatous Thalli (Order Ectocarpales)

Ectocarpus has one of the simpler thalli among the brown algae and a life cycle with isomorphic generations. It is a common filamentous brown, found growing on rocks or larger algae (fig. 4.22*a*). Its construction is heterotrichous, resembling the green alga *Stigeoclonium* (see fig. 3.26). Basal filaments grow over the substrate in a compact mass, providing attachment (fig. 4.22*b*). Arising from the base are more open branches that are free of the substrate and often end in tapering hairs. Growth is diffuse.

An irregularly shaped chloroplast wraps around each cell's periphery (fig. 4.22*c*). Fucoxanthin is the dominant pigment and produces a brown color. A vacuole is centrally located (fig. 4.22*d*). **Physodes,** which are vesicles containing

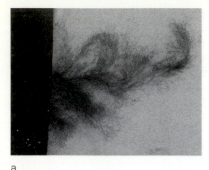

a

b

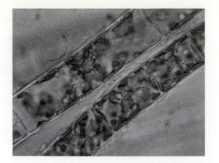

c

Figure 4.22
Ectocarpus. (*a*) Growing as
an epiphyte on *Laminaria*.
(*b*) Young plants showing
heterotrichous construction
with compact branches
attached to substrate and
erect branches, which will
have an open branching
pattern when mature.
(*c*) Detail of cells showing
irregularly shaped
chloroplast. (*d*) Cell
structures as seen with an
electron microscope.
ch=chloroplast; gb=Golgi
body; mi=mitochondrion;
nu=nucleus; ph=physode;
pl=plasmodesmata;
py=pyrenoid; va=vacuole;
wa=wall.

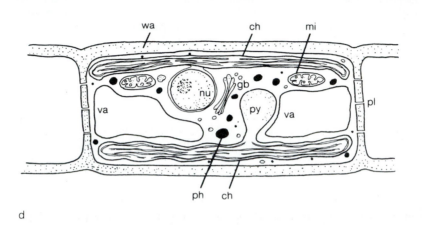

d

polyphenolic compounds, are characteristic of brown algae. Cells are uninucleate.
A prominent Golgi body is associated with the nucleus. Adjacent cells in branches
of *Ectocarpus* share a common end wall, which intercellular connections (plasmodesmata) traverse.

The *Ectocarpus* life cycle involves an alternation of generations (fig. 4.23).
Both phases are filamentous and have a similar appearance, but gametophytes may
be more restricted in their distribution and seasonality than sporophytes. Reproductive structures form at the ends of branches. Sporophytes form two types of
sporangia. At warmer temperatures, **plurilocular sporangia** (often shortened to
"plurilocs") predominate. They are multiseriate regions consisting of a large number of small cells (fig. 4.24*a*, *b*). Each of these cells develops into a zoospore
(fig. 4.24*c*), which is diploid and produces a new sporophyte. At lower temperatures, **unilocular sporangia** ("unilocs") form. Meiosis occurs in these enlarged
cells, followed by several mitotic divisions (fig. 4.24*d*). Usually, a uniloc forms and
releases thirty-two to sixty-four uninucleate spores, which are haploid and give rise
to gametophytes.

The gametophytes form **plurilocular gametangia** ("plurilocs"), which resemble plurilocular sporangia. Gametes are isogamous and, like zoospores, bear
laterally inserted flagella. Male and female gametes are similar in appearance but
functionally distinct. Female gametes settle to the bottom and secrete a chemical
called ectocarpen, which attracts male gametes. Similar pheromones controlling
reproduction have been identified in other brown algae (Müller 1988). The zygote
of *Ectocarpus* develops, without a period of dormancy, into a new sporophyte.

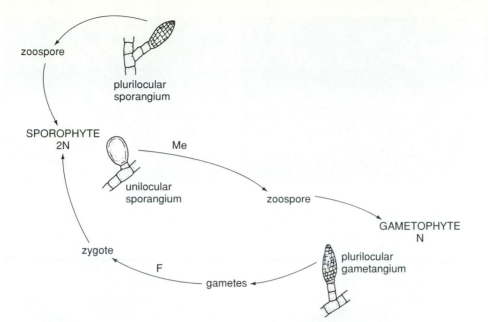

zoospore

plurilocular
sporangium

SPOROPHYTE
2N

Me

unilocular
sporangium

zoospore

GAMETOPHYTE
N

zygote

F

gametes

plurilocular
gametangium

Figure 4.23
Life cycle of *Ectocarpus*.
Sporophyte (2N) forms both
unilocular and plurilocular
sporangia. Gametophyte (N)
forms plurilocular
gametangia. Sporophytes
and gametophytes are
isomorphic. Meiosis (Me)
occurs in unilocular
sporangium. F=fusion of
gametes.

a

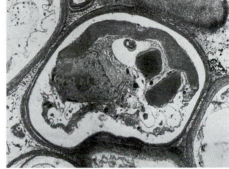

b

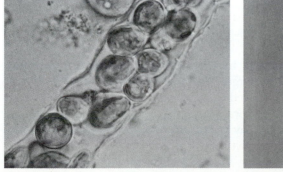

c

d

Figure 4.24
Reproductive structures of
Ectocarpus. (*a*, *b*) Plurilocular
sporangia. (*c*) Zoospores in
sporangium. (*d*) Unilocular
sporangium. (*b* courtesy S.
Carroll.)

Figure 4.25
Pilayella. (*a*) Epiphyte on
blades of *Fucus*.
(*b*) Filamentous thallus.
(*c*) Unilocular sporangia in a
series. (*d*) Intercalary
plurilocular sporangium.

Figure 4.26
(*a*) *Giffordia* showing discoid
chloroplasts and plurilocular
reproductive structure.
(*b*) Cross section of crust of
Ralfsia showing compact
arrangement of branches.

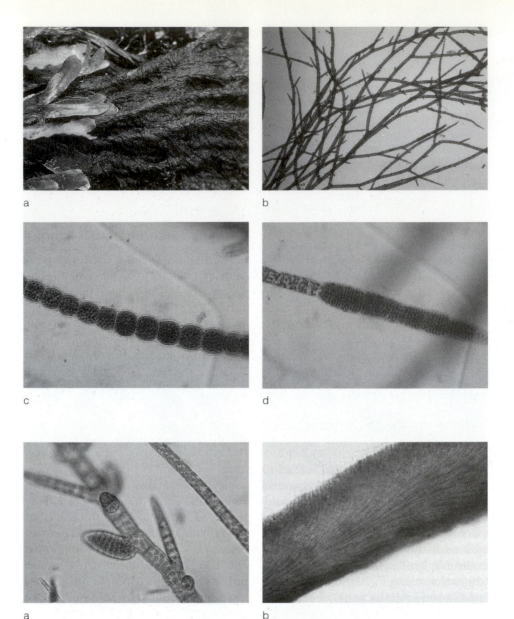

a   b

c   d

a   b

A number of other heterotrichous brown algae are related to *Ectocarpus*. *Pilayella* is a common epiphyte on larger algae (fig. 4.25*a*, *b*). It differs from *Ectocarpus* by having many small, discoid chloroplasts in each cell and intercalary reproductive structures. Unilocular sporangia form in a series between normal vegetative cells (fig. 4.25*c*) (meiosis may not occur in unilocs, Müller and Stache 1989). Plurilocs are also interspersed with vegetative cells (fig. 4.25*d*). *Giffordia* also differs from *Ectocarpus* by having small, discoid chloroplasts in each cell. *Giffordia* forms terminal reproductive structures (fig. 4.26*a*).

Ralfsia forms a dark brown crust on rocks (plate 8*a*). Its thallus has a leathery texture and sometimes resembles a crustose lichen or a tar spot. *Ralfsia* has a pseudoparenchymatous construction. Basal filaments grow along the substrate, and short, erect filaments in a compact mass arise from the basal layer (fig.

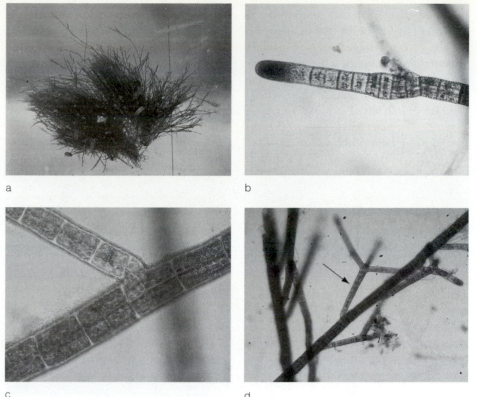

Figure 4.27
Sphacelaria. (*a*) Filaments.
(*b*) Apical cell. (*c*) Tiers of
cells. (*d*) Propagule (*arrow*).

a

b

c

d

4.26*b*). These filaments are the principal site of photosynthesis and thus are called
assimilators. Reproductive structures form in clusters (sori) on the upper surface
of the thallus. Although *Ralfsia* has isomorphic generations, other brown algae
also have ralfsoid stages.

Specialized Groups

Four other specialized orders of brown algae have isomorphic generations (see
table 4.4). Representatives of the Sphacelariales and Dictyotales are described
here.

Sphacelaria is a filamentous genus that shows a regular growth pattern
(fig. 4.27). At the end of each major axis is an apical cell with dense cytoplasm. It
divides to produce a series of axial cells. Axial cells, in turn, divide longitudinally
to produce regular tiers of cells. Branching in *Sphacelaria* may be sparse or show
a regular pinnate pattern.

Sphacelaria reproduces by forming special branches or **propagules**
(fig. 4.27*d*). Propagules detach from the parent filament and are carried by the
water until they encounter a suitable substrate. In one species of *Sphacelaria*,
propagules form during the summer, when temperatures are high and photoperiods
are long. Unilocular sporangia and plurilocular gametangia form during the winter
and spring in response to lower temperatures and shorter photoperiods (Colijn and
Hoek 1971).

The algae in the Dictyotales are more common in tropical oceans than in
colder waters. The common genus *Dictyota* has a ribbonlike thallus with a regular

Figure 4.28
Dictyota. (*a*) Ribbonlike thallus. (*b*) Apical cell.

a

b

Figure 4.29
(*a, b*) Fan-shaped thallus of *Padina.*

a

b

dichotomous branching pattern (fig. 4.28). Growth results from divisions of a dome-shaped apical cell at the end of each branch. The three-layered thallus consists of smaller pigmented cells on each surface and a layer of larger, colorless cells internally. Although *Dictyota* and its relatives have a parenchymatous construction, they are not closely related to the more complex brown algae in the Laminariales and Fucales.

Dictyota shows a number of advanced features in reproduction. Plurilocular and unilocular structures are modified, and sexual reproduction is oogamous. Gametophytes release either uniflagellated sperm or eggs. Sporophytes bear unilocular sporangia that produce four aplanospores (tetraspores). Reproduction in *Dictyota* is usually synchronized with the tides (Phillips et al. 1990 describe reproduction).

Padina is a relative of *Dictyota* with a fan-shaped thallus that a row of apical cells along its edge produces (fig. 4.29). *Padina* is one of the few brown algae with a thallus encrusted with calcium carbonate.

Heteromorphic Generations

Primitive brown algae are filaments with isomorphic generations. Brown algal evolution has shown the following trends: (1) development of an alternation of heteromorphic generations, (2) increased structural complexity of the sporophyte (gametophyte in Scytosiphonales) with either parenchymatous or pseudoparenchy-

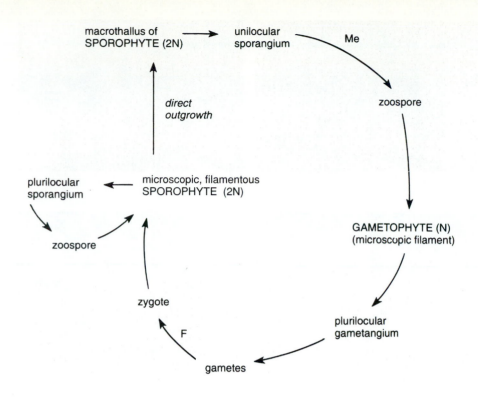

Figure 4.30
Heteromorphic life cycle.
Me=meiosis; F=fusion of
gametes.

macrothallus of
SPOROPHYTE (2N) ⟶ unilocular
sporangium   Me

direct
outgrowth

zoospore

plurilocular
sporangium ⟵ microscopic, filamentous
SPOROPHYTE (2N)

zoospore

GAMETOPHYTE (N)
(microscopic filament)

zygote

plurilocular
gametangium

F

gametes

matous construction, (3) reduction of the gametophytic phase to microscopic
filaments (encrusting sporophyte in Scytosiphonales), and (4) loss of plurilocular
sporangia on the macrothallus.

Among browns with heteromorphic generations, two major lines of development of the sporophytic phase can be distinguished. In the orders Chordariales and
Desmarestiales, sporophytes have a pseudoparenchymatous structure. In the Dictyosiphonales and Laminariales, sporophytes are parenchymatous. In the Scytosiphonales, upright macrothalli are also parenchyamatous. Heteromorphic life
cycles may be responses to seasonal changes, with different phases growing at
different times of the year (Clayton 1988).

In a typical life cycle of an alga with heteromorphic generations, the
macrothallus phase of the sporophyte bears unilocular sporangia (fig. 4.30). The
haploid zoospores produce microscopic filamentous gametophytes, resembling reduced versions of *Ectocarpus* and capable of carrying on photosynthesis. The
gametophytes form gametes. Sexual reproduction ranges from isogamy to oogamy.
The zygote from gametic fusion often gives rise to a filamentous juvenile phase
(plethysmothallus). These microscopic filaments are diploid and may reproduce by
forming zoospores in plurilocular sporangia. Under favorable conditions, they produce the macrothallus as an outgrowth. Thus, the sporophytic generation may have
both macrothallus and microthallus phases.

Some members of the Chordariales, Dictyosiphonales, and Scytosiphonales
modify the typical life cycle. Although they alternate between macrothallus and
microthallus phases in what appears to be a normal alternation of generations, both
phases have the same chromosome number, and neither sexual reproduction nor
meiosis occurs in the life cycle.

Figure 4.31
(*a, b*) *Myrionema* on *Ulva.*
(*c*) *Elachista* on *Fucus.*
(*d*) Thallus of *Elachista* spread to show assimilators (long branches), paraphyses (club-shaped branches), unilocular sporangia (dark cells), and central medullary filaments.

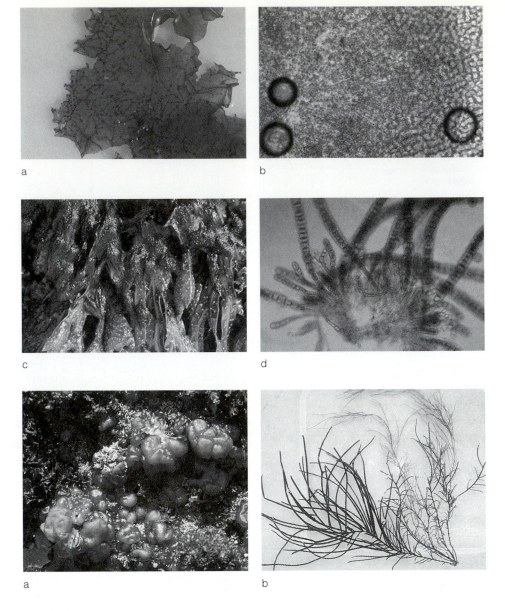

a

b

c

d

Figure 4.32
(*a*) Saccate thallus of *Leathesia.* (*b*) *Chordaria* (darker plant on lower left) with *Dictyosiphon* as an epiphyte.

a

b

Pseudoparenchymatous Sporophytes (Orders Chordariales and Desmarestiales)

Myrionema, Elachista, and *Leathesia* are epiphytes. The macrothallus of *Myrionema* is a pseudoparenchymatous disk of radiating filaments and appears as a dark spot on its host (fig. 4.31*a, b*). *Elachista* forms a tuft of filaments on larger algae, with long, assimilatory filaments radiating from a central mass of medullary filaments (fig. 4.31*c, d*). At the bases of assimilators are unilocular sporangia and associated sterile filaments called paraphyses. In the common Atlantic species *Elachista fucicola,* meiosis fails to occur in unilocs, and diploid zoospores form more typical *Elachista* macrothalli at higher temperatures but microfilaments at lower temperatures (Koeman and Cortel-Breeman 1976). The globular thallus of *Leathesia* ("Sea Potato") is composed of radiating branches (fig. 4.32*a*). In the interior, branches are separated from each other, but at the surface, terminal cells of branches form a continuous, palisade-like layer.

a   b

Figure 4.33
(a) *Desmarestia*. (b) Growing
tip with trichothallic meristem
(me) beneath terminal hair.
co=cortication. (b after
Fritsch 1945.)

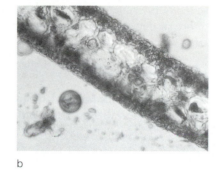

a   b   c

Chordaria has a terete thallus with wiry branches (fig. 4.32b). Its thallus is compactly organized with longitudinally oriented central filaments and side branches forming the outer layer.

Desmarestia is representative of a specialized order of pseudoparenchymatous brown algae. The four genera in the order Desmarestiales have their greatest diversity in the Southern Ocean around Antarctica. *Desmarestia*, however, is widespread in temperate oceans. It has a complex pseudoparenchymatous construction and an advanced oogamous form of sexual reproduction. Species of *Desmarestia* range in general appearance from terete to bladelike (fig. 4.33). All grow by means of a trichothallic meristem at the base of a hair terminating each axis. Thick cortication fills out the thallus around the central axis of filaments. Cells of *Desmarestia* are unusual in accumulating sulfate (SO_4^{2-}) ions from the seawater. The sulfate reacts with water to produce sulfuric acid, which protects against herbivores.

Parenchymatous Macrothalli (Orders Dictyosiphonales, Scytosiphonales)

The order Dictyosiphonales contains brown algae that are multiseriate filaments (for example, *Stictyosiphon*), terete thalli (*Dictyosiphon*, see fig. 4.32b), and blades (*Punctaria*).

Pelatonia and *Scytosiphon* belong to the order Scytosiphonales. *Petalonia* forms a ribbonlike blade (fig. 4.34a, b). Internally, a central medulla of colorless

Figure 4.34
(a) *Petalonia*. (b) Cross
section of *Petalonia* showing
differentiation into cortex and
medulla. (c) *Scytosiphon*.

cells is differentiated from outer cortical layers of smaller, photosynthetic cells. *Scytosiphon* forms a hollow, tubular thallus, often with regular constrictions (fig. 4.34c). During development, the outer cortical layers grow more rapidly than the medulla, which pulls apart except at constrictions.

Scytosiphon is widespread and forms a tubular thallus in winter. This upright thallus, bearing plurilocular reproductive structures, alternates with an encrusting phase resembling *Ralfsia,* which forms unilocular sporangia. Thus, an alternation of heteromorphic generations occurs in which the tubular thallus is the gametophyte and the encrusting thallus is the sporophyte. However, gametic fusion and meiosis only occur in a few populations (Clayton 1980, 1981). More commonly, the flagellated stages from the tubular thallus form the encrusting ralfsoid phase without gametic fusion.

Different environmental conditions in different populations influence the seasonal occurrence of the upright and encrusting thalli of *Scytosiphon*. In some, photoperiod (or photoperiod and temperature) is important in controlling seasonality (Correa, Novaczek, and McLachlan 1986). Swarmers from plurilocs form upright thalli under short photoperiods and form crusts when daylight periods are long. However, other studies report "out-of-season" growth of the upright thallus in the absence of grazing animals (Lubchenco and Cubit 1980). Since the upright thallus is more susceptible to grazers, it normally occurs in winter, when herbivores are less active.

Kelps (Order Laminariales)

The large sporophytes of the kelps with parenchymatous construction show greater differentiation into specialized tissues and overall structural complexity than any other algae. In contrast, gametophytes are reduced filaments, and sexual reproduction is oogamous.

Laminaria saccharina is a common kelp that shows the basic structure of the order (fig. 4.35). Its thallus may reach 3 meters or more in length and is divided into a blade, stipe, and holdfast. The blade grows at its base and is worn away at its apex. The **stipe** is a cylindrical, stemlike region that may vary considerably in length. Rather than the discoid holdfast typical of other brown algae, the holdfast of a kelp is composed of fingerlike extensions called **haptera** that grow into cracks in the substrate.

The kelp show a distinctive growth pattern. The outer layer of cells, called the **meristoderm,** is the principal region of new cell production. The meristoderm is especially active at the transition between the blade and stipe. To a lesser degree, the surface layer throughout the rest of the thallus also shows meristematic activity.

Internally, the thallus differentiates into several layers (fig. 4.35d). A cuticle composed of alginic acid covers the thallus surface. The outer cortex, which includes the meristodermal layer, consists of small, pigmented cells that are active in photosynthesis. The inner cortex consists of more elongate, unpigmented cells that are compactly arranged. The medulla in the center has a looser organization with long, entangled, threadlike cells that do not laterally join. Watery mucilage surrounds them. Sieve cells in the medulla join end-to-end in longitudinal rows. They have characteristic flared ends where two cells meet (fig. 4.35e). Organic material is transported through sieve cells principally as mannitol.

Laminara reproduces by forming a layer of unilocular sporangia and paraphyses on the surface of its blade (fig. 4.36a). Zoospores released from unilocs are haploid and produce filamentous gametophytes consisting of only a few cells

a

bl

st

ha

b

c

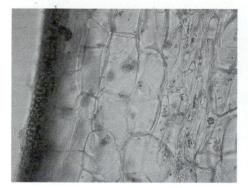

d

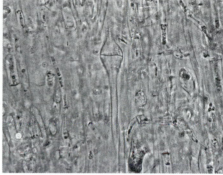

e

Figure 4.35
Laminaria saccharina.
(*a*) Growing in shallow water.
(*b*) Blade (bl), stipe (st), and
holdfast of haptera (ha).
(*c*) Haptera composing
holdfast. (*d*) Cross section
through young blade.
(*e*) Detail of medulla showing
"trumpet hyphae" with flared
ends where two sieve cells
join.

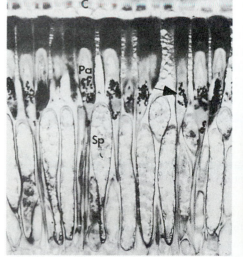

C

Pa

Sp

a

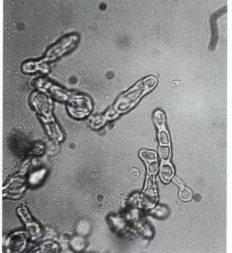

b

Figure 4.36
Laminaria. (*a*) Layer of
unilocular sporangia (Sp) and
paraphyses (Pa) on the
surface of a blade.
C=cuticle. (*b*) Microscopic
gametophytes. (*a* from Henry
and Cole 1982, courtesy
Journal of Phycology.
b courtesy R. K. Scheckler.)

Figure 4.37
Laminaria digitata with a split blade.

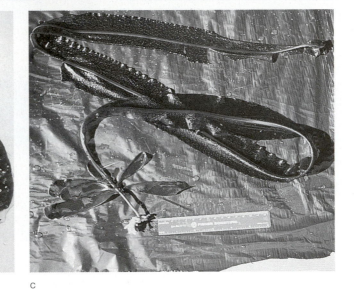

a   b   c

Figure 4.38
East Coast kelps. (*a*) *Chorda.* (*b*) *Agarum.* (*c*) *Alaria* with main blade and sporophylls at base.

(fig. 4.36*b*). Exposure to blue light stimulates gametophytes to form gametangia (Lüning 1980). Separate gametophytes bear oogonia and antheridia. One sperm forms in each antheridium and a single egg in each oogonium. After extrusion from its oogonium, the egg remains attached to the outside of the oogonium and secretes a pheromone (lamoxirene) that stimulates antheridia to release sperm and attracts sperm to the egg (Müller, Maier, and Gassmann 1985). Following fertilization, the young sporophyte overgrows the female gametophyte.

The kelps show a marked contrast between their morphologically complex sporophytic generation and their reduced gametophytic generation. Gametophytes are capable of limited independent existence supported by photosynthetic activity of their vegetative cells. Sexual reproduction by the kelps is an advanced form of oogamy.

Other species of *Laminaria* differ from *Laminaria saccharina* in having inflated stipes or digitate blades. A digitate blade results when wave action splits the blade along predetermined weakened areas (fig. 4.37). Splits extend into the stipe.

If *Laminaria saccharina* represents the basic pattern in the kelps, the following modifications occur in the sporophytes of more complex genera: (1) splitting through the blade into the stipe, (2) inflation of the stipe below the blade to

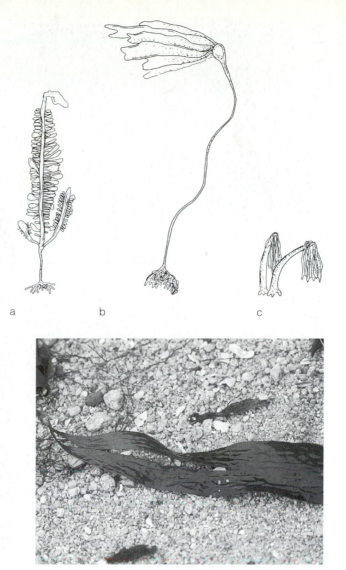

Figure 4.39
West Coast kelps.
(*a*) *Egregia*. (*b*) *Nereocystis*.
(*c*) *Postelsia* (not to scale).

a   b   c

Figure 4.40
Dictyoneurum showing split
in lower part of blade and
stipe.

form a gas-filled bladder, and (3) formation of separate reproductive blades or **sporophylls.** Figures 4.38–4.41 show examples of the kelps. *Chorda* shows a primitive condition. Its thallus is a long, unbranched cylinder without differentiation into a blade and stipe, and its holdfast is a discoid structure without haptera (fig. 4.38*a*). *Agarum* ("Devil's Apron," "Sea Colander," "Shotgun Kelp") has a single blade with distinct holes (fig. 4.38*b*). *Alaria* has a long primary blade and forms a number of smaller paddle-shaped sporophylls at its base for reproduction (fig. 4.38*c*). *Egregia* ("Boa Kelp") has numerous secondary blades along the margins of its stipe and narrow primary blade (fig. 4.39*a*).

Among the larger, more specialized kelps on the West Coast, blade splitting continues through the meristoderm into the stipe. *Dictyoneurum* is a simple example (fig. 4.40). *Nereocystis* ("Bull Kelp") is an annual plant that initially resembles *Laminaria saccharina,* but its stipe lengthens rapidly during development, and its primary blade splits into a large number of straplike secondary blades that may reach over a meter in length (fig. 4.39*b*). Much of the stipe is hollow, and a large bladder containing carbon monoxide at the top of the primary stipe provides

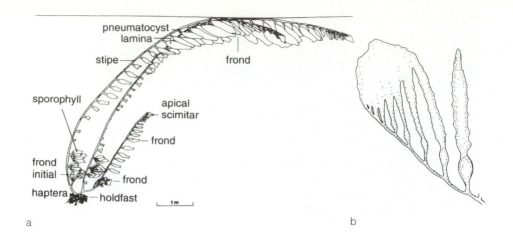

buoyancy. *Postelsia* ("Sea Palm") has a rigid primary stipe that supports the alga while out of the water during low tide. Straplike blades drape down from the top of the stipe, giving the appearance of a palm (fig. 4.39*c*).

Macrocystis ("Giant Kelp"), the largest of the kelps, develops by successive splitting of its terminal blade (fig. 4.41). As each new blade separates, it enlarges and forms a bladder at its base. A mature individual may have several hundred blades. Thalli are usually 10–20 meters long but may reach lengths of 60 meters. Blades may be 1.5 meters long and collectively produce a dense canopy on the water surface. In *Macrocystis* and other large kelps, conducting cells move organic material from blades near the sea surface, which are actively photosynthesizing, to lower parts of the thallus. The sieve cells, which are more highly developed than in *Laminaria*, form a layer between the cortex and medulla and have large pores in their end walls. They transport mannitol and amino acids. Sieve cell structure in the kelps is strikingly similar to that of conducting cells (phloem) in vascular plants, even though these groups are unrelated.

Brown Algae without a Free-Living Haploid Phase (Orders Fucales, Durvillaeales)

Representatives of the order Fucales are widespread in the oceans. The fucoids have a parenchymatous construction but are less complex than the kelps. The fucoid life cycle is more advanced, however, because of the absence of free-living gametophytes.

The genus *Fucus* is a common representative of the order. Its flattened thallus, less than a meter long, has a dichotomous branching pattern (fig. 4.42). An apical cell in a depression at the end of each branch produces new cells. Branches have a thickened midrib and thinner tissue on either side. Small cavities called cryptostomates are scattered over the surface, appearing as bumps with tufts of sterile hairs. These hairs may facilitate uptake of nutrients from the seawater (Hurd et al. 1993). Internally, the thallus differentiates into a cortex and medulla. The common species *Fucus vesiculosus* may have pairs of air bladders along its thallus, which provide buoyancy when submerged (fig. 4.42*a*). Bladder development depends on environmental conditions, being more common in algae growing in calm water.

Fucus forms special reproductive regions, called **receptacles,** at the ends of its branches. In *Fucus vesiculosus,* the receptacles are swollen with mucilage (fig. 4.42*c*), but other species, such as *Fucus distichus,* have flattened receptacles

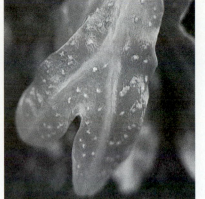

a

b

c

d

Figure 4.42
Fucus. (*a*) *Fucus vesiculosus* with paired air bladders. (*b*) Tip with midrib and cryptostomates. (*c*) Swollen receptacles of *Fucus vesiculosus.* (*d*) *Fucus distichus* with flattened receptacles.

Figure 4.42
Fucus. (*a*) *Fucus vesiculosus* with paired air bladders. (*b*) Tip with midrib and cryptostomates. (*c*) Swollen receptacles of *Fucus vesiculosus.* (*d*) *Fucus distichus* with flattened receptacles.

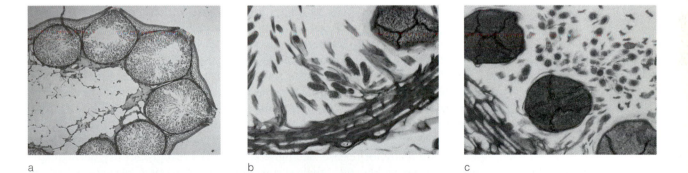

a

b

c

with little mucilage (fig. 4.42*d*). Just below the surface of a receptacle are spherical cavities called **conceptacles** that produce gametes (fig. 4.43*a*). Colorless hairs extend from an opening in each conceptacle. Within a conceptacle, special branches produce gametangia. Depending on the species, male and female gametes may form in the same conceptacle or in conceptacles in different algae. In each antheridium (male gametangium), meiosis is followed by several mitotic divisions to form packets of sixty-four sperm (fig. 4.43*b*). Each oogonium (female gametangium) produces a packet of eight eggs as a result of meiosis and a mitotic division

Figure 4.43
Reproduction in *Fucus.* (*a*) Conceptacles in section through a receptacle. (*b*) Antheridial branch. (*c*) Detail of oogonium with eight eggs.

Figure 4.44
Ascophyllum. (*a*) Common in intertidal region of the North Atlantic. (*b*) Air bladders. (*c*) Receptacles on special branches.

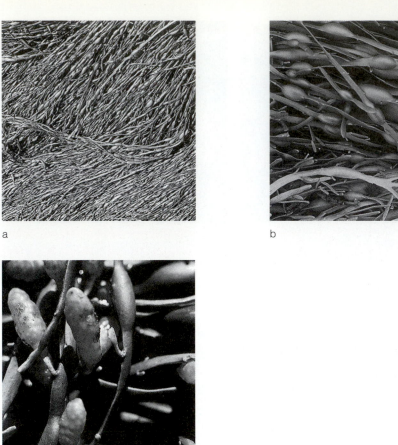

a

b

c

(fig. 4.43*c*). Gamete packets are released through the conceptacle opening into the surrounding seawater, where the packets come apart. Restricting gamete release to periods of calm water leads to a high level of fertilization (Serrão et al. 1996). Sperm are chemically attracted to the much larger eggs. Fertilized zygotes secrete an adhesive to aid in attachment to the substrate. Initial zygote development has been studied extensively (Evans, Callow, and Callow 1982).

The presence of cryptostomates on the surface of *Fucus* (see fig. 4.42*b*) suggests that, at one time, reproduction was not confined to special receptacles. Cryptostomates are homologous to conceptacles but do not form gametangial branches. Receptacles may have evolved to control gamete release. In intertidal plants, such as *Fucus vesiculosus,* the receptacles become swollen and turgid when gametes are ready for release (see fig. 4.42*c*). As the receptacles dry during exposure to the air at low tide, they shrink, forcing the gamete packets out of the conceptacles. The packets stick to the hairs protruding from the conceptacles. The incoming tide dissolves the mucilage around the packets, releasing the gametes into the seawater. By synchronizing gamete release, receptacles maximize fertilization chances and give zygotes time to attach to the substrate before the tide goes out.

Ascophyllum is a common fucoid in the intertidal region of the North Atlantic (fig. 4.44). It may reach 3 meters in length and varies greatly in appearance, depending on the amount of wave activity. Main axes of the thallus are compressed

Figure 4.45
Pelvetia growing high in
intertidal region (in Iceland).

cylinders and bear numerous secondary branches of limited growth. At intervals along the axes, swollen gas bladders provide sufficient buoyancy to support the thallus while submerged. The annual formation of an air bladder along an axis provides a means for aging a thallus (Cousens 1985). *Ascophyllum* commonly has a life span of eight to ten years, compared to the two- to three-year life span of other perennial algae. Some *Ascophyllum* are reported to have lived twenty years. The ascomycete *Mycosphaerella* grows within *Ascophyllum* tissues. The relationship of this fungus to its host is unclear, but *Ascophyllum* sporelings grow better when infected with *Mycosphaerella* (Fries 1988; Garbary and MacDonald 1995).

Ascophyllum forms receptacles on special side branches in response to short photoperiods (fig. 4.44c) (Terry and Moss 1980). Male and female gametes occur in conceptacles on different thalli. The formation of four eggs per oogonium is a more advanced condition than in *Fucus*. Usually, the entire receptacle is shed into the water, aiding in dispersal. The orange-yellow or olive-colored receptacles look like berries as they are carried by the seawater. Temperature controls their gamete release (Bacon and Vadas 1991).

Pelvetia grows in the upper intertidal region along the European Atlantic coast and along the West Coast of North America, but is absent from the western Atlantic Coast (fig. 4.45). Its thallus resembles *Fucus,* but its blades fold to form a channel. Two eggs form per oogonium.

Sargassum is a widespread genus with a large number of species. Its tan thallus resembles a branch of holly. Narrow, stemlike regions connect flattened, leaflike regions with toothed edges, and gas bladders at the ends of branches resemble berries (fig. 4.46). Receptacles form on short branches. The formation of oogonia with only a single egg is considered an advanced condition among fucoids. Each egg has eight nuclei, but only one is fertilized by a sperm.

The genus *Durvillaea* is placed in a separate order on the basis of a diffuse growth pattern. *Durvillaea* is common on rocky shores in colder oceans of the Southern Hemisphere, where it resembles a kelp with a long (up to 10 meters) blade arising from a cylindrical stipe and holdfast. Eggs and sperm form in conceptacles scattered over the thallus.

Figure 4.46
Sargassum. (*a*) Attached species. (*b*) Pelagic species from Sargasso Sea.

a   b

Brown Algal Ecology and Commercial Uses

Most brown algae are members of marine benthic communities. Larger species usually require a firm substrate, such as rock, while smaller species may attach to floating objects or grow as epiphytes. Only a few species normally live free-floating. Brown algal diversity is greatest in colder oceans of both the Northern and Southern Hemispheres, especially on rocky shores. However, members of the order Dictyotales and many fucoids are common in tropical waters.

In the intertidal region on temperate rocky shores, fucoids ("rockweeds") are often abundant and may form distinct zones (review by Chapman 1995). Common genera include *Fucus, Ascophyllum,* and *Pelvetia.* The fucoids are important primary producers of organic material, and protect other intertidal inhabitants by maintaining a moist environment under their fronds during low tide and reducing the force of waves. Their fronds also are surfaces to which epiphytic algae and sessile invertebrates attach. Periodic exposure to the air may increase growth of some intertidal algae (Strömgren 1983), but only *Pelvetia* cannot tolerate continual submergence.

Submarine forests of kelps usually grow in the subtidal region of waters that remain below 20° C throughout the year. *Laminaria* is the major kelp in the Atlantic, while *Macrocystis, Nereocystis,* and a variety of understory genera are important in the Pacific. In the Arctic, *Laminaria* and *Agarum* are common. The ecologic role of the kelps is similar to that of trees in a terrestrial forest in creating a structural framework for the community. In the Antarctic, members of the Desmarestiales replace the kelps as the dominant subtidal browns (Moe and Silva 1977; Amsler et al. 1995).

In tropical and subtropical waters, brown algae commonly grow in lagoons behind reefs. Common genera include *Sargassum, Turbinaria,* and *Padina.* Most species of *Sargassum* are benthic, but two species grow free-floating in the central North Atlantic, where they form extensive rafts. This region enclosed by the major current system is called the *Sargasso Sea.* Although this pelagic *Sargassum* probably originated from coastal populations, *Sargassum natans* and *Sargassum fluitans* are considered separate species. The floating algae do not form receptacles but propagate vegetatively by fragmentation. A unique community of fishes and invertebrates, many of them camouflaged to resemble *Sargassum,* inhabit *Sargassum* rafts.

Sargassum muticum was introduced accidentally into Europe, probably on Japanese oysters (Critchley et al. 1983). Since first reported in 1973, *Sargassum muticum* has spread aggressively along the coasts of France and Great Britain. *Sargassum muticum* is bisexual, allowing the alga to self-fertilize and its zygotes to be carried in detached algae over long distances before release.

Only a few brown algae grow in freshwater habitats. *Sphacelaria* occasionally grows in lakes, and the encrusting brown *Heribaudiella* is found in streams.

Some brown algae are commercially important. In Japan and other Asian countries, the kelps *Laminaria (kombu)* and *Undaria (wakame)* are cultivated for food. Although not a major crop, the reproductive blades of *Alaria* are edible and make an interesting snack. Some brown algae are also sources of alginates (=algin), which are mucopolysaccharides in brown algal walls. These phycocolloids are used as stabilizers and gels in a number of commercial products, including cosmetics, foods, and paints. The principal alginate sources are *Macrocystis* in the North Pacific and *Ascophyllum* in the North Atlantic. Off the U.S. West Coast, boats with rotating blades cut and harvest the tops of *Macrocystis*. Other kelps and fucoids are secondary sources of alginates. Minor uses of brown algae are as animal feed and crop fertilizer.

Summary

1. Algae in the division Chromophyta are characterized by chlorophylls *a* and *c* and heterokontous flagellation. They also have a carbohydrate reserve of β-1,3-linked glucose units and chloroplasts containing thylakoids arranged in groups of three and surrounded by chloroplast endoplasmic reticulum. Chromophyte chloroplasts may be derived from a eukaryotic endosymbiont.

2. In the Chrysophyceae and Synurophyceae, fucoxanthin is an important photosynthetic pigment and gives cells a golden brown color. Flagellated cells and colonies of flagellated cells are sometimes common in freshwaters. Statocysts form as dormant stages.

3. Members of the classes Tribophyceae (yellow-green algae), Eustigmatophyceae, and Raphidophyceae usually are green and are rarely found in abundance. Unicellular and filamentous tribophytes usually grow in freshwater habitats. The class Eustigmatophyceae is separated from the Tribophyceae largely on the basis of ultrastructural differences of its flagellated cells. The class Raphidophyceae contains unicellular flagellates.

4. The diatoms (class Bacillariophyceae) are important in freshwater and marine environments as members of both planktonic and benthic communities. Vegetative cells of the diatoms are nonflagellated and occur as solitary cells, chains of cells, or members of colonies. The diatoms have a distinctive siliceous covering called a frustule that consists of two valves and girdle bands. Markings on valve surfaces and other decorations of the frustule are distinctive for different species. Symmetry of the valve markings is used to divide diatoms into centric and pennate diatoms. Cell divisions for asexual reproduction result in the diatoms decreasing in cell size. Sexual reproduction leading to the formation of an auxospore restores maximum cell size.

5. The brown algae (class Phaeophyceae) are common in coastal environments, especially in colder oceans. Simpler brown algae are heterotrichous filaments with an alternation of isomorphic generations. More advanced browns usually have pseudoparenchymatous or parenchymatous diploid phases and reduced

filamentous haploid phases. The sporophytes of the kelps show the greatest morphological complexity among algae. The fucoids have no free-living haploid phase. The brown algae are common members of marine benthic communities. Alginic acid from some brown algae is used in a variety of commercial products. A few brown algae are cultivated for food in Asia.

Further Reading

Green, J. C., B. S. C. Leadbeater, and W. L. Diver, eds. 1989. *The Chromophyte Algae: Problems and Perspectives*. Clarendon Press.

Round, F. E., R. M. Crawford, and D. G. Mann. 1990. *The Diatoms*. Cambridge University Press.

Sandgren, C. D., ed. 1988. *Growth and Reproductive Strategies of Freshwater Phytoplankton*. Cambridge University Press.

Sandgren, C. D., J. P. Smol, and J. Kristiansen, eds. 1995. *Chrysophyte Algae*. Cambridge University Press.

5

Haptophytes, Dinoflagellates, Cryptomonads, and Euglenophytes

The four divisions of algae discussed in this chapter—Haptophyta, Dinophyta, Cryptophyta, and Euglenophyta—are not closely related to each other. Most of their representatives are unicellular flagellates, and table 5.1 summarizes their characteristics. Each group shows evidence of having independently acquired its photosynthetic system as a eukaryotic endosymbiont (another alga) through a secondary symbiotic event. Molecular comparisons of ribosomal RNA support separating these four groups from other algal divisions (Douglas et al. 1991). At the end of this chapter is a brief consideration of a fifth "odd" algal division, the chlorarachniophytes.

Division Haptophyta

The haptophytes resemble the chromophytes in their photosynthetic pigments and carbohydrate reserve. Their pigments are chlorophylls a, c_1, c_2 (c_3 in some) and the carotenoids fucoxanthin and β-carotene. Their carbohydrate reserve is chrysolaminarin. Haptophyte cells either lack external coverings or have organic scales, which calcify in some species. A typical haptophyte cell has two smooth flagella and a third hairlike appendage, the haptonema, which superficially resembles a flagellum but is used to capture food. Separation of the haptophytes from the chromophytes is primarily on the basis of flagellar structure and the haptonema, but is supported by molecular comparisons (Bhattacharya et al. 1992; Medlin, Lange, and Baumann 1994). The haptophytes are placed in a single class Prymnesiophyceae (or Haptophyceae).

Most haptophytes are solitary, flagellated cells (fig. 5.1). Two smooth flagella arise at a cell's anterior end and are usually of equal length. The flagellar apparatus also consists of two basal bodies, fibrous roots, and three or four microtubular roots. The haptonema also arises from the cell's anterior end and may be straight or coiled. In contrast to the 9+2 microtubular arrangement of flagella, the haptonema contains a ring of three to seven microtubules and has a sticky tip for collecting food particles and possibly for attachment. In some species, the haptonema is reduced or absent.

The haptophyte cell surface may be naked or covered by several layers of scales. Organic scales, composed of carbohydrates, form internally in Golgi vesicles and are then layered on the cell surface. Scales gradually are shed into the surrounding water. Cellulose is a component of the scales in some species.

Table 5.1 *Characteristics of Divisions Haptophyta, Dinophyta, Cryptophyta, Euglenophyta, and Chlorarachniophyta*

	Haptophyta	Dinophyta	Cryptophyta	Euglenophyta	Chlorarachniophyta
Major photosynthetic pigments	Chlorophylls a, c_1, c_2; fucoxanthin	Chlorophylls a, c_2; peridinin (some chlorophylls a, c_1, c_2; fucoxanthin)[1]	Chlorophylls a, c_2; phycobiliproteins	Chlorophylls a, b	Chlorophylls a, b
Carbohydrate reserve	Chrysolaminarin (β-1,3-linked glucan)	Starch (α-1,4-linked glucan)	Starch (α-1,4-linked glucan)	Paramylon (β-1,3-linked glucan)	Unknown
Chloroplast structure					
Thylakoid associations	Three	Three	Two (one to four)	Three	One to three
Envelope	Two membranes	Usually three membranes	Two membranes	Three membranes	Two membranes
Chloroplast ER	Present	Absent	Present (extensive periplastidal space)	Absent	Present
Cell covering	Scales	Theca	Periplast	Pellicle	None
Flagellar appendages	Smooth	Smooth	Mastigonemes (tubular hairs)	Nontubular hairs	Smooth

[1] A few dinoflagellates have chlorophyll a and phycobilins, or chlorophylls a and b.

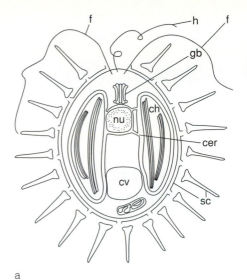

a

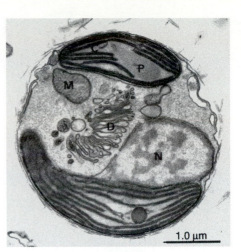

b

Figure 5.1
Haptophytes. (*a*) Typical cell
with a chloroplast (ch)
surrounded by endoplasmic
reticulum (cer),
chrysolaminarin vacuole (cv),
flagella (f), Golgi body (gb),
haptonema (h), nucleus (nu),
and scales (sc). (*b*) Electron
micrograph of
Chrysochromulina with
chloroplasts (C), pyrenoid
(P), nucleus (N), Golgi body
(D), and mitochrondrion (M).
(*b* from Eikrem 1996,
courtesy *Phycologia*.)

a

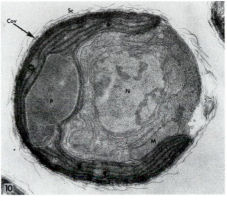

b

Figure 5.2
The coccolithophore
Emiliania. (*a*) Scanning
electron micrograph showing
scales covering cell surface.
(*b*) Electron micrograph
showing cell structures.
N=nucleus; P=pyrenoid;
M=mitochrondrion;
Sc=scales. (*a* courtesy E.
Venrick. *b* from Kaveness
1972, courtesy the author
and the British Phycological
Society.)

The coccolithophores, a distinct group within the haptophytes, have calcified scales
called coccoliths, formed of calcite crystals (fig. 5.2*a*). The characteristic forms of
coccoliths distinguish different species. Larger scales may be visible at high
magnification (1000×) with a light microscope, but only a scanning electron
microscope shows structural details. Functions suggested for coccoliths and other
scales are (1) protect the cell membrane from physical damage, (2) act as ballast
for flotation, (3) concentrate light for photosynthesis, and (4) aid in nutrient uptake
by trapping a layer of water between the scales and cell membrane.

Cells contain two saucer-shaped chloroplasts extending along each side
(fig. 5.1). Chloroplasts have thylakoids organized into groups of three, contain
pyrenoids, and are surrounded by chloroplast endoplasmic reticulum (CER). The
presence of CER suggests that the chloroplast is derived from a eukaryotic endosymbiont. Most haptophytes lack eyespots. In the cell anterior near the base of
the flagella is a Golgi body. A large vacuole with chrysolaminarin occupies the cell
posterior.

Figure 5.3
Phaeocystis occurs (*a*) in large colonies with nonflagellated cells embedded in mucilage or (*b*) as solitary, flagellated cells with a short haptonema.

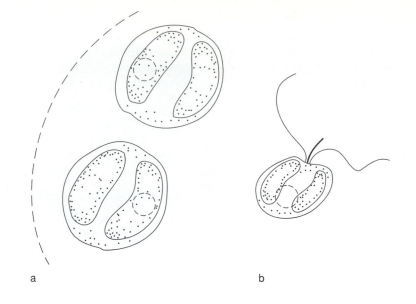

a   b

Cells reproduce by division into two daughter cells. The mitotic spindle may be open or closed. Reported occurrences of sexual reproduction and meiosis are few, and life cycles are poorly known. Many haptophytes have distinct morphologic phases in their life cycles, sometimes with flagellated stages alternating with benthic phases.

Representative Haptophyte Genera

Chrysochromulina is a unicellular flagellate that grows in both freshwater and marine environments. Its ovoid cells appear to have three flagella, but the third "flagellum" is actually a long haptonema. Organic scales but not coccoliths cover cells. Some species have scales bearing long spines. *Chrysochromulina*, like many haptophytes, is a mixotroph, using its haptonema to capture food (Kawachi et al. 1991; Kawachi and Inouye 1995). The sticky-tipped haptonema extends to capture suspended particles (bacteria, detritus) and then bends to deliver captured particles to a site on the side of the cell, where they are ingested by phagocytosis.

Phaeocystis is a colonial haptophyte that is widespread in the oceans, sometimes becoming abundant in polar waters (review by Davidson and Marchant 1992). Typically, it occurs in large colonies, with thousands of spherical cells embedded in extensive mucilage (palmelloid condition) (fig. 5.3*a*). In this stage, cells are nonflagellated and do not have scales or a haptonema. The large colony size and surrounding mucilage may protect against herbivory. *Phaeocystis* also occurs as solitary biflagellated cells with organic scales and a short haptonema (fig. 5.3*b*). When abundant, *Phaeocystis* may discolor the water brown and cause foaming by releasing large quantities of organic material. New colonies form by division of solitary cells and splitting of colonies.

Typical coccolithophores are biflagellated cells with a haptonema. Several layers of scales cover cells, including an outer layer formed of one or several types of coccoliths (see fig. 5.2). Adjacent coccoliths may overlap or abut each other. Underneath the coccoliths are several layers of organic scales. Coccolith structure distinguishes different species and requires viewing with a scanning electron microscope. Most coccolithophores show different morphologic stages in their life cycles, sometimes with a coccolith-bearing stage alternating with a benthic stage without coccoliths. Different stages may have been described as different species. The common coccolithophore *Emiliania* has three types of cells: nonflagellated cells with coccoliths, flagellated cells with organic scales, and nonflagellated coccoid cells without scales.

Haptophyte Ecology

The haptophytes are common in the marine plankton but also occur in freshwater. They are smaller than many diatoms and dinoflagellates. The haptophytes are most diverse in the tropics, where their small size and large surface area, ability to move vertically by flagellar action, and mixtotrophy may allow them to grow when concentrations of inorganic nutrients are low. Some haptophytes are common at higher latitudes, including species that produce dense growths and discolor seawater. At times, *Emiliania* is responsible for extensive blooms in the North Atlantic, while *Phaeocystis* is abundant at high latitudes in both hemispheres, sometimes blooming in the North Sea and in the ice-edge zone around the Antarctic. In temperate oceans, the haptophytes often increase in abundance after a spring diatom bloom. When abundant, coccoliths shed into the water may cause a milky discoloration that satellites can detect (Balch et al. 1991). Dense growths of *Chrysochromulina* may damage fish gills and cause fish mortality (Hallegraeff 1993; Hansen, Nielsen, and Kaas 1995), and *Prymnesium* produces a toxin responsible for killing fish. Both are a threat to fish farms in northern Europe.

As major components of the marine phytoplankton, the haptophytes play an important role in the global carbon cycle by their photosynthetic production and by their uptake of inorganic carbon for coccolith formation. The coccolithophores use bicarbonate to form coccoliths:

$$Ca^{++} + 2\,HCO_3^- \rightarrow CaCO_3 + CO_2 + H_2O$$

The coccoliths of calcium carbonate on the cell surface are continually replaced. Older coccoliths are shed and may accumulate in aggregates of floating particles in the water. When these aggregates sink, they transport the carbon in coccoliths from the surface waters into the deep oceans. Coccoliths may dissolve when they reach deeper, colder water that is undersaturated with calcium carbonate, or they may accumulate on the bottom as calcareous sediments or oozes, which eventually may become chalk deposits. Relatively intact coccoliths in deposits are important microfossils and date back to the early Mesozoic era.

The haptophytes also play an important role in the marine sulfur cycle (Charlson et al. 1987; Iverson, Nearhoof, and Andreae 1989; Ayers, Ivey, and Gillett 1991). Because of their abundance at times, *Phaeocystis* and coccolithophores such as *Emiliania* are important sources of dimethylsulfide (DMS). Their cells use dimethylsulfonium propionate (DMSP) in osmoregulation. After release into the seawater, usually when zooplankton lyse cells, DMSP is converted to DMS and acrylic acid. DMS is a volatile compound that diffuses into the atmosphere, where it is oxidized. These sulfur-oxides contribute to acid precipitation, and sulfur-aerosols seed cloud formation over the oceans, where the haptophytes are a significant source of atmospheric sulfur (fig. 5.4). (Human sources are more

Figure 5.4
Dimethylsulfide.
Dimethylsulfonium propionate
(DMSP), used for
osmoregulation in a cell, is
converted to dimethylsulfide
(DMS) after release into
seawater. DMS is volatile and
in the atmosphere is oxidized
to form sulfur-aerosol that
seeds cloud formation and
sulfur-oxides that contribute
to acid rain.

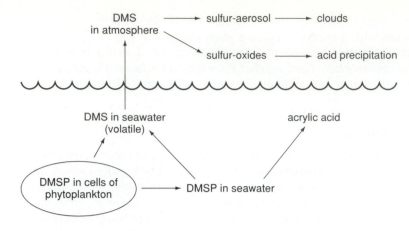

important over land and in coastal areas.) The dinoflagellates and some benthic algae may also release significant quantities of DMS. Acrylic acid, also from the breakdown of DMSP, has antibiotic properties and may inhibit bacterial growth around haptophyte cells.

Several haptophytes (*Pavlova* and *Isochrysis*) are used as food in the mariculture of oysters and mussels.

Division Dinophyta

Most dinoflagellates are unicellular flagellates and include approximately equal numbers of photosynthetic and nonphotosynthetic species. The dinoflagellates with photosynthetic systems are the focus here. Many have chlorophylls a and c_2 and peridinin as their principal pigments, but others have chlorophylls a, c_1, and c_2 and fucoxanthin, or other pigment combinations. Carbohydrates are stored as starch, but lipids are also important reserves. A dinoflagellate cell is not surrounded by a wall but has a theca underlying the cell membrane, which may consist of distinct plates of cellulose. The nucleus and chloroplasts have unusual features. Table 5.1 summarizes dinoflagellate characteristics. Molecular comparisons using ribosomal RNA suggest that the dinoflagellate cell is more closely related to ciliates and apicomplexans than to other algal groups (see fig. 1.11) (Bhattacharya et al. 1992; McNally et al. 1994), and variation in pigmentation suggests that dinoflagellates may have acquired their photosynthetic systems from several different sources. The dinoflagellates also have an unusual form of ribulose bisphosphate not found in other eukaryotes, consisting of only one type of subunit instead of the usual two (Morse et al. 1995).

Biflagellated dinoflagellate cells are of two types. The **desmokonts** have two anterior flagella (fig. 5.5*a*). One flagellum may be coiled over the cell surface. In contrast, the **dinokonts** have laterally inserted flagella. One flagellum is ribbonlike and encircles the cell in a groove, and the other flagellum extends back (fig. 5.5*b*, *c*). An equatorial groove or girdle divides a typical dinokont cell into an **epicone** and a **hypocone.** The posterior flagellum extends through a depression called the sulcus. The name *dinoflagellate* comes from the whirling motion of swimming cells.

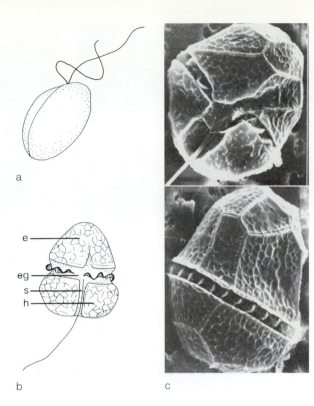

Figure 5.5
Flagellar arrangements of
dinoflagellates.
(*a*) Desmokont with apical
flagella. (*b*) Dinokont with
transverse and posterior
flagella. e=epicone;
eg=equatorial groove with
transverse flagellum;
h=hypocone; s=suclus.
(*c*) Flagella. (*c* from Berdach
1977, courtesy *Journal of
Phycology*.)

Although most dinoflagellates are unicellular flagellates (fig. 5.6*a–f*), colonies of flagellated cells, nonflagellated cells (fig. 5.6*g*), and filaments are also known. Nonflagellated vegetative cells show their dinoflagellate nature when they form dinokont reproductive stages.

Typical Dinoflagellate Cell

Dinoflagellate cells have several unusual features, including: (1) the **theca** and associated structures, (2) the nucleus, and (3) chloroplasts. Figure 5.7 shows typical cells.

Thecal vesicles are in a layer beneath the cell membrane (fig. 5.8*a*). They are flattened vesicles that may enclose distinct cellulose plates or may lack any obvious contents. The size, number, and arrangement of thecal plates varies among dinoflagellates and is important in their taxonomy. The desmokonts typically have two large plates, while thecal plates in the dinokonts vary considerably. Some dinokonts have numerous, indistinct thecal plates, while others with larger plates are described as "armored" (fig. 5.8*b*). Phycologists use the number of thecal plates to try to identify evolutionary patterns. They do not agree, however, on whether the primitive condition had many small plates, suggesting that plates have enlarged and decreased in number, or whether a few plates was the primitive condition, suggesting that the number of plates has increased.

Microtubules and additional membranes (sometimes considered to include the cell membrane) may underlie thecal vesicles. Also associated with the theca may

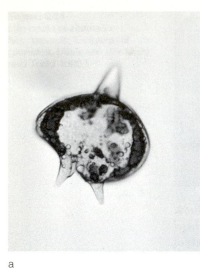

a

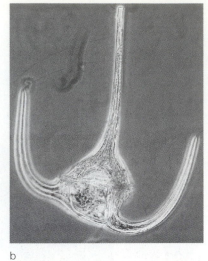

b

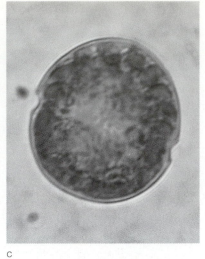

c

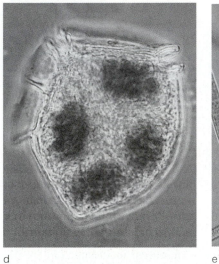

d

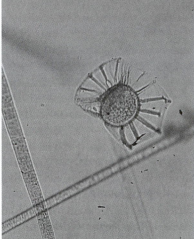

e

Figure 5.6
Representative
dinoflagellates.
(*a*) *Peridinium.* (*b*) *Ceratium.*
(*c*) *Glenodinium.*
(*d*) *Dinophysis.*
(*e*) *Ornithocercus.*
(*f*) *Prorocentrum.*
(*g*) *Pyrocystis.*

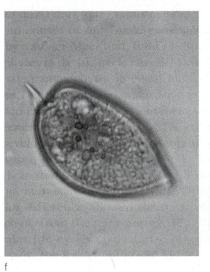

f

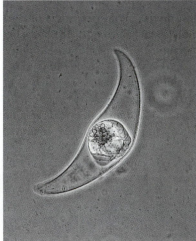

g

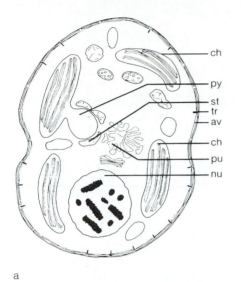

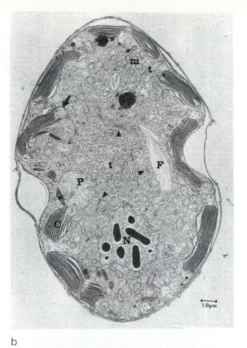

a

b

Figure 5.7

Typical dinoflagellate cell. (*a*) Nucleus (nu) with condensed chromosomes, chloroplast (ch) with protruding pyrenoid (py), thecal vesicles (av), trichocyst (tr), pusule (pu), and starch grains (st).

(*b*) Electron micrograph showing chloroplast (C), mitochondrion (m), nucleus (N), pyrenoid (P), and trichocyst (t). (*b* from Herman and Sweeney 1976, courtesy *Journal of Phycology*.)

a

b

Figure 5.8

Theca. (*a*) Thecal vesicles in section. 1=cell membrane; 2,3=membranes of thecal vesicle; 4=inner membrane. (*b*) Surface view of *Gonyaulax* showing thecal plates. (*a* from Wilcox, Wedemayer, and Graham 1982, courtesy *Journal of Phycology*. *b* from Dodge 1983, courtesy the author and the British Phycological Society.)

be trichocysts. **Trichocysts** are vesicles containing crystalline rods that can be discharged and, presumably, function in defense (fig. 5.9*a*).

The dinoflagellate nucleus differs from the usual condition in eukaryotic cells in a number of features. An envelope surrounds the nucleus, but the chromosomes as seen in electron micrographs are distinct, rod-shaped structures that remain condensed throughout the cell cycle and attached to the nuclear envelope (see fig. 5.7). They also lack histone proteins. The dinoflagellate nucleus is often described as **dinokaryotic** (or mesokaryotic) to distinguish it from the more typical eukaryotic condition.

Mitosis in the dinoflagellates is also unusual. The spindle forms outside the nuclear envelope, which remains intact during mitosis. Channels containing bun-

Figure 5.9
Unusual cell structures.
(*a*) Trichocyst. (*b*) Pusule.
(*a,b* from Wilcox,
Wedemayer, and Graham
1982, courtesy *Journal of
Phycology.*)

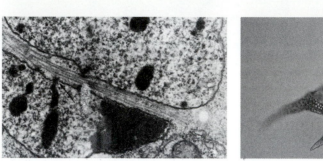

a   b

Figure 5.10
Mitosis and cell divisions.
(*a*) Channel with microtubules
through nucleus during
mitosis. (*b*) *Ceratium* dividing
to form two daughter cells.
(*a* from Dodge 1983,
courtesy the author and the
British Phycological Society.)

a   b

dles of microtubules develop through the nucleus (fig. 5.10*a*), and other micro-tubules attach to the nuclear envelope adjacent to chromosomes. Each chromosome splits into a pair of daughter chromosomes, which separate as the nucleus elongates and eventually "pulls apart" into two daughter nuclei.

Cells contain a number of small chloroplasts, which are often brown but may be red, green, or blue-green. An envelope consisting of three closely associated membranes usually surrounds each chloroplast, and thylakoids within the chloro-plast are in groups of three. Pyrenoids may be embedded in the chloroplast or protrude. Starch does not accumulate in chloroplasts but in the cytosol, sometimes forming a cap on a protruding pyrenoid.

Strong evidence indicates that dinoflagellate chloroplasts are derived relatively recently from endosymbionts that probably were other eukaryotic algae. Support for this origin are: (1) the large number of dinoflagellates without chloroplasts, (2) the chloroplast's triple envelope, and (3) a binucleate condition in a few dinoflagellates. In binucleate cells, one nucleus is a typical mesokaryotic nucleus, and the other is eukaryotic (fig. 5.11). The eukaryotic nucleus is associated with the chloroplast, and a membrane, presumably representing the symbiont's cell membrane, separates both the chloroplast and the nucleus from the remainder of the cell. This membrane becomes the third membrane in the envelope of other dinoflagellates.

Some dinoflagellates have chlorophylls a, c_1, and c_2 and the carotenoid fucox-anthin, which suggests that their photosynthetic system originated from a member of the Chromophyta by the following sequence of events (fig. 5.12): (1) uptake by a dinoflagellate lacking chloroplasts of an endosymbiont with fucoxanthin and an embedded pyrenoid to produce a binucleate condition, (2) loss of the endosym-biont's eukaryotic nucleus, and (3) replacement of fucoxanthin by peridinin, loss

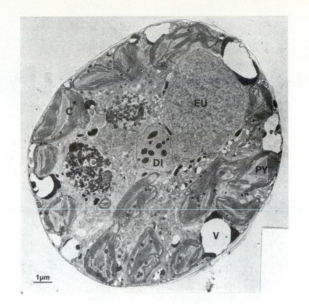

Figure 5.11
Peridinium balticum with
mesokaryotic (DI) and
eukaryotic (EU) nuclei. (From
Tomas and Cox 1973,
courtesy *Journal of
Phycology*.)

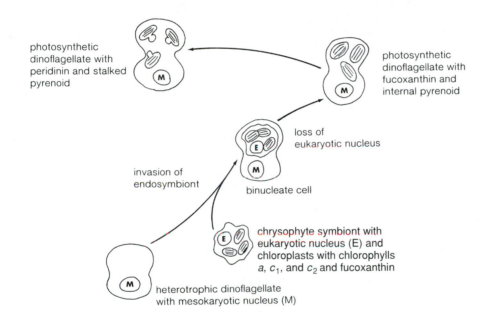

photosynthetic
dinoflagellate with
peridinin and stalked
pyrenoid

photosynthetic
dinoflagellate with
fucoxanthin and
internal pyrenoid

loss of
eukaryotic nucleus

invasion of
endosymbiont

binucleate cell

chrysophyte symbiont with
eukaryotic nucleus (E) and
chloroplasts with chlorophylls
a, c_1, and c_2 and fucoxanthin

heterotrophic dinoflagellate
with mesokaryotic nucleus (M)

Figure 5.12
Possible evolution of
chloroplast in dinoflagellates.
(After Bujak and Williams
1981.)

of chlorophyll c_1, and development of a pyrenoid protruding from the chloroplast
in advanced dinoflagellates.

Chloroplasts of other dinoflagellates may have arisen from different endosymbionts. The blue-green or pink chloroplasts with phycobilins of a few
dinoflagellates may be derived from cryptomonad endosymbionts (Wilcox and
Wedemayer 1985; Geider and Gunter 1989; Farmer and Roberts 1990). A green
dinoflagellate with chloroplasts containing chlorophylls a and b may have arisen
from a green algal endosymbiont (Watanabe et al. 1987).

Other organelles in dinoflagellate cells include Golgi bodies near the nucleus.
The **pusule** is a distinctive dinoflagellate organelle consisting of a series of vesicles near the base of the flagella (see fig. 5.9b). It functions in osmoregulation,

Figure 5.13
Sexual reproduction in *Peridinium*. (*a*) Gametes. (*b*) Fusion. (*c*) Planozygote. (*d*) Resting cyst. 1=spore wall; 2=oil droplet; 3=two nuclei. (*a–d* from Pfiester 1975, courtesy *Journal of Phycology*.)

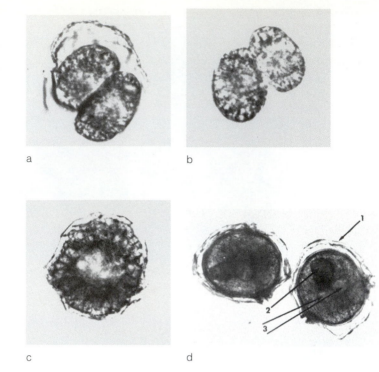

a   b

c   d

macromolecule uptake, and secretion (Klut, Bisalputra, and Antia 1987). A few dinoflagellates have eyespots in the cytoplasm independent of the chloroplasts. Eyespots vary in complexity, with some having lenslike structures. Reserves are stored in the cytosol as starch or lipids. Heterotrophic dinoflagellates may have a tubular protrusion or peduncle near the base of their flagella that is used to suck out the contents of prey.

The dinoflagellates reproduce either by division into two daughter cells (see fig. 5.10*b*) or by zoospore formation. Most dinoflagellates are haploid. (Some heterotrophic dinoflagellates have diploid vegetative cells.) Low nitrogen induces gamete formation (Pfiester 1984), and isogametes are usually produced. Compatible gametes fuse, forming a planozygote, which develops a thick wall and becomes dormant (fig. 5.13). Thus, the zygote functions as a resting cyst and settles to the bottom. The zygote is the only diploid stage, with meiosis occurring when the zygote germinates. Vegetative cells may also differentiate directly into cysts.

Dinoflagellate Ecology

Most dinoflagellates live in the oceans as part of the plankton, but they are also found in freshwaters. A few species are benthic or occur in symbiotic associations. The dinoflagellates vary greatly in their nutrition, ranging from autotrophic to heterotrophic forms that may parasitize invertebrates and fishes or phagocytize other algae. Even photosynthetic dinoflagellates require vitamins (auxotrophy) and may supplement their energy requirements as mixotrophs.

Dense growths of planktonic dinoflagellates produce brown or red water discolorations called **red tides.** Red tides most often occur in coastal waters and estuaries. Some dinoflagellates producing red tides are luminescent, and some contain toxins that are released into the water or accumulate in food chains. In

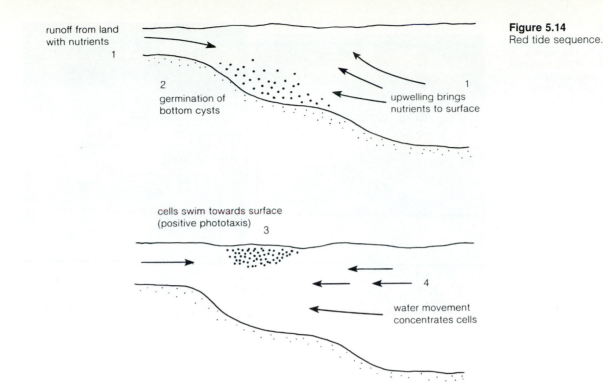

Figure 5.14
Red tide sequence.

runoff from land
with nutrients
1

2
germination of
bottom cysts

1
upwelling brings
nutrients to surface

cells swim towards surface
(positive phototaxis)
3

4
water movement
concentrates cells

severe cases, the toxins may cause fish-kills or lead to human poisonings from eating contaminated mollusks or fishes.

Causes of dinoflagellate blooms are diverse and often related to local conditions (see Anderson, White, and Baden 1985; Granéli et al. 1990). However, some common patterns exist. Growths are normally "seeded" by germination of cysts that have remained dormant on the bottom until conditions are favorable for growth. Water enrichment by upwelling or runoff is often followed by dense cell concentrations that produce red tides. A typical red tide sequence is as follows (fig. 5.14):

1. Germination of cysts on the bottom inoculates cells into the overlying water.

2. The population of cells increases by asexual reproduction.

3. Cells accumulate near the surface as a result of their positive phototaxis.

4. Cells may concentrate further as a result of water flow (produced by onshore winds or tides).

5. After sexual reproduction, zygotes become cysts that sink, maintaining a reserve of dormant stages on the bottom.

Red tides sometimes start in estuaries and spread to coastal waters. The impact of a red tide on marine communities depends on the species responsible. Dinoflagellate respiration at night and cell decomposition when a bloom ends may deplete oxygen supplies. Effects may be more severe when the abundant species contain toxins.

Only a few dinoflagellates (approximately twenty species) are toxic (Hallegraeff 1993). Ordinarily, each species forms several different toxins. The major toxins are **saxitoxin** and its derivatives produced by *Alexandrium*, **brevetoxin**

Figure 5.15
(*a*) *Alexandrium tamarense*, responsible for red tides in New England (*b*) Sign posted to indicate shellfishing is closed because of bloom. (*a* courtesy L. Fritz.)

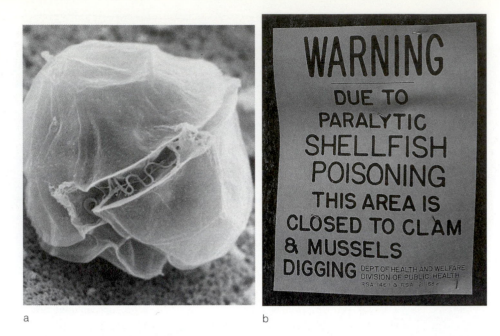

a

b

produced by *Gymnodinium* (=*Ptychodiscus*), and **ciguatoxin** produced by *Gambierdiscus* (see table 7.2). Toxins are deterrents against herbivorous zooplankton, but in some cases, herbivores tolerate and accumulate toxins. When toxins become concentrated in food chains, they can poison humans who eat contaminated fish or bivalves.

Since 1972, blooms of *Alexandrium tamarense* have occurred during the summer in New England (fig. 5.15*a*). "Paralytic shellfish poisoning" (or PSP) in humans may result from eating contaminated bivalves that have fed on *Alexandrium*. The principal toxin, saxitoxin, is a neurotoxin that blocks sodium channels and produces the following symptoms:

> The first symptoms are usually a prickling sensation of the lips, a numbness and swollen feeling of the tongue, then tingling of toes and fingers, a dizziness and headache usually in the back of the head, progressive numbness of arms and legs, staggering and helplessness, shortness of breath, and finally, failure of respiration. (Medcof 1985)

Death may ultimately result from respiratory paralysis. A standard bioassay with mice is used to monitor toxin levels in shellfish, and when a critical level is exceeded, shellfish beds are closed (fig. 5.15*b*).

Gymnodinium causes red tides along the Florida coast and in the Gulf of Mexico and Caribbean Sea. It produces brevetoxin, a neurotoxin. Blooms of *Gymnodinium* usually start on the west coast of Florida and may be carried in currents into the Atlantic. In high concentrations, *Gymnodinium* may kill fish and marine mammals, and may poison humans who eat contaminated bivalves.

Diarrhetic shellfish poisoning is associated with blooms of *Dinophysis* and results from eating contaminated mussels and other bivalves. The resulting digestive tract disturbance lasts for several days but is not fatal. The principal toxin is **okadaic acid.**

Ciguatera refers to poisoning from eating contaminated fishes and is widespread in tropical oceans (Anderson and Lobel 1987). Toxin sources are benthic dinoflagellates, such as *Gambierdiscus* and *Prorocentrum,* that grow on coral

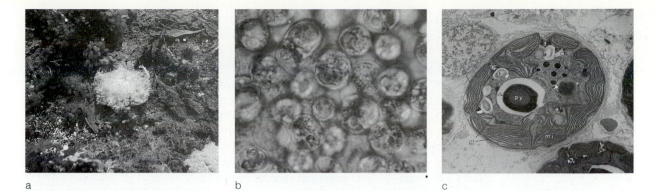

a   b   c

Figure 5.16
Zooxanthellae. (*a*) Mangrove jellyfish (*Cassiopea*) "upside down" to expose zooxanthellae to sunlight. (*b*) Zooxanthellae from *Cassiopea*. (*c*) Electron micrograph showing dinoflagellate within host tissue. C=chloroplast; Ch=chromosomes; mi=mitochondrion; py=pyrenoid; st and ss=starch. (*c* courtesy A. Sam.)

rubble and macroalgae, and that grazing fishes eat. The principal toxins in ciguatera are ciguatoxin and maitotoxin, and unfortunately, there is no easy and reliable way of detecting their presence. Cooking does not destroy the toxins. Typical symptoms of ciguatera resemble those of a number of other illnesses and include gastrointestinal disturbance and numbness around the mouth. Symptoms develop several hours after eating contaminated fish and may last for several days.

A recently discovered nonphotosynthetic dinoflagellate kills fish and feeds on their remains (Burkholder et al. 1992; Boyle 1996). *Pfiesteria*, described as an "ambush predator" and a "phantom fish killer," has been reported in estuaries along the southeastern United States. Although it may occur in a typical dinokont form, it usually exists in an amoeboid stage (Steidinger, Burkholder, and Glasgow 1996). This stage secretes a neurotoxin that kills fish and then phagocytizes their flesh, resulting in a "backward" food chain in which an alga kills and eats fish, rather than vice versa.

A few dinoflagellates, including species of *Gonyaulax, Pyrodinium, Pyrocystis,* and the nonphotosynthetic *Noctiluca,* are luminescent. They usually emit flashes of blue light in response to mechanical disturbance of the water. The light is enzymatically produced (luciferin-luciferase reaction) in small vesicles (scintillons) in their cells. Bioluminescence may protect the dinoflagellates against herbivores by startling attackers (Esaias and Curl 1972) or by acting as a "burglar alarm" to attract fish that prey on the herbivores (Mensinger and Case 1992; Abrahams and Townsend 1993). Some luminescent dinoflagellates are responsible for spectacular displays. In Oyster Bay, Jamaica, migration of *Pyrodinium* toward the surface during the day and concentration of cells by tidal flow into the bay produce impressive luminescent displays at night (Seliger et al. 1970).

Many dinoflagellates show daily vertical migrations, ascending to the surface during the day and descending into deeper water at night. This movement allows them to take advantage of higher nutrient concentrations in deeper waters when surface waters are nutrient limited (Lieberman, Shilo, and Rijn 1994).

Symbiotic, photosynthetic dinoflagellates, called **zooxanthellae,** belong to *Symbiodinium* and several other genera, and are found in cells of a variety of hosts, including protozoa, giant clams (*Tridacna*), flatworms, sea anemones, jellyfish (fig. 5.16*a*), and reef-forming corals. The zooxanthellae are present in a non-flagellated condition in vesicles of host cells (fig. 5.16*b, c*). The alga provides its host with oxygen and organic material, primarily in the form of glycerol but also glucose, alanine, and fatty acids in lesser amounts, and in turn, receives protection and inorganic nutrients from its host. More than 50% of algal production may be transferred to the host. Chemical "host factors" regulate symbionts' release of organic material.

Division Cryptophyta

The cryptomonads or cryptophytes are unicellar flagellates containing chlorophylls a and c_2 and phycobiliproteins as their photosynthetic pigments. (Some cryptomonads have chlorophyll c_1 [Schimek et al. 1994].) Phycobiliproteins are also present in the cyanobacteria and red algae, where they organize into granular phycobilisomes on thylakoid surfaces. In contrast, the phycobiliproteins of the cryptomonads are in the space the thylakoids enclose and only a single type is present at a time. The cryptomonads have either phycoerythrin or phycocyanin; allophycocyanin is never present (Spear-Berstein and Miller 1989). Some cryptomonads are nonphotosynthetic. Cryptomonad cells have several unusual structural features, including a cell covering called a **periplast** and an extensive **periplastidal space** that CER encloses. The principal reserve is starch.

Typical Cryptomonad Cell

A typical cryptomonad has an ovoid cell with two flagella arising slightly below the anterior end (fig. 5.17). Each flagellum usually has a single row of mastigonemes. Some cryptomonads have a furrow (sometimes erroneously called a gullet) extending along one side of their cell, with the flagella arising near its anterior end (Kugrens, Lee, and Andersen 1986). The flagellar apparatus consists of microtubular and fibrous roots. A periplast composed of protein plates associated with the inner side or both sides of the cell membrane surrounds the cells (fig. 5.18). The periplast maintains cell shape. Usually associated with the periplast are **ejectosomes,** organelles with tightly coiled ribbons that can be discharged. Presumably, they have a defensive function like the trichocysts of dinoflagellates. Cells are uninucleate. The cryptomonads usually lack an eyespot.

A chloroplast extends the length of a cryptomonad cell. The chloroplast has two lobes with a pyrenoid in the cross connection (see figs. 1.9d, 5.17b) and contains thylakoids in groups ranging from one to four, with pairs common. The thylakoids contain either phycocyanin or phycoerythrin. The cryptomonads vary greatly in color, with some being brown, blue, or red.

In addition to an envelope of two closely associated membranes, a layer of CER surrounds the chloroplast. Unlike the chromophytes and haptophytes, where space between the CER and chloroplast envelope is limited, the cryptomonads have an extensive periplastidal space containing distinct structures, such as ribosomes (80S), vesicles, microtubules, and a **nucleomorph** (fig. 5.19). Like a nucleus, the nucleomorph has a double-membrane envelope and contains DNA (Ludwig and Gibbs 1985). The pyrenoid projects from the chloroplast into the periplastidal space, where starch grains may form a cap on the pyrenoid.

The CER and structures in the periplastidal space strongly suggest that the cryptomonad chloroplast originated from a eukaryotic endosymbiont. The presence of starch and phycobiliproteins, and the nucleotide sequences in ribosomal RNA (see fig. 1.11) indicate that the symbiont was a red alga (Penny and O'Kelly 1991; Douglas et al. 1991). The periplastidal space contains the remains of other cellular components of this symbiont, including the nucleomorph as the remnant of its nucleus. Molecular comparisons show that the host cell is related to rhizopodal protozoa (McFadden, Gilson, and Hill 1994).

The cryptomonads reproduce by cell division, preceded by mitosis. During mitosis, the nuclear envelope partially breaks down. Division occurs by furrowing. Sexual reproduction among the cryptomonads is rare (see Kugrens and Lee 1988).

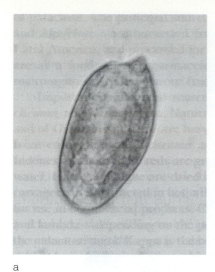

a

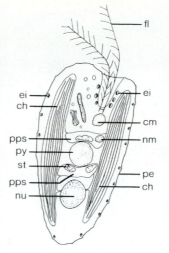

b

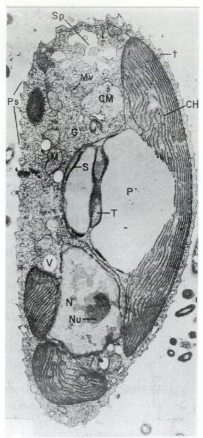

c

Figure 5.17
Typical cryptomonad structures. (a) *Cryptomonas* as seen with a light microscope. (b) Cell showing chloroplast (ch), vesicle (cm, Corps de Maupas) ejectosome (ei), flagellum (fl) with mastigonemes, nucleomorph (nm), nucleus (nu), periplast (pe), periplastidal space (pps), pyrenoid (py), and starch grain (st). (c) Electron micrograph showing section through *Cryptomonas*. Ch=chloroplast; P=pyrenoid; S=starch; CM=Corps de Maupus; N=nucleus; Ps=periplast; t=ejectosome. (c from Lucas 1970, courtesy *Journal of Phycology*.)

Figure 5.18
(a) Periplast of *Chroomonas*. (b) Section with periplast and ejectosomes. (a from Kugrens, Lee, and Andersen 1986, courtesy *Journal of Phycology*. b from Gantt 1971, courtesy *Journal of Phycology*.)

a

b

Figure 5.19
Periplastidal space between chloroplast (C) and chloroplast endoplasmic reticulum (CER).
E=ejectosome;
Fg=fibrillogranular body;
G=globules;
M=mitochondrion;
N=nucleus;
NM=nucleomorph;
Pc=periplastidal space;
Py=pyrenoid; S=starch.
(From Gillott and Gibbs 1980, courtesy *Journal of Phycology*.)

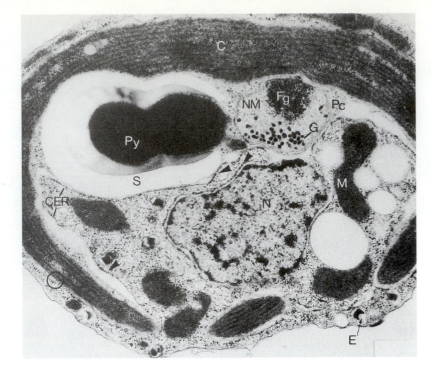

Cryptomonad Ecology

The cryptomonads grow in both freshwater and marine habitats. In lakes, they are often important members of the phytoplankton. Because of their small size and delicate nature, however, they are often overlooked in studies and their ecologic importance underestimated.

The cryptomonads are also endosymbionts in invertebrates, protozoa, and dinoflagellates. *Mesodinium,* which sometimes discolors seawater reddish brown, is a ciliate containing a cryptomonad symbiont.

Division Euglenophyta

Many euglenophytes lack photosynthetic systems and are strictly heterotrophic. Only photosynthetic species are considered here. The euglenophytes resemble the chlorophytes in having chlorophylls *a* and *b* as their major photosynthetic pigments, but their carbohydrate reserve, called **paramylon,** is a β-1,3-linked polymer of glucose. With one exception, all euglenophytes are unicellular flagellates. They lack a cell wall but have a pellicle underlying the cell membrane. Table 5.1 summarizes euglenophyte characteristics.

Typical Euglenophyte Cell

A typical euglenophyte has one or two flagella arising from an indentation or reservoir slightly below the cell's anterior end (fig. 5.20*a, c*). Two flagella, as seen in *Eutreptia,* is the primitive condition. However, common genera, including *Euglena,* have only a single emergent flagellum. A second rudimentary one may be present in the reservoir. Most euglenophytes do not have any external covering

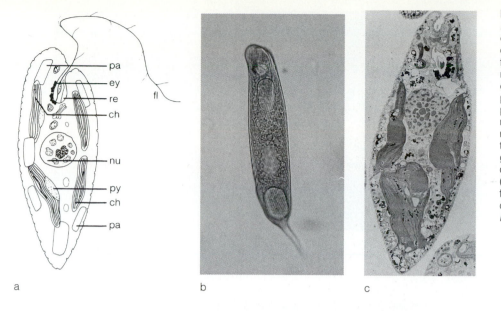

Figure 5.20
Euglena. (*a*) Cell showing chloroplast (ch), eyespot (ey) on reservoir (re), emergent flagellum (fl) with swelling in reservoir, nucleus (nu) with condensed chromosomes, paramylon granule (pa), pyrenoid (py). (*b*) Light micrograph showing numerous small chloroplasts, two large paramylon granules, and pellicle (strips on surface of cell.)
(*c*) Electron micrograph. (*c* from Walne and Arnott 1967, courtesy the authors and *Planta* Springer-Verlag.)

outside of the cell membrane, but a **pellicle** composed largely of protein is associated with the cell membrane's inner surface. The strips of the pellicle spiral around the cell (fig. 5.20*b*). Some euglenophytes have a plastic form that changes as they swim. As the cell expands and contracts, the strips of the pellicle slide relative to each other (Suzaki and Williamson 1985). Other euglenophytes have rigid pellicles. In the genus *Trachelomonas,* a lorica surrounds the cell. Minerals give the lorica a dark brown color.

Within the reservoir of a euglenophyte cell, a swelling on one flagellum contains flavin and acts as a light receptor. Associated with the inner surface of the reservoir membrane is an eyespot composed of pigmented granules. The eyespot interacts with a flagellar swelling to sense the direction of light. The cell rotates as it swims, causing the eyespot to shade the swelling during each rotation. Orienting to minimize this shading causes the cell to swim toward the light. Mutants without a flagellar swelling do not respond to light, even when an eyespot is present. The euglenophytes also respond to gravity (Lebert and Häder 1996).

The euglenophytes are uninucleate (fig. 5.20*a,c*). Like the dinoflagellates, the euglenophytes have chromosomes that remain condensed. Granules of the reserve paramylon occur in the cytoplasm as crystals surrounded by a membrane (Kiss, Vasconcelos, and Triemer 1987). Cells have a number of small chloroplasts. Each chloroplast is surrounded by an envelope consisting of three closely associated membranes and contains thylakoids associated in triplets.

As in the other algae considered in this chapter, the euglenophyte chloroplast probably originated from an endosymbiotic eukaryotic ancestor. Supporting evidence are the large number of colorless species and the presence of a triple envelope. The extra membrane of the chloroplast envelope is interpreted as the symbiont's cell membrane. The presence of chlorophyll *b* indicates that the endosymbiont was a green alga, but nucleotide sequences for ribosomal RNA in the chloroplast suggest a closer relationship with chromophytes (Douglas and Turner 1991).

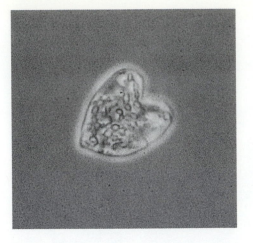

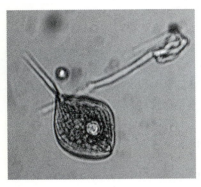
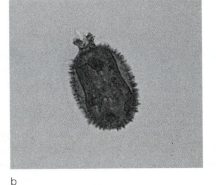
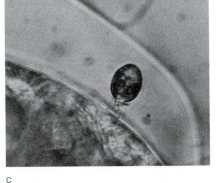

a
b
c

Figure 5.22
Representative
euglenophytes. (a) *Phacus.*
(b) *Trachelomonas* (see
fig. 1.12a), (c) *Colacium.*

Reproduction in the euglenophytes occurs by longitudinal division of cells (fig. 5.21). During mitosis preceding cell division, the nuclear envelope remains intact, and the nucleolus persists. Under unfavorable conditions, the euglenophytes may form thick-walled cysts and become dormant. Sexual reproduction is unknown in the group.

Figure 5.22 shows representative euglenophytes.

Euglenophyte Ecology

The euglenophytes are found in freshwater and seawater, and tend to be abundant in waters with a high organic content. Blooms of a red species of *Euglena* sometimes turn pond water the color of blood. Even though many euglenophytes are photosynthetic, they can also assimilate organic material (mixotrophy) and require at least one vitamin (cyanocobalamin and sometimes thiamine).

Colacium is the only euglenophyte normally occurring as a nonflagellated cell (fig. 5.22c). When swimming, it resembles *Euglena,* but it loses its flagellum when it attaches by a mucilaginous stalk to submerged substrates. *Colacium* often grows on the exoskeletons of zooplankton.

Chlorarachnion has an amoeboid cell containing green chloroplasts but does not fit into any recognized algal group (Hibberd and Norris 1984). Its cells lack walls and form delicate pseudopods that sometimes link cells in a network. Cells contain a central nucleus and several chloroplasts with chlorophylls *a* and *b*. Starch has not been detected as a reserve. Four membranes, interpreted as the chloroplast envelope and CER, surround each chloroplast. Pyrenoids protrude from the chloroplasts into the periplastidal space between the two sets of membranes. The periplastidal space also contains a nucleomorph with DNA. The cells are photosynthetic but also obtain nourishment by ingesting organic particles. Reproduction is by formation of zoospores with a single lateral flagellum, and a dormant cyst stage occurs.

The chlorarachniophytes may have arisen through a secondary endosymbiotic event in which an amoeboid cell phagocytized a eukaryotic alga, possibly a green alga, which became its chloroplasts (McFadden et al. 1994).

Division Chlorarachniophyta

Summary

1. The algal divisions discussed in this chapter consist largely of unicellular flagellates that are not closely related to each other. Each shows a different combination of characteristics that suggests that its chloroplast was derived from a eukaryotic endosymbiont.

2. The haptophytes (division Haptophyta) are common in the marine plankton. Most are unicellular flagellates bearing two smooth flagella and a haptonema for capturing food particles. Chlorophylls *a* and *c* and fucoxanthin are their principal photosynthetic pigments. Chloroplast endoplasmic reticulum surrounds their chloroplasts. Scales cover many haptophytes, including the coccolithophores with calcified scales. Some haptophytes become abundant at higher latitudes. The sinking of their calcareous scales transports carbon into deep water, and their release of dimethylsulfide seeds cloud formation over the oceans.

3. The dinoflagellates (division Dinophyta) usually have chlorophylls *a* and *c* and peridinin and a starch reserve. (Some have fucoxanthin in place of peridinin, or less commonly, other pigment combinations.) Most species are biflagellated, with either two apical flagella (desmokonts) or one flagellum encircling the equator of the cell and the other posteriorly directed (dinokonts). Distinctive features of dinoflagellate cells include a theca composed of vesicles associated with the cell membrane, a nucleus with condensed chromosomes attached to the nuclear envelope, and an unusual form of mitosis. The chloroplast of most dinoflagellates probably is derived from an endosymbiotic chromophyte. The dinoflagellates are often abundant in the oceans and occasionally in lakes. Dense growths of some species produce red tides and luminescent waters. A few species are toxic. Some dinoflagellates enter into symbiotic associations with invertebrates, including reef-forming corals.

4. The cryptomonads (division Cryptophyta) have chlorophylls *a* and *c* and a phycobiliprotein as their principal photosynthetic pigments. Two flagella with mastigonemes arise near the anterior end of cryptomonad cells. Distinctive cell features include a periplast and an extensive periplastidal space that the chloroplast endoplasmic reticulum surrounds and that contains a nucleomorph. The chloroplast may be derived from a red algal ancestor. Cryptomonads are probably more widespread than reported.

5. The euglenophytes (division Euglenophyta) contain chlorophylls *a* and *b* and a reserve called paramylon. One or two flagella arise from a reservoir near the cell's anterior end. Distinctive cell features include a pellicle of proteinaceous strips, an eyespot associated with the reservoir, and a nucleus with condensed chromosomes. The chloroplast may be derived from endosymbiotic green algae. Euglenophytes are widespread, especially in organically enriched waters.

6. Chlorarachniophytes with an amoeboid form contain chloroplasts derived from a green algal symbiont (as indicated by their pigments), a nucleomorph, and chloroplast endoplasmic reticulum.

Further Reading

Gibbs, S. P. 1990. The evolution of algal chloroplasts. In W. Wiessner, D. G. Robinson, and R. C. Starr, eds., *Experimental Phycology 1*, 145–57. Springer-Verlag

Granéli, E., B. Sunström, L. Edler, and D. M. Anderson, eds. 1990. *Toxic Marine Phytoplankton*. Elsevier.

Green, J. C., and B. S. C. Leadbeater, eds. 1994. *The Haptophyte Algae*. Clarendon Press.

McFadden, G. I. 1993. Second-hand chloroplasts: Evolution of cryptomonad algae. *Advances in Botanical Research* 19:189–230.

Santore, U. J. 1992. The Cryptophyceae or: History repeats itself. In W. Reisser, ed., *Algae and Symbioses,* 443–69. Biopress.

Winter, A., and W. G. Siesser, eds. 1994. *Coccolithophores*. Cambridge University Press.

6

Red Algae (Division Rhodophyta)

The red algae (division Rhodophyta) are distinguished from other eukaryotic algae by their lack of flagellated stages and their photosynthetic systems with chlorophyll *a* (and carotenoids) and phycobilisomes (table 6.1). Phycobilisomes are composed of three types of phycobiliproteins: phycoerythrin, phycocyanin, and allophyco-cyanin (see table 2.2). Sometimes, chlorophyll *d* is reported as a second form of chlorophyll in red algae, but no evidence indicates that it has a functional role in photosynthesis. The red algae have a carbohydrate reserve of α-1,4-linked units of glucose with branches at the sixth carbon, which is similar to glycogen or the amylopectin form of starch.

A wall surrounds red algal cells. In many species, the principal wall component is cellulose, but other reds have mannans (polymers of mannose) or xylans (polymers of xylose). Mucopolysaccharides associated with the walls include **agar** and **carrageenan.** Both are polymers of galactose and commercially important as stabilizers and gels.

A typical red algal cell has one or more nuclei (fig. 6.1). A large vacuole may fill the central region of a cell. One or more chloroplasts are in the surrounding cytoplasmic layer. Unlike in the chloroplasts of other eukaryotic algae, the thylakoids are separated from each other (see figs. 1.9*a*, *e*, 6.1*d*) and have granular phycobilisomes on their outer surfaces, which are the principal light-harvesting structures. Light energy transfers to chlorophyll *a*, embedded in the thylakoid membrane, by the following pathway:

Table 6.1 *Characteristics of Division Rhodophyta*

Major photosynthetic pigments	Chlorophyll *a*; phycobiliproteins
Carbohydrate reserve	Starch
Chloroplast structure	
Thylakoids	Unassociated
Envelope	Two membranes
Chloroplast ER	Absent
Cell covering	Wall
Flagella	Absent

Figure 6.1
Representative red algal cells. (*a*) *Audouinella* showing central vacuole surrounded by cytoplasmic layer with chloroplasts and nucleus (*middle right*). (*b*) Electron micrograph of *Polysiphonia*. c=chloroplast; n=nucleus; v=central vacuole. (*c*) Three cells in a filament of *Clathromorphum* showing intercellular connections with plugs. (*d*) Electron micrograph of chloroplasts and mitochondria of *Polysiphonia*. (*e*) Pit connection with plug. CM=cap membrane; IC=inner cap; OC=outer cap; P=cell membrane; W=wall between cells. (*b* from Scott, Phillips, and Thomas 1981, courtesy *Phycologia. c* from Pueschel and Miller 1996, courtesy *Journal of Phycology. d* courtesy J. Z. Kiss. *e* from Pueschel 1987, courtesy *Journal of Phycology.*)

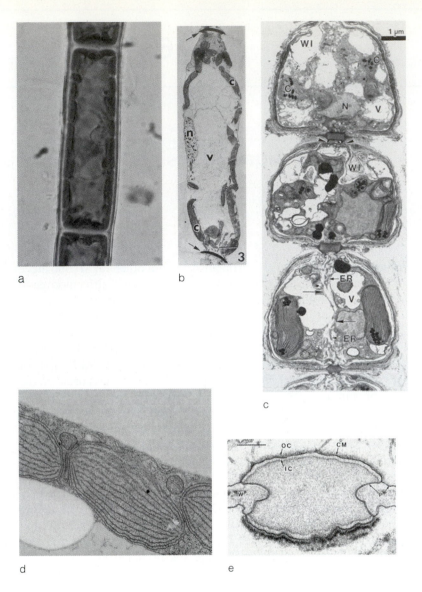

phycoerythrin → phycocyanin → allophycocyanin → chlorophyll *a* (in photosystem II)

When phycoerythrin is the dominant pigment, cells have a red color, but the overall color of red algae varies greatly. The rhodophyte chloroplast with a double-membrane envelope probably derives from a cyanobacterial symbiont. The chloroplasts of some red algae have pyrenoids. Grains of starch occur in the cytoplasm.

Most red algae are multicellular and grow by vegetative divisions. During mitosis, the nuclear envelope persists. Centrioles are lacking, but the mitotic spindle radiates from "nuclear associated organelles," which are often ring-shaped (Scott and Broadwater 1990). Cell division is by furrowing.

Although some red algae are solitary cells or parenchymatous blades, most are filaments. Some filamentous reds have openly branching filaments, but many have a compact thallus with a pseudoparenchymatous construction in which individual branches are difficult to distinguish.

a

b

Plate 1
(*a*) A bloom of cyanobacteria on Onondaga Lake, New York.
(*b*) Cyanobacteria and other bacteria forming colored bands around a hot spring in Yellowstone National Park.

Plate 2
(*a*) The flagellated green alga *Haematococcus* with a layer of mucilage separating the cell wall from the cell membrane. A red pigment (astaxanthin) is present in the cytoplasm. Flagella are not visible.
(*b*) The colonial green alga *Pediastrum.*
(*c*) *Ulva* growing submerged in a tidepool.

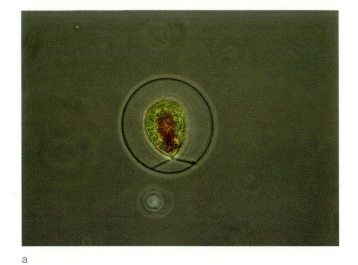

a

b

c

a

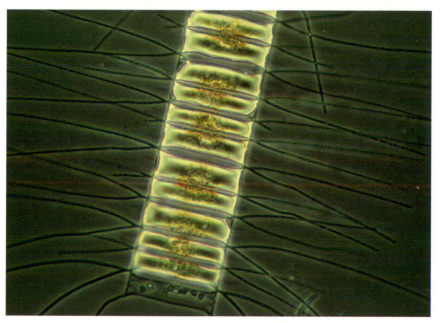

b

Plate 3
(*a*) Colony of *Synura* showing typical golden brown color of a chrysophyte alga.
(*b*) The marine diatom *Chaetoceros* in a ribbonlike chain.

Plate 4
(*a*) The red alga *Gracilaria* (Indian Ocean).
(*b*) Red algae in a New England tidepool: *Porphyra* (sheetlike, *left*), *Corallina* (jointed pink, *lower center*), encrusting coralline with limpet on surface (*center*), *Chondrus* (brown, *right*).

a

b

a

b

Plate 5
(*a*) The cyanobacterium *Gloeocapsa* forming a purple discoloration on a rock.
(*b*) Snow discolored by algal growth.

Plate 6
(*a*) Intertidal zonation on Maine coast, showing upper spray zone with dark film of cyanobacteria, barnacle zone (light band), broad fucoid zone (mostly *Ascophyllum*), and lower red algal zone (tan). (*b*) Lower intertidal region in New England with edge of fucoid zone of *Ascophyllum* (left), *Mastocarpus* (brown), and *Chondrus* (yellow).

a

b

a

b

c

Plate 7
(*a*) Underwater view of kelp zone with *Laminaria* on a rocky shore in New England. (*b*) Subtidal red algae *Callophyllis* between *Phycodrys*. (*c*) Urchin barren (mussels visible beyond urchins).

Plate 8

(*a*) High tidepool with red crusts of *Hildenbrandia*, brown crusts of *Ralfsia*, green filaments of *Rhizoclonium*, and herbivorous snails (*Littorina*).

(*b*) Tidepool with kelps and green algae, and surrounded by fucoids (*Ascophyllum*, *Fucus*).

(*c*) *Littorina littorea*, grazing on the green alga *Enteromorpha*.

a

b

c

Table 6.2   *Comparison of Bangiophycidae and Florideophycidae*

	Bangiophycidae	Florideophycidae
Thallus form	Filament, blade, unicell	Filament, pseudoparenchymatous thallus
Growth	Usually diffuse	Usually apical
Chloroplasts per cell	One stellate	Many discoid
Wall fibrils	Mannans, xylans, cellulose	Cellulose
Life cycle	Monophasic or biphasic	Usually triphasic, biphasic in some
Reproductive structures	Undifferentiated	Differentiated

The red algae are placed in a single class Rhodophyceae, which is divided into two subclasses, Bangiophycidae and Florideophycidae. Formerly, a distinct pit connection and plug between cells distinguished the floridean red algae from bangean reds (fig. 6.1*e*). Cell division by furrowing is incomplete in floridean algae, and a pit connection is left in the center of the cross wall between adjacent cells. A grooved plug, which fits into the edges of the furrow, fills the opening. Secondary pit connections may develop between nonsister cells. All floridean algae have this characteristic plug. However, plugs have been found in stages of some bangean algae. A combination of characteristics, including the presence of a plug, now is used to distinguish the two subclasses of red algae (table 6.2). Floridean algae normally show apical growth, their cells have numerous small chloroplasts, and their life cycles are complex. In contrast, bangean red algae show diffuse growth, usually have a single central chloroplast, and have simpler life cycles without special reproductive structures.

Subclass Florideophycidae

Compared to other algal groups, floridean reds, which are all branching filaments, vary relatively little in basic structure. In some species, branching is open and the filamentous nature apparent. In many species, however, filaments grow in a compact pseudoparenchymatous mass. Thallus structure is of relatively little value for taxonomic classification.

Floridean red algae have complex life cycles. In contrast to the two phases in an alternation of generations of other algae, most species of floridean red algae have three phases: the gametophyte, the carposporophyte, and the tetrasporophyte (fig. 6.2). Gametophytes are haploid and produce nonflagellated gametes. Sexual reproduction is a specialized form of oogamy. The carposporophyte is a diploid stage that develops from a zygote on the female gametophyte and produces spores that develop into tetrasporophytes. Meiosis occurs during spore formation on tetrasporophytes, and the spores develop into gametophytes. Thus, the life cycle has two diploid phases: the carposporophyte and the tetrasporophyte. The tetrasporophyte and gametophyte are independent (free-living) phases, while the carposporophyte is associated with the gametophyte. Life cycle details, particularly regarding carposporophyte development, and characteristics of pit plugs are important in distinguishing taxonomic groups.

We shall discuss four aspects of floridean diversity: (1) thallus structure, (2) life cycles, (3) postfertilization events during carposporophyte formation, and (4) plugs in pit connections.

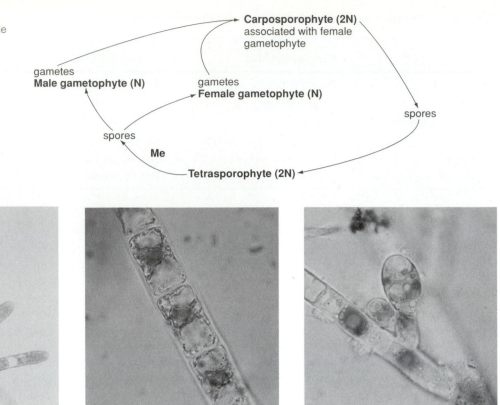

Figure 6.2
Typical three-phase life cycle of floridean red algae. F=fusion of gametes; Me=meiosis.

Carposporophyte (2N) associated with female gametophyte

gametes
Male gametophyte (N)

gametes
Female gametophyte (N)

spores

spores

Me

Tetrasporophyte (2N)

a

b

c

Figure 6.3
Audouinella. (*a*) Filament of cells with parietal chloroplasts. (*b*) Stellate chloroplasts. (*c*) Monospore formation for asexual reproduction.

Thallus Structure

The floridean reds have four general structural types: (1) openly branching filaments, (2) fleshy and foliose erect thalli, (3) encrusting thalli (including calcareous forms), and (4) calcareous erect thalli. The last three categories have pseudoparenchymatous construction.

Openly Branching Filaments

In its morphology and life cycle, *Audouinella* shows a primitive condition among floridean algae. Its thallus is a heterotrichous filament, which grows on rocks or other algae (fig. 6.3*a*). As in other heterotrichous forms, basal branches grow over the substrate, providing attachment. Sometimes, the holdfast is reduced to a single cell. The erect branches have a more open pattern and may have terminal hairs. *Audouinella* grows from apical cell divisions at the end of each branch. The number and shape of chloroplasts in *Audouinella* cells vary (and formerly were used to distinguish several genera) (see figs. 6.1*a*, 6.3*a*, *b*).

Batrachospermum is a freshwater red alga found in streams. Its thallus is soft and gelatinous, and may be brown, olive, gray, violet, or blue-green. *Batrachospermum* shows greater specialization among its branches than *Audouinella*. Growth results from an apical cell dividing at the end of each axis (fig. 6.4). The resulting cells form a central axial filament of cylindrical or barrel-shaped cells. At the upper end of each axial cell, a ring of four to six small pericentral cells forms. These cells initiate three different types of branches. Most branches are of limited or

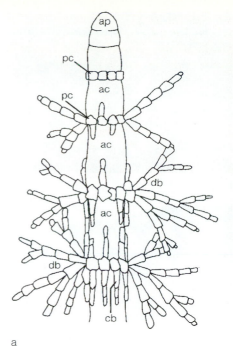

a

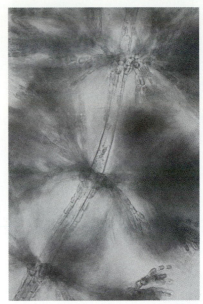

b

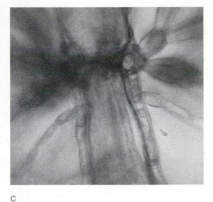

c

Figure 6.4
Batrachospermum.
(*a*) Growing axis. Divisions of apical cell (ap) produce axial cells. Each axial cell (ac) forms a ring of pericentral cells (pc), which give rise to branches. Determinate branches (db) radiate out, while corticating branches (cb) grow over axial cells. (*b*) Uniaxial construction with whorls of lateral branches. (*c*) Cortication on axial cell.

Figure 6.5
Callithamnion, a simple filamentous red alga.

determinate growth and are composed of beadlike cells arising in dense whorls from the nodal regions. Occasionally, a new axis, showing indeterminate growth, grows in place of a determinate lateral. Also arising from pericentral cells are **branches** of limited growth that form a layer around the axis called **cortication.** Instead of growing outward, they grow over the adjacent axial cells, forming a covering of smaller cells.

Simple uniseriate filaments, such as *Callithamnion* (see figs. 1.13*c*, 6.5), are also found in the order Ceramiales, which is considered an advanced group on the basis of its life cycle. Other ceramialean algae form complex filaments with

Figure 6.6
Ceramium. (*a*) General appearance. (*b*) Branch tip showing apical cell and "lobster claw" tips. (*c*) Complete cortication. (*d*) Incomplete cortication.

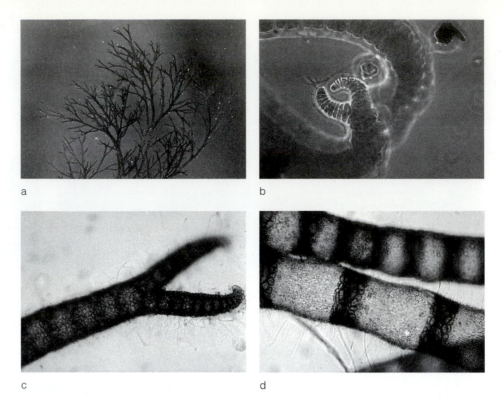

a

b

c

d

Figure 6.7
Corticated genera. (*a*) *Plumaria* with a pinnate branching pattern. Branch tips are uncorticated. (*b*) *Centroceras.* (*c*) *Spyridia.*

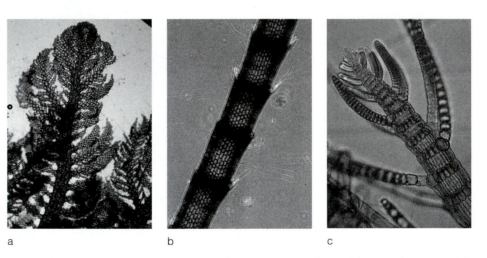

a

b

c

cortication or polysiphonous construction. *Ceramium* is a widespread genus with branching filaments (fig. 6.6). Several layers of smaller corticating cells cover its axial cells. Cortication may be limited to the nodal regions between axial cells or extend completely over axial cells. As in *Batrachospermum,* cortication results from special branches that arise from pericentral cells at the nodes and grow over the adjacent axial cells. Cells of the corticating branches are pigmented, while the axial cells containing large vacuoles may be unpigmented. Figure 6.7 shows other corticated genera.

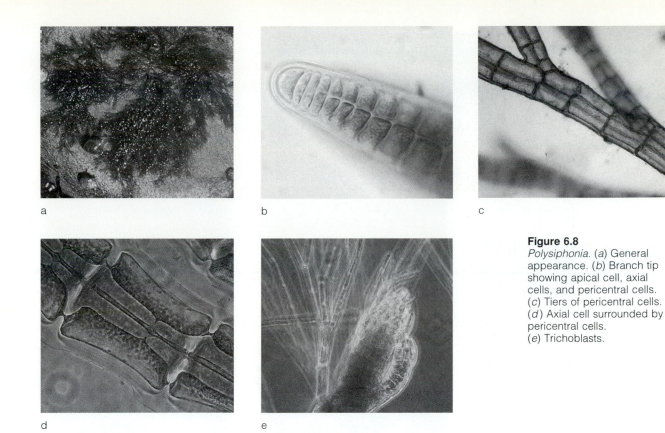

a

b

c

d

e

Figure 6.8
Polysiphonia. (*a*) General appearance. (*b*) Branch tip showing apical cell, axial cells, and pericentral cells. (*c*) Tiers of pericentral cells. (*d*) Axial cell surrounded by pericentral cells. (*e*) Trichoblasts.

Polysiphonia is a common filamentous red alga composed of cells arranged in regular tiers, a condition described as **polysiphonous** (fig. 6.8). The number of cells per tier characterizes a species. Each major branch grows from transverse divisions of an apical cell. The resulting axial cells lengthen and divide longitudinally to form pericentral cells. The pericentral cells, surrounding the axial cell, are the same length as the axial cell, which can be seen in the center in cross section. *Polysiphonia* and other red algae may have colorless hair cells (trichoblasts) (fig. 6.8*e*). The hair cells form in response to nutrient deficiency and may increase the surface area for nutrient uptake (Oates and Cole 1994).

Fleshy and Foliose Erect Thalli

Pseudoparenchymatous algae normally attach by discoid holdfasts. From the holdfast center arises the upright thallus, which may be a blade, a system of cylindrical branches, or hollow tubes. Each upright axis is composed of one or more central axial filaments and lateral branches of determinate growth. The ends of lateral branches form a palisade-like layer, which creates the outer surface of the thallus (fig. 6.9).

Nemalion has a relatively loose pseudoparenchymatous construction. Its dark purple thallus consists of soft, cylindrical branches, which feel like earthworms (fig. 6.10). In contrast to *Batrachospermum* with a single central axial filament, *Nemalion* has many axial filaments (multiaxial). Side branches form a compact outer surface. The considerable space among the branches within the thallus is

Figure 6.9
Pseudoparenchymatous
construction in the erect axis
of *Crouania*. Note central
axial filament and lateral
branches. Terminal cells of
branches form continuous
outer surface of thallus.

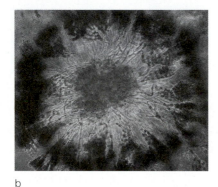

a

b

c

Figure 6.10
Nemalion. (*a*) General
appearance of gametophyte.
(*b*) Cross section of thallus
showing central axial
filaments (multiaxial) and
lateral branches. Pigmented
cells form outer dark layer.
(*c*) Two compact
carposporophytes
(cystocarps).

filled with mucilage. The outer layer of the thallus differentiates as a cortex where photosynthesis occurs, while the colorless inner region is a medulla.

Dumontia has hollow branches (usually collapsed tubes) (fig. 6.11*a*). Each branch is organized around a central axial filament (uniaxial), as in *Batrachospermum*. Space between the side branches is extensive, except where the terminal cells form a compact, palisade-like layer at the outer surface.

Agardhiella, Gracilaria, and *Chondrus* have compact constructions. Branches of *Agardhiella* and some species of *Gracilaria* are cylindrical (fig. 6.11*b*; plate 4*a*), while *Chondrus* and other species of *Gracilaria* have flattened branches (fig. 6.11*c*). *Chondrus* usually is brown, but upper parts of intertidal algae may bleach to a light yellow.

In foliose reds, such as *Phycodrys,* lateral branches arising from the central axial filament are largely confined to one plane (fig. 6.12). When the branches are compact, they produce a continuous surface.

Encrusting Thalli

Encrusting reds grow appressed to the substrate. Some are calcified. *Hildenbrandia* is widespread in the oceans and occasionally found in freshwaters. Its crusts are usually so thin that they appear to be simply dark red discolorations on rocks

a

b

c

Figure 6.11
(*a*) *Dumontia.* (*b*) *Agardhiella.*
(*c*) *Chondrus,* a common
intertidal plant in the North
Atlantic.

a

b

Figure 6.12
Foliose red algae.
(*a*) *Phycodrys* with leaflike
blade. (*b*) Foliose
construction in *Hypoglossum*
resulting from branching in
one plane. (*b* redrawn from
Fritsch 1945.)

(plate 8*a*; fig. 6.13). The thallus is composed of basal filaments that grow over the substrate and of upright branches that arise from the basal layer, forming a compact mass.

Crustose corallines belonging to the order Corallinales, are common in the oceans, often appearing as pink spots on submerged surfaces (fig. 6.14; plate 4*b*). Instead of forming flat crusts, a few corallines grow as nodules, called rhodoliths. Coralline algae have a thick covering of calcium carbonate (and magnesium carbonate), which may represent over 70% of their dry weight. Calcification gives the thallus a pink color and a stony texture. Like *Hildenbrandia,* the coralline thallus consists of a basal layer of filaments that produce short, upright branches in a compact mass. Calcium carbonate, in the form of calcite, deposits in the walls of these cells. A decalcified and sectioned thallus reveals several layers (fig. 6.14*c*). The basal layer is called the hypothallium. In erect branches, the meristematic region, where new cells are produced, is normally intercalary, rather than terminal as in most reds. Meristematic cells form an epithallium above and a perithallium below. The epithallium is thin and continually sloughs off, or it may be lacking.

Figure 6.13
Hildenbrandia. (*a*) Producing dark red spots on a rock. (*b*) Cross section of thallus with a conceptacle containing tetrasporangia.

a

b

Figure 6.14
(*a, b*) Encrusting coralline algae. Limpet in (*a*) keeps surface of crust free of epiphytes. (*c*) Cross section of thallus showing epithallium (ep), perithallium (pe), hypothallium (hy), meristematic layer (m), and a conceptacle (co) with carpogonia.

a

b

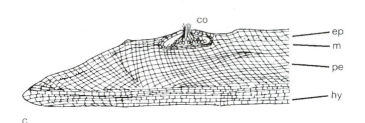

c

Calcareous Erect Thalli

Besides the encrusting forms just described, the order Corallinales contains upright, articulate corallines. In these forms, the thallus has distinct, calcified segments that alternate with flexible, uncalcified joints (fig. 6.15; plate 4*b*). *Corallina* is a common representative. Its erect axes have multiaxial construction similar to *Nemalion,* with several hundred axial filaments. The segments are of the lateral branches surrounding the axial filaments. The joints consist only of the axial filaments.

Life Cycles

Most floridean red algae have separate, free-living, gametophytic and tetrasporophytic phases, while carposporophytes develop on gametophytes. We will consider a typical life cycle before discussing the variations in specific groups (fig. 6.16).

Male and female gametophytes are often separate. The male gametophytes produce male gametes, called spermatia, on spermatangial branches (figs. 6.17*a,*

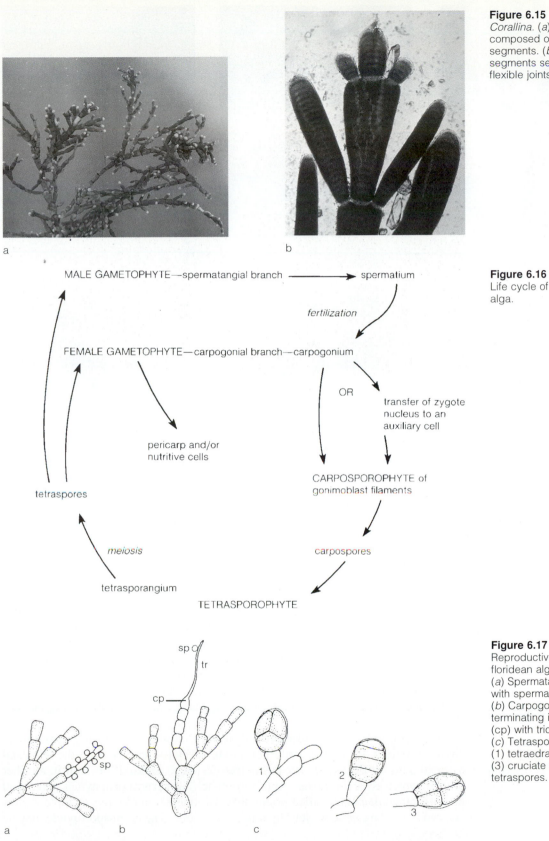

Figure 6.15
Corallina. (*a*) Thallus composed of calcified segments. (*b*) Closeup of segments separated by flexible joints.

MALE GAMETOPHYTE—spermatangial branch ⟶ spermatium

fertilization

FEMALE GAMETOPHYTE—carpogonial branch—carpogonium

OR

transfer of zygote nucleus to an auxiliary cell

pericarp and/or nutritive cells

CARPOSPOROPHYTE of gonimoblast filaments

tetraspores

carpospores

meiosis

tetrasporangium

TETRASPOROPHYTE

Figure 6.16
Life cycle of a floridean red alga.

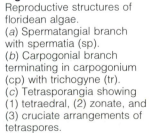

Figure 6.17
Reproductive structures of floridean algae.
(*a*) Spermatangial branch with spermatia (sp).
(*b*) Carpogonial branch terminating in carpogonium (cp) with trichogyne (tr).
(*c*) Tetrasporangia showing (1) tetraedral, (2) zonate, and (3) cruciate arrangements of tetraspores.

Red Algae (Division Rhodophyta)

a

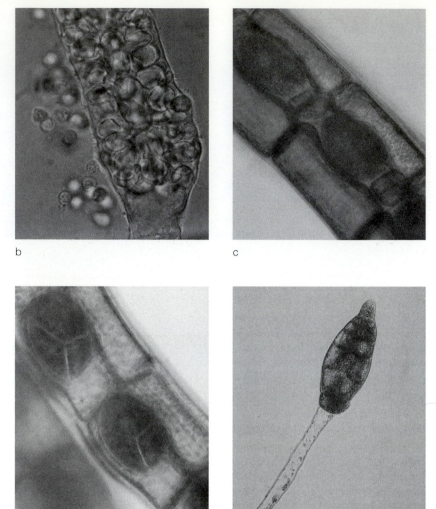

b

c

d

e

Figure 6.18
Stages in the life cycle of
Polysiphonia.
(*a*) Spermatangial branches.
(*b*) Spermatia.
(*c*) Tetrasporangia
surrounded by pericentral
cells. (*d*) Tetraspores.
(*e*) Germling.

6.18*a*, *b*). The spermatia are unpigmented. When released, they are carried by the water. (Remember that red algae lack flagellated stages.) Female gametophytes produce a special branch, the carpogonial branch, in which the terminal cell functions as an oogonium (figs. 6.17*b*, 6.19*a*, *b*). In red algae, the term **carpogonium** is used in place of *oogonium*. The carpogonium is a flask-shaped structure with a long extension called a **trichogyne** (fig. 6.19*c*) that normally extends beyond the surface of the female gametophyte, while the rest of the carpogonial branch is buried within the thallus. The contents of the carpogonium function as an egg. After a spermatium attaches to the trichogyne, the nucleus of the spermatium enters the trichogyne, migrates down into the carpogonium, and fuses with the egg nucleus (fig. 6.19*d*).

Different groups of floridean red algae vary considerably during carposporophyte formation. Either the carposporophyte develops directly from the fertilized carpogonium, or the zygote nucleus is transferred to an **auxiliary cell** in another branch, which gives rise to the carposporophyte. The carposporophyte is composed of branching filaments, called gonimoblast filaments, consisting of diploid cells derived from the zygote nucleus by mitotic divisions. The carposporophyte may be

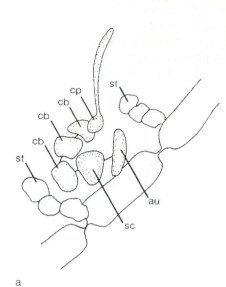

a

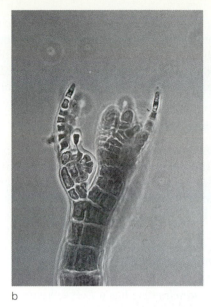

b

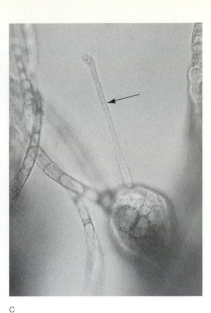

c

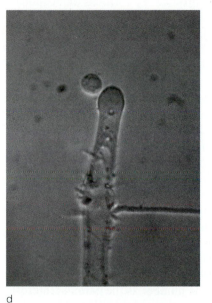

d

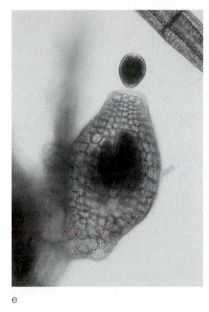

e

Figure 6.19
Stages in the life cycle of
Polysiphonia. (*a*) Procarp
consisting of four-celled
carpogonial branch (cb) with
terminal carpogonium (cp)
and auxiliary cell (au) formed
after fertilization from
supporting cell (sc). Sterile
cells (st) develop around
procarp to form pericarp.
(*b*) Carpogonial branch.
(*c*) Trichogyne (*arrow*).
(*d*) Spermatium on
tip of trichogyne.
(*e*) Carposporophyte
surrounded by pericarp, with
release of a carpospore.
(*a* modified from Broadwater
and Scott 1982.)

a compact cluster of filaments recognizable as a distinct structure (cystocarp)
within (or on) the female gametophyte, or the gonimoblast filaments may be
diffuse, growing interspersed with branches of the gametophyte. Additional cells
of the gametophyte may form a layer (pericarp) around the developing carposporophyte and transfer nutrients to it. At the ends of the gonimoblast filaments,
carpospores are produced and released into the water (fig. 6.19*e*).

Tetrasporophytes develop from carpospores, and like the carposporophytes,
are diploid. When mature, tetrasporophytes form single-celled tetrasporangia in
which meiosis occurs to form four tetraspores (fig. 6.18*c, d*) in a characteristic
arrangement (fig. 6.17*c*). When released into the water, tetraspores separate from
each other and, after settling on a suitable substrate, produce new gametophytes
(fig. 6.18*e*).

The triphasic life cycle of most red algae may have arisen to compensate for the infrequency of fertilization resulting from nonflagellated gametes (Searles 1980). Retention of the zygote and proliferation of diploid tissue on the gametophyte allow the production of a larger number of diploid spores.

Four different types of life cycles are found among floridean red algae. The first two described are the most common and consist of the typical three phases.

Polysiphonia Type

Polysiphonia is openly filamentous, so reproductive structures are easily observed. In the life cycle of *Polysiphonia,* both gametophytes and tetrasporophytes are similar in form. Spermatangial branches are colorless and produce a cluster of spermatia (fig. 6.18*a, b*). Female gametophytes produce a four-celled carpogonial branch in which the terminal cell is the carpogonium (fig. 6.19*a, b*). The zygote nucleus is transferred to an auxiliary cell near the carpogonium, and a compact carposporophyte surrounded by a pericarp develops (fig. 6.19*e*). The carposporophyte forms carpospores. Tetrasporophytes are similar in appearance to gametophytes but bear tetrasporangia at the upper ends of branches (fig. 6.18*c, d*).

Audouinella also has a *Polysiphonia* type of life cycle. Its carpogonium is the only cell of the carpogonial branch. In addition to the typical reproductive stages, both tetrasporophytes and gametophytes can reproduce by forming monospores. **Monospores** form singly in sporangial cells (see fig. 6.3*c*) and develop into the same phase as the parent. They are nonflagellated and are normally carried by water movement.

In *Chondrus,* tetrasporophytes and gametophytes have a similar form. However, tetrasporophytes tend to grow in deeper water than gametophytes, and in some populations, male gametophytes are rare or lacking. The type of carrageenan in the wall can distinguish gametophytes from tetrasporophytes, even in the absence of reproductive stages. Gametophytes form kappa carrageenan, while tetrasporophytes (and carposporophytes) have lambda carrageenan.

In the coralline algae, reproductive structures form in cavities called **conceptacles** beneath the thallus surface (see fig. 6.14*c*). Only one type of reproductive structure forms in a conceptacle.

The absence of sexual fusion and meiosis sometimes modifies a life cycle. For example, in the encrusting algae *Hildenbrandia,* tetrasporophytes produce tetraspores, which give rise to more tetrasporophytes without a gametophyte phase (DeCew and West 1977). Presumably, meiosis does not occur. In the North Atlantic, some populations of *Plumaria* reproduce by releasing spores (paraspores) that develop into thalli with the same chromosome number as the parents. Interestingly, these algae are all triploid.

Bonnemaisonia Type

In the life cycle of *Bonnemaisonia,* tetrasporophytes and gametophytes are dissimilar in form (fig. 6.20). Usually, gametophytes have well-developed pseudoparenchymatous thalli, while tetrasporophytes are simple filaments or crusts. When two phases differ in appearance, they may initially have been described as different genera, and only studies in culture have shown their relationship as different phases of one species. In *Nemalion,* the pseudoparenchymatous thallus described earlier is the gametophyte. Side branches bear reproductive structures, and a compact carposporophyte is produced (see fig. 6.10*c*). Tetrasporophytes are filaments resembling *Audouinella.* In the life cycle of *Ahnfeltia,* gametophytes have an upright thallus with wiry branches, while tetrasporophytes are crusts (fig. 6.21) (Maggs and Pueschel 1989).

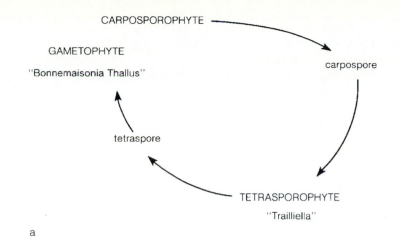

CARPOSPOROPHYTE

GAMETOPHYTE

"Bonnemaisonia Thallus"

carpospore

tetraspore

TETRASPOROPHYTE

"Trailliella"

a

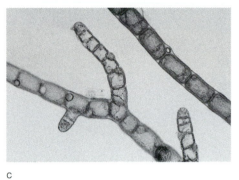

b

c

Figure 6.20
Life cycle of *Bonnemaisonia*.
(*a*) Summary diagram.
(*b*) Gametophyte with
distinctive hooked branches.
(*c*) Tetrasporophyte originally
described as genus
Trailliella, with small gland
cells alternating with normal
vegetative cells. (*e* courtesy
J. S. Prince.)

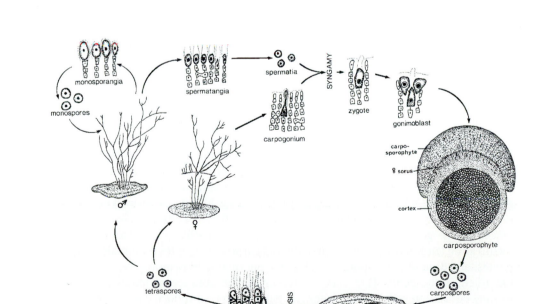

monosporangia

monospores

spermatangia

spermatia

SYNGAMY

carpogonium

zygote

gonimoblast

carpo-
sporophyte

♀ sorus

cortex

carposporophyte

carpospores

tetraspores

MEIOSIS

tetrasporangia

tetrasporophyte

Figure 6.21
Life cycle of *Ahnfeltia*. (From
Maggs and Pueschel 1989,
courtesy *Journal of
Phycology*.)

Figure 6.22
Life cycle of *Mastocarpus*.
(*a*) Summary diagram.
(*b*) Gametophyte of
Mastocarpus stellatus found
in North Atlantic. (*c*) Dark
purple-brown crusts of
Petrocelis stage
(tetrasporophyte).

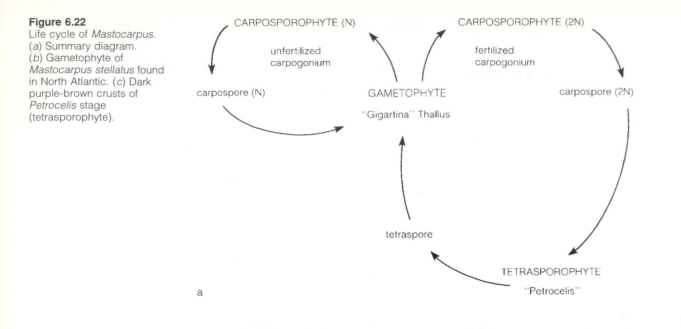

b

c

Studies by John West and his students showed two types of life cycles among species formerly in the genus *Gigartina* (Polanshek and West 1977; West, Polanshek, and Shelvin 1978; Zupan and West 1988). Some species have isomorphic free-living phases, as in the related genus *Chondrus,* and have been retained in the genus *Gigartina*. In other species, the upright thallus alternates with an encrusting tetrasporophyte, or no alternate phase forms (fig. 6.22). These species have been placed in the genus *Mastocarpus*. When male gametophytes provide spermatia, fertilization leads to the production of diploid carposporophytes on female gametophytes, and diploid carpospores develop into encrusting tetrasporophytes, formerly identified as the genus *Petrocelis*. However, in the absence of males (males are lacking entirely in some species), female gametophytes produce carposporophytes without fertilization, and the carpospores (presumably haploid) develop into more female gametophytes.

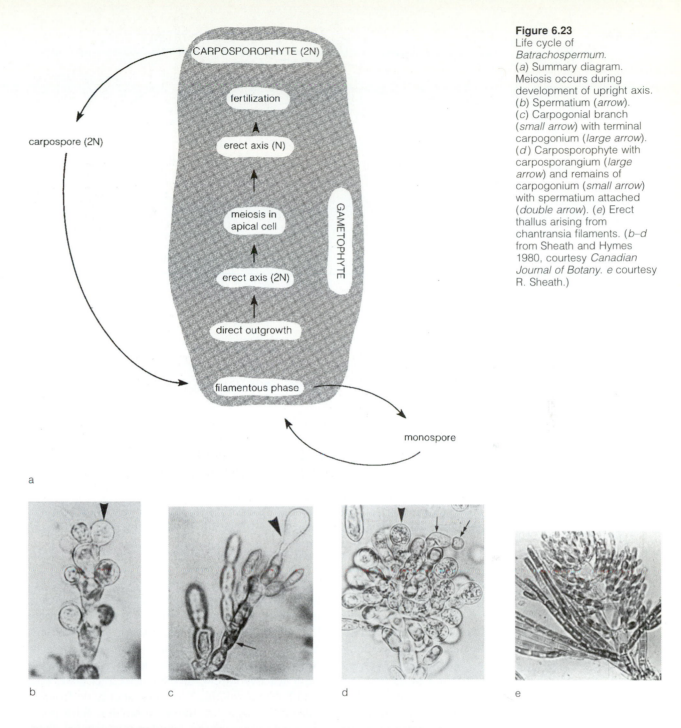

Figure 6.23
Life cycle of *Batrachospermum.* (*a*) Summary diagram. Meiosis occurs during development of upright axis. (*b*) Spermatium (*arrow*). (*c*) Carpogonial branch (*small arrow*) with terminal carpogonium (*large arrow*). (*d*) Carposporophyte with carposporangium (*large arrow*) and remains of carpogonium (*small arrow*) with spermatium attached (*double arrow*). (*e*) Erect thallus arising from chantransia filaments. (*b–d* from Sheath and Hymes 1980, courtesy *Canadian Journal of Botany*. *e* courtesy R. Sheath.)

Batrachospermum Type

The freshwater red algae *Batrachospermum* and *Lemanea* (order Batracho-spermales) lack tetrasporophytes and do not have triphasic life cycles (fig. 6.23). Spermatia and carpogonia are produced on gametophytes, and fertilization occurs as in other red algae. The carposporophyte, which develops from the carpogonium, is composed of diploid cells and forms diploid carpospores. Carpospores give rise to reduced filaments (chantransia stage), which may overwinter and reproduce themselves by monospores. Under favorable conditions, the erect macrothallus is

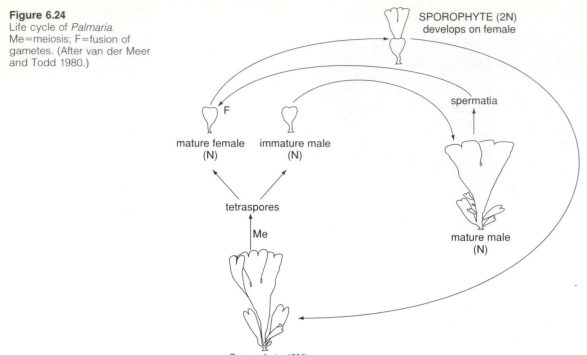

Figure 6.24
Life cycle of *Palmaria*.
Me=meiosis; F=fusion of
gametes. (After van der Meer
and Todd 1980.)

SPOROPHYTE (2N)
develops on female

spermatia

F

mature female
(N)

immature male
(N)

mature male
(N)

tetraspores

Me

Sporophyte (2N)

produced as an outgrowth. During development of the erect thallus, the apical cell undergoes meiosis. Therefore, the lower part of the thallus is diploid, and the upper part, where reproductive structures form, is haploid. Since meiosis is not associated with the formation of reproductive stages, it is **somatic meiosis.**

Palmaria **Type**

Palmaria has a dark red thallus of bladelike branches. Natural populations were thought to consist of only male gametophytes and tetrasporophytes until culture studies by van der Meer and Todd (1980) explained this apparent lack of female gametophytes in the life cycle (fig. 6.24). They found that tetraspores produce male and female gametophytes in equal numbers, but that females are much smaller (less than 1 millimeter) and mature in a few days, while the much larger males take nine to ten months to develop and form spermatia. Therefore, spermatia from an earlier generation of males fertilize the current generation of females. Tetrasporophytes develop directly from carpogonia and overgrow female gametophytes. Thus, a carpospore-producing stage is absent.

In their study, van der Meer and Todd used a color mutant. In the haploid condition, mutants were green, while diploid cells had normal red pigmentation. This color difference allowed the researchers to distinguish tissue of the female gametophyte from the tetrasporophyte.

Similar life cycles characterize other members of the order Palmariales. In *Halosaccion,* the female gametophyte is reduced, consisting only of a vegetative cell and a carpogonium (Mitman and Phinney 1985).

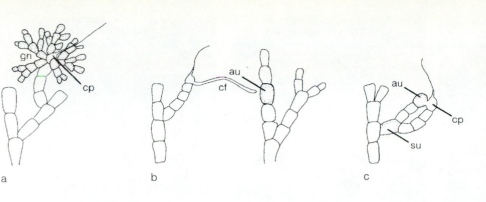

Figure 6.25
Postfertilization events in floridean red algae. (*a*) Carposporophyte (gn=gonimoblast filament) developing from carpogonium (cp). (*b*) Transfer of zygote nucleus by a connecting filament (cf) to an auxiliary cell (au) in an accessory branch distant from the carpogonial branch. Carposporophyte will develop from auxiliary cell. (*c*) Procarp condition with accessory branch and carpogonial branch arising from same supporting cell (su). After fertilization of the carpogonium, the zygote nucleus is transferred to the auxiliary (au) cell in the accessory branch.

Table 6.3 *Orders of the Red Algae*

Subclass / Order	Pit Plugs Number of Cap Layers[1]	Cap Membrane[1]	Life Cycle Auxiliary Cell	Phases[2]
Bangiophycidae				
Porphyridiales	(unicells without plugs)		Absent	Monophasic
Compsopogonales	0	−	Absent	Monophasic (?)
Bangiales	1 (conchocelis phase)	−	Absent	Biphasic
Rhodochaetales	0	−	Absent	Biphasic
Florideophycidae				
Acrochaetiales	2 (d in some)	+	Absent	Triphasic (P)
Gelidiales	1	+	Absent	Triphasic (P)
Nemaliales	2	+	Absent	Triphasic (B)
Batrachospermales	2d	−	Absent	Biphasic (Bat)
Bonnemaisoniales	0	+	Absent	Triphasic (B)
Palmariales	2	+	Absent	Biphasic (Pal)
Corallinales	2d	−	Present before fertilization	Triphasic (P)
Hildenbrandiales	1	+	Present before fertilization	Triphasic (P)
Gigartinales	0	+	Present before fertilization	Triphasic (P, B)
Gracilariales	0 (?)	+	Present before fertilization	Triphasic (P)
Ahnfeltiales	0	−	Present before fertilization	Triphasic (B)
Rhodymeniales	0	+	Present before fertilization	Triphasic (P)
Ceramiales	0	+	Formed after fertilization	Triphasic (P)

Source: After Garbary and Gabrielson 1990; Pueschel 1990, 1994.
[1] d = dome-shaped second layer; presence(+) or absence(−) of cap membrane.
[2] Typical life cycles (but may be exceptions): P = *Polysiphonia* type; B = *Bonnemaisonia* type; Bat = *Batrachospermum* type; Pal = *Palmaria* type.

Postfertilization Events and Development of the Carposporophyte

Table 6.3 summarizes the orders of red algae. Formerly, taxonomic positions of red algae in the Florideophycidae were based almost entirely on reproductive events following fertilization of the carpogonium and leading to development of the carpo-

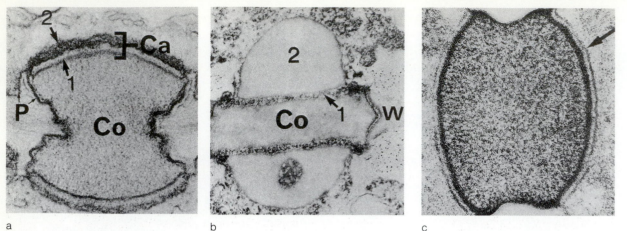

a   b   c

Figure 6.26
Types of plugs in pit
connections of floridean red
algae, showing variation in
number of cap layers. (*a*) *Palmaria* (Palmariales) with two
cap layers. Co=plug core;
Ca=cap; P=cell membrane;
1=inner cap layer; 2=outer
cap layer. (*b*) *Clathromorphum*
(Corallinales) with enlarged
second layer. W=wall.
(*c*) *Gracilaria* (Gracilariales)
with no cap (arrow indicates
cap membrane). (*a–c* from
Pueschel and Cole 1982,
courtesy *American Journal of
Botany.*)

sporophyte. More recent studies with pit plugs (described in the next section) have led to a reevaluation of the orders based primarily on life-cycle criteria.

As described in the previous section, the members of Batrachospermales and Palmariales have biphasic life cycles with distinct features. Some of the variation in postfertilization events of the remaining orders of floridean red algae, with triphasic life cycles of either the *Polysiphonia* or *Bonnemaisonia* type, will be discussed further. These orders can be divided into two groups based on whether the carposporophyte develops directly from the carpogonium (or the zygote nucleus is transferred to a nearby cell in the carpogonial branch) (fig. 6.25*a*), or the zygote nucleus is transferred to an auxiliary cell, from which the carposporophyte develops. When the auxiliary cell is in a branch distant from the carpogonium, a long connecting filament without cross walls forms (fig. 6.25*b*). For example, following fertilization in *Dumontia,* the carpogonium fuses with a cell lower in the carpogonial branch, which may provide nutrients, and several connecting filaments develop from this fusion cell. Accessory branches with auxiliary cells are usually more numerous than carpogonial branches. Each connecting filament grows considerable distance through the tissue of the thallus to reach an auxiliary cell, and then a derivative of the zygote nucleus migrates through the connecting filament to the auxiliary cell (its nucleus does not participate in carposporophyte development).

In other species, carpogonia and auxiliary cells associate more closely. In the advanced **procarp** condition, the carpogonial branch and the accessory branch with the auxiliary cell arise from a common supporting cell, or the supporting cell functions as the auxiliary cell (fig. 6.25*c*). The number of cells in the two branches is also reduced.

Red algae in the orders Rhodymeniales and Ceramiales form distinctive procarps. Only in the Ceramiales does the auxiliary cell form after fertilization. The procarp is very uniform in this group. Supporting cells bear four-celled carpogonial branches (see fig. 6.19*a*). After fertilization, division of the supporting cell forms the auxiliary cell. The zygote nucleus transfers by direct fusion or through a small connecting cell.

Pit Plugs

As described earlier, the intercellular connection between cells of a floridean red alga and some bangean algae is filled with a large lenticular plug, which may be visible with high magnification through a light microscope but requires an electron microscope to see details of taxonomic importance. All plugs have a central protein core, and some species have one or two layers, or caps, of polysaccharides (fig. 6.26). Ordinarily, a membrane surrounds the plug and passes between the inner and outer caps when both are present. A plug forms following cell division. As cells grow, the size and shape of their plugs may change, but the number of cap layers and the presence of a cap membrane do not.

Pueschel and Cole (1982) proposed using pit-plug structure as a taxonomic characteristic and started a reevaluation of the orders of floridean red algae. The orders of red algae in table 6.3 are based on this and subsequent studies that have supported and expanded the use of the number of cap layers and the presence or absence of a plug membrane. In many instances, this reexamination has lead to a splitting of previously recognized orders. For example, the traditionally recognized Nemaliales, containing floridean algae without an auxiliary cell, has been split into five orders based in part on pit-plug differences. Similarly, two orders formerly distinguished on the basis of their auxiliary cells (the Gigartinales and Crypto-nemiales) were combined (as Gigartinales), with other species being separated into the Corallinales, Hildenbrandiales, Gracilariales, and Ahnfeltiales, based on plug characteristics.

Subclass Bangiophycidae

The bangean red algae show a greater range of morphologic types than the floridean reds. The simplest bangean reds are solitary cells, while multicellular thalli include filaments and parenchymatous blades. Macrothalli usually show diffuse growth and lack pit connections with plugs. Cells typically have a single stellate chloroplast. Reproduction in multicellular forms is commonly by monospore formation. The bangean reds do not form distinct carpogonia, lack carposporophyte stages, and do not have differentiated tetrasporangia. Some genera, however, have alternate phases that resemble simple filamentous members of the Florideophycidae. As table 6.3 shows, the Bangiophycidae are divided into four orders (Garbary and Gabrielson 1990).

Porphyridium is a "little round red thing" (fig. 6.27). Its spherical cells grow in aerial habitats and are surrounded by a mucilaginous sheath but lack a wall. Cells contain a central, stellate chloroplast, which may occupy much of the cell (no large vacuole is present). To reproduce, *Porphyridium* divides by a furrow into two cells.

Erythrotrichia, *Rhodochaete*, and *Bangia* are filaments. *Erythrotrichia* is a simple filament that often grows epiphytically on larger algae (fig. 6.28). *Rhodochaete*, a branching filament, resembles floridean algae in its apical growth and pit plugs. The lack of caps and a membrane around plugs may be a primitive condition. *Bangia* grows in both freshwater and seawater, sometimes forming a feltlike covering on rocks wetted by spray (fig. 6.29). Its thallus is an unbranched filament in which older regions become multiseriate. Each cell has a stellate chloroplast and is surrounded by a wall composed of polymers of mannose. The macrothallus of *Bangia* is present during the winter, while an alternate conchocelis phase of microscopic filaments occurs throughout the year.

Porphyra has a parenchymatous thallus in the form of a delicate blade consisting of one or two layers of cells (fig. 6.30*a*; plate 4*b*). Blade color varies greatly, often being yellow, olive, pink, or purple. As in *Bangia,* cell walls are composed of mannans. In the life cycle of *Porphyra,* the foliose macrothallus alternates with a microscopic filamentous phase called the conchocelis phase (fig. 6.31). Some populations show a true alternation of generations, in which the foliose phase is

Figure 6.27
Porphyridium. (*a*) Electron
micrograph showing cell with
large chloroplast, nucleus,
and starch grains in
cytoplasm. (*b*) Light
micrograph. (*a* from
Schornstein and Scott 1982,
courtesy *Canadian Journal of
Botany*.)

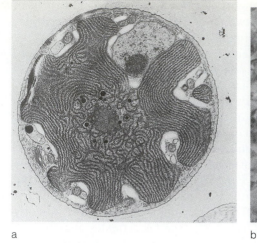

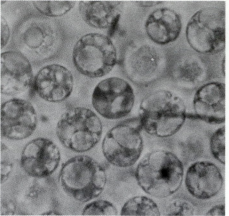

a

b

Figure 6.28
Unbranched filaments of
Erythrotrichia.

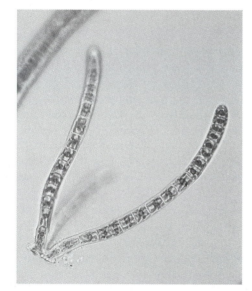

Figure 6.29
Bangia. (*a*) Filaments matted
to rock in spray zone of New
England. (*b*) Unbranched
filament with multiseriate
arrangement of cells.

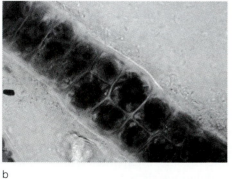

a

b

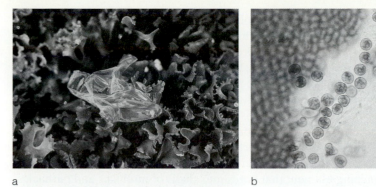
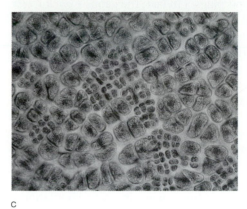
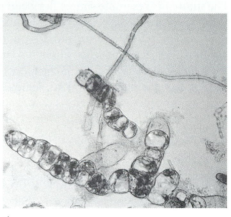

a

b

c

d

Figure 6.30
Porphyra. (*a*) Delicate blade of macrothallus.
(*b*) Monospores being released from edge of blade.
(*c*) Formation of carpospores.
(*d*) Conchocelis phase.
(*d* courtesy J. S. Prince.)

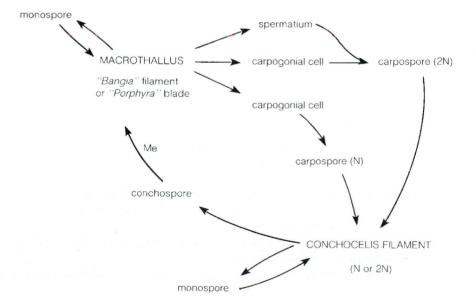

Figure 6.31
Summary of life cycle of *Porphyra* and *Bangia.* When the conchocelis stage is diploid, meiosis (Me) occurs when conchospores germinate.

haploid and the conchocelis phase is diploid. In others, sexual reproduction and meiosis have been lost, and both phases have the same chromosome number. The macrothallus forms three types of reproductive stages in unspecialized cells: monospores, carpospores, and spermatia. Monospores form singly by release of the cell contents as a spore and produce more macrothalli (fig. 6.30b). Sixteen or more spermatia form at a time by repeated divisions of the cell contents. In sexually reproducing species, spermatia fertilize unspecialized carpogonial cells, which divide several times to form diploid carpospores (fig. 6.30c). In nonsexual species, carpogonial cells divide without being fertilized and form haploid carpospores. In either case, carpospores produce the microscopic **conchocelis phase** (fig. 6.30d). This phase is a branching filament that bores into shells of mollusks and barnacles. The filaments resemble simple floridean reds such as *Audouinella,* with cells having parietal chloroplasts, cellulose walls, and pit plugs. (In fact, the genus *Conchocelis* was considered a floridean red alga until its relationship with bangean reds was established.) Conchocelis filaments may reproduce themselves by releasing monospores or may form conchospores that give rise to the macrothallus. When the conchocelis phase is diploid, meiosis occurs during conchospore germination, and all four products of meiosis contribute through cell divisions to the new haploid thallus (Mitman and van der Meer 1994).

Red Algal Ecology and Commercial Uses

Red algae are important members of many benthic communities, from tropical to polar oceans. A few species also grow in freshwater and terrestrial habitats. Despite the wide occurrence of red algae as a group, most species are restricted to suitable habitats within a more limited temperature range. In some populations, tetrasporophytes and gametophytes have different seasonal or spatial distributions (Dyck and DeWreede 1995).

On many rocky shores, a broad intertidal region is exposed to the air for part of the tidal cycle and submerged the rest of the time, and a subtidal region remains submerged but receives sufficient light to support benthic algal growth. Many red algae have adapted to living in the intertidal region. Among these are species that grow in the spray zone near or slightly above the limit reached by seawater at high tide. An example is *Bangia,* which grows high on shore during the winter but survives as a subtidal conchocelis phase in summer. Many red algae grow lower in the intertidal region, often producing a distinct red algal zone. While exposed to the air during low tide, intertidal reds must tolerate periods of desiccation, exposure to high light levels, and a greater temperature range than when submerged. Less tolerant species are restricted to the lower part of the shore, with shorter periods of air exposure, to tidepools that retain seawater during low tide, or to the understory below a canopy of brown algae, where moist conditions are maintained. Many intertidal red algae are brown or dark purple because carotenoids act as light shields (and do not function in photosynthesis) and mask the red photosynthetic pigment phycoerythrin.

Epiphytism is well developed among red algae, especially in the Ceramiales and Acrochaetiales. Epiphytes grow on the surfaces of larger algae. Some species may live on a wide range of host species, while others require a specific host. In the North Atlantic, *Polysiphonia lanosa* grows only on the fucoid *Ascophyllum* (see fig. 9.14b). Its holdfast penetrates the tissues of its host, but host and epiphyte exchange little or no organic material, since both are photosynthetic (Harlin and Craigie 1975).

A few red algae with reduced filaments grow as endophytes in other algae. Some are true parasites. Most parasitic species are very specific as to the host species they infect and are often closely related taxonomically to their hosts (Goff et al. 1996). Parasitic species have reduced structures, and may lack or have reduced levels of pigments. Growing within host tissues, parasitic reds develop pit connections with their hosts, and parasite nuclei may enter host cells (Goff and Coleman 1985).

Subtidal red algae, especially in deeper water, are often calcified as a defense against herbivores or are foliose with a large surface for collecting light and a rich red color due to phycoerythrin. Their pigmentation is an adaptation to a light environment that provides only low levels of blue-green light. Since red is the chromatic complement of blue-green, phycoerythrin absorbs most of the light for photosynthesis. In some species, tetrasporophytes grow in deeper water than gametophytes (Mathieson 1979). Some crustose corallines are the deepest growing algae, reported at a depth of 268 meters in very clear water (Littler et al. 1985).

Encrusting corallines are widespread and common in coastal environments. On tropical coral reefs, they may cover a large portion of the reef surface, sometimes forming a ridge where waves break on the reef front. Corallines help bind the reef together, and their calcareous structures also contribute to sediments. Recently, a bacterial pathogen, recognized by its bright orange spots, has been killing coralline algae on Pacific reefs (Littler and Littler 1995).

Because of their habit and slow growth, corallines are readily overgrown by other organisms, especially more rapidly growing algae. In fact, some invertebrate larvae are chemically attracted to encrusting algae, where they settle and metamorphose into adults (Pearce and Scheibling 1991; Morse 1991). Corallines may clear their surfaces of other organisms by sloughing their outer epithallial layer (Johnson and Mann 1986), or they may rely on herbivores to remove settling species. In New England tidepools, the limpet *Acmaea* grazes on the outer layer of the coralline *Clathromorphum,* cleaning it of epiphytes, and thus both limpet and alga benefit (Steneck 1982). Similarly, chitons grazing on the surface of tropical corallines stimulate coralline growth and remove sporlings of other algae (Littler, Littler, and Taylor 1995).

Although most red alga are marine, a few (approximately 3%) grow in freshwater (Sheath and Hambrook 1990). Freshwater reds are usually found in rivers and streams. (Cyanobacteria with phycoerythrin also produce red growths.) Common genera include *Batrachospermum, Audouinella,* and *Lemanea.* They vary greatly in color—from yellow-green and blue-green to brown and blackish. Several red algae, such as *Bangia,* have spread in the twentieth century from the Atlantic Oceans into the Great Lakes (Sheath and Cole 1980; Sheath and Morison 1982).

In flowing freshwater environments, red algae may show modifications in their pigments and life cycles, compared to their marine relatives. Many freshwater red algae are not red; the blue pigment phycocyanin is dominant, possibly as an adaptation for absorbing light filtered through a tree canopy. Instead of two free-living phases as in the life cycles of many reds, freshwater red algae often have a single free-living phase (*Batrachospermum* life cycle), perhaps to reduce loss of reproductive stages in flowing water.

Cyanidium, a unicellular red algae, grows in the outflow from hot springs. It has a bluish green color from chlorophyll *a* and phycocyanin. *Cyanidium* occurs only in very acidic waters (pH 2–4) at temperatures as high as 55° C.

Figure 6.32
Summary of commercial growth and processing of *Porphyra*.

Stock culture of conchocelis
on shells or artificial substrates
↓
Induction of conchospore formation
(low temperature, increased irradiance,
short photoperiods)
↓
Attachment of spores to nets
↓
Transfer of nets to fixed poles or rafts
in estuaries (natural and artificial fertilization)
↓
Growth of blades to harvestable size
(nets may be removed and frozen after growth
starts and later returned to estuaries)
↓
Harvesting of blades
↓
Processing: washing and chopping of blades,
slurry spread on mats to form sheets,
sheets dried and packaged
↓
Nori sheets for sale

Humans use red algae as foods and as sources of phycocolloids (hydrocolloids) in a wide range of commercial products. Seaweed consumption in Europe and the Americas is limited. The two most widely used red algae in North America are *Palmaria,* called dulse, and *Porphyra,* called laver. In the Pacific and Asia, humans eat a much greater variety and quantity of seaweeds (see Abbott 1988). Most are collected by foraging along the shore. *Porphyra,* however, is an important food crop and is farmed on a large scale in Japan and China (see Mumford and Miura 1988; Lobban and Harrison 1994). *Porphyra* is used for flavoring or as a wrapper for *sushi*, or may be eaten alone. It is known as *nori* in Japan and as *zicai* in China. Use of *Porphyra* in China dates back to the sixth century. Kathleen Drew's discovery of the conchocelis stage in 1949 significantly advanced cultivation techniques since "seed" stocks of the conchocelis phase derived from quality blades could now be maintained. With the approach of the winter growing season, conchospore release is stimulated by changing the culture conditions. After spores attach to nets, the nets are transferred to estuaries, where the foliose macrothallus receives both natural and artificial enrichment as it grows to harvestable size. Then mechanical harvesters collect the blades, which are processed for sale as black, paperlike sheets (fig. 6.32).

In North America, seaweeds are sources of phycocolloids. Phycocolloids, derived from the mucilaginous polysaccharides of cell walls, are used as thickeners, gels, and stabilizers in foods, cosmetics, and a wide range of other commercial products (see Lewis, Stanley, and Guist 1988). Two important phycocolloids (or gums) derived from red algae are carrageenan and agar, which are both polymers of galactose. The principal sources of agar—*Gelidium, Gracelaria, Pterocladia,* and *Ahnfeltia*—are harvested from natural populations, primarily in Asia and Latin America, and processed for commercial use in Asia. Important uses of agar are as a food gel, in pharmaceutical capsules, and as a medium for culturing microorganisms. The agarose fraction is used for gel electrophoresis.

Important commercial sources of carrageenan are *Chondrus, Gigartina, Eucheuma,* and *Kappaphycus.* Natural populations of *Chondrus* in the North Atlantic and of *Gigartina* in Chile are harvested. However, most carrageenan today comes from cultivation of *Eucheuma* and *Kappaphycus* in the Philippine Islands and Indonesia, where these reds are grown on nets raised off the sea bottom in shallow water. Harvested algae are dried before shipment to processing plants, where the carrageenan is extracted in hot, alkaline water, cleared, concentrated, and blended for use in commercial products. Carrageenan comes in three forms—kappa, iota, and lambda—depending on the position and number of sulfate groups attached to the galactose units. Kappa is the best gel, iota is a weaker gel, and lambda does not gel but has a high viscosity, making it valuable as a thickener. The three forms are derived from different algae or different phases of one species (see *Chondrus*) and are blended to produce the desired gel strength and viscosity. Carrageenan is widely used in toothpaste, cosmetics, and foods, such as chocolate milk, ice cream, dessert gels, and pet foods. Because carrageenan forms weak bonds with milk proteins (casein), it is often used with milk products.

Summary

1. In the red algae (division Rhodophyta), the principal photosynthetic pigments are chlorophyll *a* and phycobiliproteins, the reserve is a branched form of starch, the thylakoids in the chloroplasts are unassociated and have phycobilisomes on their surfaces, and no flagellated stages form. The red algae are divided into two subclasses.

2. The floridean reds (subclass Florideophycidae) have a characteristic grooved plug in the pit between adjacent cells. Algae are filamentous in construction and usually show apical growth. They range from openly branching filaments to pseudoparenchymatous thalli. The latter may be erect or encrusting, and sometimes are covered with calcium carbonate. The life cycle of a typical floridean red has three phases. In the sexual or gametophytic phase, a spermatium fertilizes a carpogonium. The resulting diploid cells form a carposporophyte attached to the female gametophyte. The carposporophyte produces carpospores, which give rise to the tetrasporophytic phase. Meiosis occurs during tetraspore formation by tetrasporophytes, and the tetraspores form new gametophytes. Traditionally, the orders of floridean red algae have been distinguished by events after fertilization leading to the development of the carposporophyte, but more recently, the orders have been revised on the basis of the pit-plug structure.

3. The bangean red algae (subclass Bangiophycidae) include unicells, filaments, and simple parenchymatous blades with diffuse growth. In some genera, a macroscopic phase alternates with a microscopic filamentous phase, with pit plugs similar to floridean red algae.

4. Red algae are common in coastal marine environments, and a few species grow in freshwater and aerial habitats. Some reds are grown for food or harvested for their phycocolloids (carrageenan or agar), which are used in a variety of commercial products.

Further Reading

Cole, K. M., and R. G. Sheath, eds. 1990. *Biology of the Red Algae*. Cambridge University Press.

Murray, S. N., and P. S. Dixon. 1992. The Rhodophyta: Some aspects of their biology. III. *Oceanography and Marine Biology, An Annual Review* 30: 1–148.

Woelkerling, W. J. 1988. *The Coralline Red Algae*. Oxford University Press.

7

Phytoplankton

Members of the plankton live free-floating in the water, without association with submerged substrates. The algal component of the plankton is called the **phytoplankton,** to distinguish it from the zooplankton and the bacterioplankton. For the most part, planktonic algae move with water flow. Some have flagella that allow them to swim short distances but lack sufficient power to swim against the water. Phytoplankton communities are well developed in oceans and lakes. Although the phytoplankton generally grow poorly in flowing waters, they may be present in slowly flowing stretches of rivers.

In aquatic environments, light rapidly decreases as it penetrates the water column. The availability of light limits phytoplankton growth to a layer of water called the **photic zone** near the surface. Growth of a planktonic population depends on the rate at which new cells are produced in the photic zone and the rate at which cells are lost:

population growth (+ or −) = rate of new cell production − rate of cell loss

Light and nutrient availability are the major factors determining new cell production. Consumption by grazing animals and sinking (sedimentation) are the principal ways cells are lost from a population. Most phytoplankton cells are denser than water and therefore tend to sink. Water movements and viscous drag help keep cells suspended. Cells also may show adaptations to slow their sinking and to aid in resuspension.

Dividing the phytoplankton according to size into three categories is sometimes convenient. Each category requires different collection methods to obtain a representative sample. The larger phytoplankton, called the net plankton or **microplankton,** range in size from 20–200 micrometers and are readily collected with a fine mesh net. The **nanoplankton** range in size from 2–20 micrometers and are not retained by a standard plankton net. Ordinarily, the nanoplankton are concentrated by allowing samples to settle or by filtration. The smallest members of the phytoplankton are the **picoplankton,** which are less than 2 micrometers in diameter. Chlorophyll fluorescence commonly measures their abundance.

The phytoplankton community in a lake or ocean often consists of a relatively large number of different species. Called the "paradox of the plankton" by Hutchinson (1961), this high diversity in a relatively uniform environment appears to contradict the ecologic principle of competitive exclusion, which predicts that one or a few species should outcompete the others and dominate. Explanations proposed for the diversity of phytoplankton assemblages include: (1) insufficient time for the better competitors to eliminate the poorer competitors because of changing environmental conditions or frequent disturbances; (2) selective grazing on potentially dominant species, preventing them from becoming abundant enough to displace other species; (3) different nutrients or other resources limiting different species; and (4) patchy distributions of phytoplankton populations, so that potential competitors are separated in microhabitats, and the apparent occurrence of many species together results from mixing patches during sampling.

In the first part of this chapter, we will examine important environmental factors influencing phytoplankton growth. Then we will consider marine and freshwater phytoplankton communities.

Environmental Factors Influencing Phytoplankton Growth

Light

A photosynthesis/irradiance (P/I) curve expresses an alga's requirement for light by showing photosynthesis at different levels of light, or irradiances (fig. 7.1a). This curve has three parts. At low irradiance (photon flux densities), the rate of photosynthesis is directly related to the amount of irradiance, and light is limiting. Where the curve levels off at higher irradiance, light is saturating. At very high irradiances, photosynthetic activity may be inhibited. The lines produced by extending the first two parts of the P/I curve intersect at an irradiance designated I_k. For an algal species, the initial slope of the response curve (α), the saturation level (P_{max}), and I_k characterize the response of photosynthesis to light. Figure 7.1b shows P/I curves for different species. For a species adapted to relatively low irradiance, the initial slope of the curve is steep and the I_k value is low. Algae adapted to low light often have lower saturation levels than algae adapted to higher irradiances.

Irradiance decreases exponentially with water depth because water and suspended particles absorb and scatter light (fig. 7.2a). Near the surface, light is usually more than sufficient for photosynthesis, but nutrient availability may be limited. Deeper in the water column, light determines the level of photosynthetic production. The phytoplankton as a whole (as well as each species) have a **compensation depth,** where the amount of light allows only enough production to meet the requirements for cellular maintenance (respiration) without production of new biomass (fig. 7.2b). Above the compensation depth, where light is more than sufficient to meet cellular requirements, excess production allows growth and reproduction. Cells below the compensation depth are unable to grow and gradually deplete any reserves previously accumulated. However, in natural bodies of water, phytoplankton cells are mixed about in the water column and exposed to a range of light levels. Even though they may spend part of the time below their compensation depth, they will still show net production if they also spend sufficient time at higher irradiances. Although each species has its own compensation depth, the photic zone normally refers to the layer where the phytoplankton as a whole show net production. Its limit is often approximated as the depth where surface irradiance is 1%.

Density Stratification

Temperature and salinity control water density. In aquatic environments, **stratification,** which is the development of layers of water with different densities, influences the amount of time cells spend in the photic zone and nutrient availability.

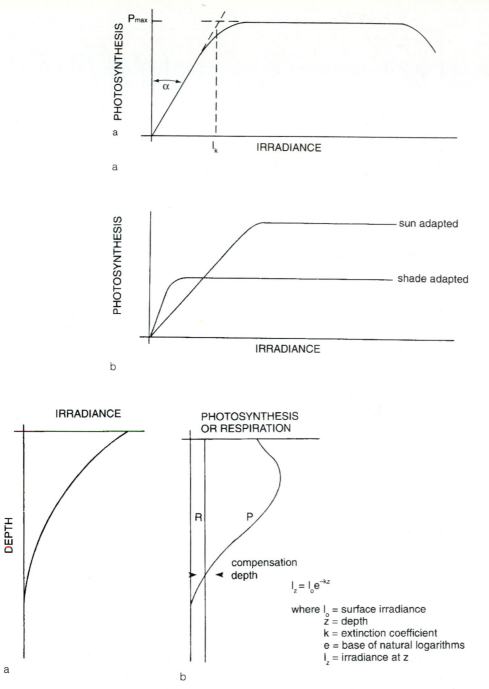

a

a

b

Figure 7.1
(*a*) P/I curve: photosynthetic production (per unit of chlorophyll) in response to different levels of irradiance. (*b*) Comparison of P/I curves of sun- and shade-adapted populations.

Figure 7.2
(*a*) Irradiance decreases exponentially with water depth. In the equation describing this relationship, the extinction coefficient (k) is characteristic of a particular body of water. (*b*) Photosynthetis (P) and respiration (R) at different depths. Photosynthesis depends on irradiance and shows a maximum at or slightly below the surface, while respiration is assumed not to change with depth. The compensation depth is where photosynthesis and respiration are equal.

Pure water has its greatest density slightly below 4° C. Above 4° C, density is inversely related to temperature. In a typical temperate lake in early spring, the water column has a uniform temperature of 4° C when the water mixes vertically in the absence of any density differences (fig. 7.3*a*). As the water absorbs solar radiation at the surface, the upper layer of water warms, while the deeper water remains cool. Between these layers is a region of rapid temperature change with depth, called a **thermocline** (fig. 7.3*b*). The temperature decrease in the

Figure 7.3

Seasonal temperature changes in a temperate lake. (*a*) In spring, the water is initially isothermal, and heating occurs at the surface. (*b*) In summer, thermal stratification occurs, with the thermocline separating the upper epilimnion (or mixed layer) from the deep layer, or hypolimnion. (*c*) In fall, the epilimnion cools and the thermocline decreases until an isothermal condition occurs. (*d*) In winter, cooler, less dense water at the surface produces inverse stratification.

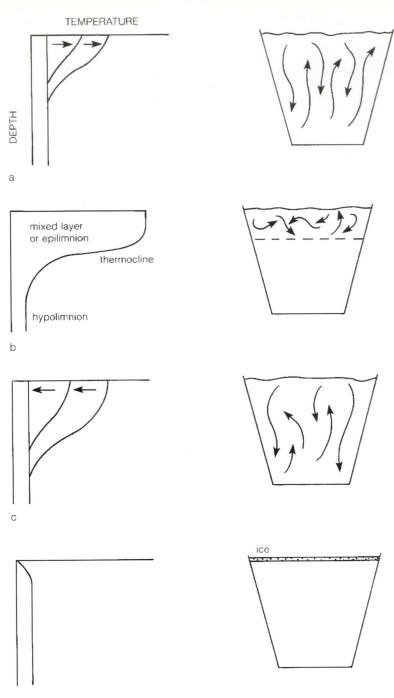

thermocline is associated with a rapid increase in water density (fig. 7.4). This density gradient, called a **pycnocline,** is a barrier to vertical mixing, preventing exchange between the upper and lower layers of a lake. Thus, development of the seasonal thermocline in the spring causes density stratification and confines mixing of phytoplankton to a region above the thermocline, called the **mixed layer** (or epilimnion in lakes). Stratification persists through summer and is strengthened by continued heating of the mixed layer. In fall, the mixed layer gradually cools until

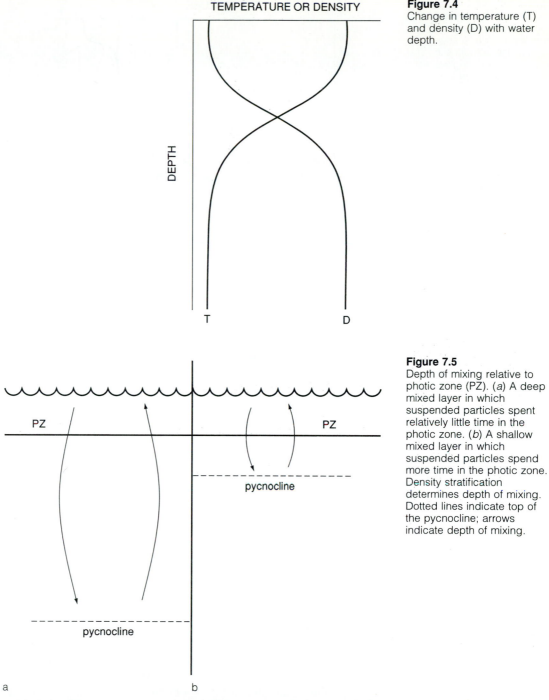

DEPTH

T   D

Figure 7.4
Change in temperature (T) and density (D) with water depth.

PZ   PZ

pycnocline

pycnocline

a   b

Figure 7.5
Depth of mixing relative to photic zone (PZ). (*a*) A deep mixed layer in which suspended particles spent relatively little time in the photic zone. (*b*) A shallow mixed layer in which suspended particles spend more time in the photic zone. Density stratification determines depth of mixing. Dotted lines indicate top of the pycnocline; arrows indicate depth of mixing.

the lake has a uniform temperature (fig. 7.3*c*). Vertical mixing is associated with this isothermal period. Continued cooling below 4° C produces inverse stratification, with cold, less dense water near the surface and denser water of approximately 4° C below (fig. 7.3*d*). Ice is even less dense and floats on the surface.

The depth of mixing, as determined by the presence or absence of a pycnocline, influences the time a cell spends in the photic zone (fig. 7.5). Although the photic zone is often shallower than the mixed layer, cells circulating in the mixed layer are

Figure 7.6
Stratification in temperate oceans showing a permanent thermocline (pt). A seasonal thermocline (st) forms in upper waters during warmer months. The mixed layer in winter (wm) extends to the permanent thermocline. The mixed layer in summer (sm) is much shallower, extending only to the seasonal thermocline.

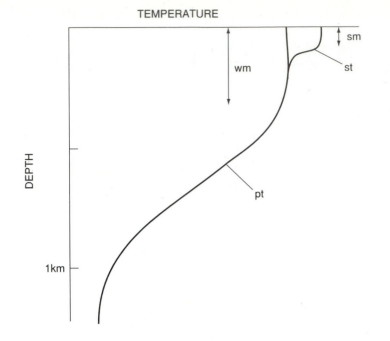

continually brought into the photic zone. If they spend sufficient time there, growth can occur. Cells sinking into the pycnocline or below it, however, cannot be resuspended into the mixed layer and are lost from the community.

The greater depth of offshore ocean waters means that solar heating and annual temperature changes are confined to the upper 1,000 meters. In tropical and temperate oceans, the thermocline from approximately 200 to 1,000 meters is permanent. In temperate oceans, a seasonal thermocline forms in spring in the layer above the permanent thermocline and persists through summer, restricting phytoplankton activity to an even shallower mixed layer (fig. 7.6).

In addition to temperature, salinity influences water density. The relationship is direct: Density increases as salinity increases. A layer showing a rapid increase in salinity with depth (analogous to the thermocline) is a **halocline** (or chemocline) and can produce or reinforce a pycnocline. In lakes, seasonal temperature changes normally have a greater effect on water density. The halocline is important, however, in lakes with a high salt content from evaporation or inflow of saline water. In seawater, the presence of salts also lowers the water's freezing point to almost $-2°$ C and lowers the temperature of maximum density below the freezing point of water. As a result, seawater density is inversely related to temperature throughout the range in which seawater is a liquid, and inverse stratification never occurs.

The salinity of seawater is expressed as grams of salt per 1,000 grams of water (parts per thousand), abbreviated ‰. In the open ocean, salinity normally ranges from 34–36‰, and in coastal regions from 28–32‰, but is lower near freshwater discharges. Salinities above 40‰ represent hypersaline conditions.

Nutrients

The availability of particular inorganic nutrients often determines the level of phytoplankton growth. Overall, nutrient concentrations vary greatly in different bodies of water. The terms **eutrophy** and **oligotrophy** respectively describe conditions of high and low nutrient levels. Nutrient sources to the mixed layer/photic zone are: (1) discharges from rivers and streams, and runoff from land; (2) upwelling or vertical mixing that brings deep-water nutrients to the surface when a shallow pycnocline is absent; (3) recycling within the mixed layer from grazers breaking cells during "sloppy" feeding, from animal wastes, and from bacterial decomposition; and (4) the atmosphere, for such gases as carbon dioxide and for dust particles. Pollutants directly discharged into some waters or precipitated from the atmosphere are sometimes significant nutrient sources. Sinking of cells is the principal way nutrients are removed from the photic zone. In this section, we will consider first the general nutritional requirements of algae before discussing the phytoplankton specifically.

Macroelements that all algae require in relatively large quantities are carbon, hydrogen, oxygen, sulfur, potassium, calcium, magnesium, phosphorus, and nitrogen. Microelements that algae require in much lower quantities, often as cofactors in enzyme systems, include iron, manganese, copper, zinc, and molybdenum (see Vymazal 1995 for information on specific nutrients). Other elements are required by only some algae, such as silicon by the diatoms. Most elements are available in excess amounts, and only a few limit or determine the level of algal growth.

The ions of many metals are relatively insoluble in water but are maintained in the water by forming weak bonds with organic materials. This interaction, called chelation, reduces the toxicity of some ions and makes them available to algae. Humic substances from the partial decomposition of plant material and mucilage that algae secrete are naturally occurring chelators.

Although algae are described as phototrophic or autotrophic, implying that they require only light and inorganic material to grow successfully, many also use organic material either as a supplementary energy source (heterotrophy) or as a necessary growth factor or vitamin (auxotrophy). Vitamins that some algae require include biotin, thiamine, and cyanocobalamin (B_{12}). In some algal groups, **auxotrophy,** or the requirement for a specific vitamin, is common, while other groups have relatively few auxotrophic species. In nature, bacteria and other algae are vitamin sources.

Algal growths below the photic zone or under thick covers of ice and snow suggest that some populations obtain energy from sources other than sunlight, probably using either dissolved or particulate organic material in the water. Heterotrophy by algae is the ability to oxidize organic compounds for energy. Since these algae may obtain energy either by photosynthesis or by heterotrophy, they are "mixed-feeders" or mixotrophs. **Mixotrophy** may involve either uptake of dissolved organic material from the surrounding water or ingestion of organic particles (phagocytosis). Some mixotrophic algae take up acetate, simple sugars, or other simple organic compounds dissolved in the water. Phagotrophic algae ingest bacteria or detrital particles that enter cells in food vacuoles. Besides allowing algae to grow at low light levels, mixotrophy may also support growth when inorganic nutrients, such as phosphates, are unavailable. Algal groups with mixotrophic members include the dinoflagellates, chrysophytes, synurophytes, haptophytes, cryptomonads, and euglenophytes. Some species, classified as algae because of their similarities to photosynthetic species, lack photosynthetic pigments and depend entirely on heterotrophic metabolism.

A phytoplankton's requirement for a specific nutrient is expressed by the **Monod equation,** derived from the Michaelis-Menten model for enzyme kinetics:

$$\mu = \mu_{max} \frac{S}{K + S}$$

where μ = growth rate based on cell production, μ_{max} = maximum growth rate, S = nutrient concentration, and K is the value of S at half the maximum growth

Figure 7.7
Growth in response to the external concentration of a nutrient (S). In this hyperbolic relationship, the half-saturation constant (K) and the maximum growth rate (μ_{max}) characterize an alga's response to a specific nutrient.

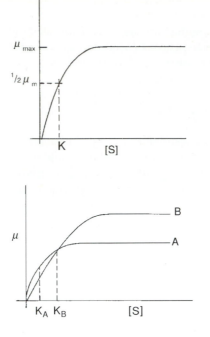

Figure 7.8
Growth of species A and B in response to the concentration of a nutrient (S).

Figure 7.9
Model describing growth in response to the internal concentration of a nutrient. (After Droop 1973.)

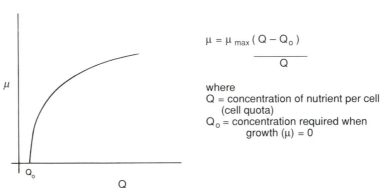

$$\mu = \mu_{max} \frac{(Q - Q_o)}{Q}$$

where
Q = concentration of nutrient per cell (cell quota)
Q_o = concentration required when growth (μ) = 0

rate (fig. 7.7). This relationship can describe the utilization of either an inorganic or organic nutrient. The half-saturation constant (K) is a measure of a population's ability to use a nutrient. A low K value indicates that a population can grow at low nutrient concentrations, while a high K value indicates the high nutrient concentrations are required for good growth. Both K and μ_{max} vary with temperature (Tilman, Mattson, and Langer 1981; van Donk and Kilham 1990).

K and μ_{max} can be used to predict the outcome of competition between two species. As shown in figure 7.8, species A will be more successful at low nutrient levels. At high concentrations, however, species B will achieve a much higher rate of growth than A.

The Monod equation assumes that nutrient uptake is reflected in growth. However, growth is not always proportional to uptake. Phosphorus is stored when it is available in excess of requirements and later used for growth. In such cases, growth is related to cellular levels of phosphorus. The Droop equation in figure 7.9 describes the relationship between growth and internal stores.

Because of the importance of nitrogen, phosphorus, carbon, and silicon in controlling growth of phytoplankton populations, these nutrients will be consid-

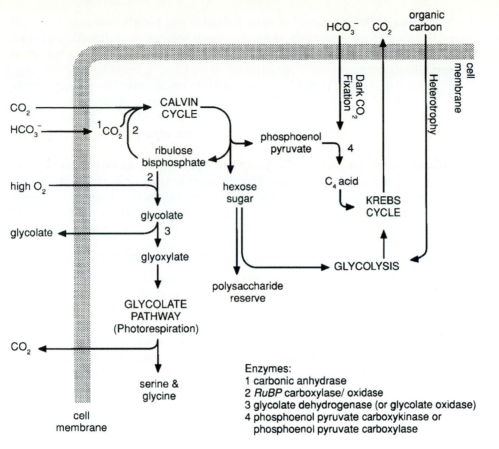

Figure 7.10
Summary of carbon
metabolism in algae.

Enzymes:
1 carbonic anhydrase
2 *RuBP* carboxylase/ oxidase
3 glycolate dehydrogenase (or glycolate oxidase)
4 phosphoenol pyruvate carboxykinase or
   phosphoenol pyruvate carboxylase

ered in more detail. In many aquatic environments, either phosphorus or nitrogen limits overall phytoplankton growth, while carbon may be depleted in the vicinity of actively photosynthesizing cells. The diatoms require silicon in large quantities, which may limit their growth. Nitrogen, phosphorus, and silicon usually increase in concentration with increasing depth in the water column.

Carbon

The total inorganic carbon (TIC) in a body of water consists of carbon dioxide (CO_2), carbonic acid (H_2CO_3), bicarbonate ions (HCO_3^-), and carbonate ions (CO_3^{2-}). TIC is derived from the atmosphere and from dissolution of carbonate rocks. The relative amounts of the different forms are related to pH. With increasing pH, the following equilibrium shifts to the right:

$$CO_2 + H_2O$$
$$H_2CO_3 \rightleftharpoons H^+ + HCO_3^- \rightleftharpoons 2H^+ + CO_3^{2-}$$

The ionic forms produce the primary buffering system in the water. Freshwater systems normally have a pH between 6 and 9, while seawater has a pH between 7.8 and 8.2.

For photosynthesis, most algae can use either carbon dioxide or bicarbonate. Bicarbonate is the predominant form of TIC at the pH of seawater and in many lakes. Before entering the Calvin cycle, carbonic anhydrase must convert bicarbonate to carbon dioxide (fig. 7.10). If photosynthetic activity is high in a confined body of water, cellular removal of carbon dioxide and bicarbonate may cause pH to rise. Under extreme conditions, the carbon equilibrium may shift sufficiently that carbonate is the dominant form of TIC. Algae are unable to use carbonate in photosynthesis.

Respiration in algae is similar to that in other organisms. The principal pathways are glycolysis in the cytosol, and the Krebs cycle and electron transport system in mitochondria. Ordinarily, respiration is assumed to be the same in the light and dark, but evidence indicates that, in some algae, this is not always the case.

Photorespiration is a potentially important pathway in the carbon metabolism of algae. It is a completely separate process from normal or dark respiration. In cells exposed to high irradiance and high levels of dissolved oxygen, ribulose bisphosphate carboxylase, which normally fixes carbon dioxide at the first step of the Calvin cycle, acts instead as an oxidase by splitting ribulose bisphosphate to form glycolate (fig. 7.10). This process is called photorespiration because oxygen is consumed. However, unlike regular respiration, no usable chemical energy is produced, and much of the glycolate may be lost from a cell. Perhaps this pathway helps to remove excess carbon from the photosynthetic system or simply reflects the early evolution of ribulose bisphosphate before an oxygen-rich atmosphere developed.

Although some algae show a significant level of photorespiration, other algae minimize its occurrence by maintaining a high CO_2/O_2 ratio. In contrast to the C_4 photosynthetic pathway in some vascular plants, algae have an alternative way of maintaining a ratio that is favorable for the Calvin cycle. Since algae normally use bicarbonate in the water as their primary source of inorganic carbon, uptake of bicarbonate and its conversion to carbon dioxide by carbonic anhydrase keep the carbon dioxide level high in chloroplasts favoring photosynthesis rather than photorespiration. Carbonic anhydrase may be in the cell membrane, cytosol, or chloroplast (Badger and Price 1994).

Many algae release large quantities of organic carbon from their cells, including glycolate from photorespiration. Sometimes, as much as 50% of the carbon fixed in photosynthesis is lost in this fashion. The dissolved organic material from algal cells may be an important nutritional source for heterotrophic bacteria.

Inorganic carbon also is used in the formation of calcareous structures, such as the scales of coccolithophores (Haptophyta):

$$Ca^{++} + 2HCO_3^- \rightarrow CaCO_3 + CO_2 + H_2O$$

Ultimately, these calcareous structures may sink, transporting carbon from the surface waters into the deep oceans, where the calcium carbonate may accumulate in sediments or be released as the structures dissolve.

Nitrogen

Nitrogen occurs in a number of different forms in water, including diatomic nitrogen (N_2), ammonium (NH_4^+), nitrate (NO_3^-), nitrite (NO_2^-), other nitrogen oxides, and nitrogen in organic compounds. Ammonium is normally low in oxygenated waters, where bacteria oxidize it to nitrate. Nitrogen oxides present as atmospheric pollutants may wash out by rain, enriching some aquatic environments.

Nitrate and ammonium are the primary forms that algae use, but most algae can also use nitrite, urea, and other forms of organic nitrogen. Although abundant, nitrogen gas (N_2) is available only to cyanobacteria capable of nitrogen fixation.

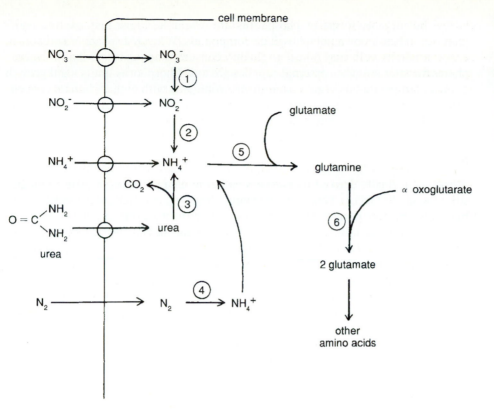

Figure 7.11
Nitrogen uptake and incorporation into amino acids. Enzymes: (1) nitrate reductase; (2) nitrite reductase; (3) urease or urea amidolase; (4) nitrogenase (nitrogen fixation); (5) glutamine synthetase; (6) glutamine-oxoglutarate aminotransferase.

When other forms of inorganic nitrogen are low, these cyanobacteria are often important in converting N_2 gas to forms usable by non-nitrogen–fixing algae (see p. 26). Nitrogen fixation is inhibited in the presence of nitrate or ammonium.

Both nitrate and nitrite must be reduced to ammonium before incorporation into amino acids and other organic compounds (fig. 7.11). Nitrate uptake is followed by reducing enzymes converting it to nitrite and then to ammonium. The energy required for these reduction steps makes it more favorable for a cell to use ammonium when it is available. Urea also must be converted enzymatically to ammonium. Algae vary in their use of free amino acids and other forms of organic nitrogen (see Antia, Harrison, and Oliveira 1991). The phytoplankton also may release amino acids and other forms of dissolved nitrogen into the water, making them available to other organisms (Bronk, Gilbert, and Ward 1994).

Protein synthesis normally depends on an adequate nitrogen supply in the surrounding water. Cellular accumulation of nitrogen in an inactive storage form is limited in comparison to phosphorus, and when nitrogen is deficient, nitrogen compounds in cellular structures may degrade.

Nitrogen often limits the phytoplankton in the open oceans (Howarth 1988) but is less commonly limiting in freshwater. In tropical and subtropical oceans with very low nutrient levels, nitrogen fixation by cyanobacteria is important.

Phosphorus

The total phosphorus content of water consists of dissolved inorganic phosphorus, dissolved organic compounds with phosphorus, and organic phosphorus in suspended particles. While algae may utilize several different forms of nitrogen, phosphorus uptake is almost exclusively as phosphate. When inorganic phosphorus is low but organic forms of phosphorus are available, algae may excrete phosphatases to break down polyphosphate compounds. When phosphate is available in excess amounts, cells may take it up (luxury consumption) and store it as polyphosphates for later use when external supplies are low. Phosphorus limits algal growth in many temperate lakes and in marine environments with high carbonate concentrations, such as coral reefs. Natural phosphorus is derived from rock weathering and biological sources, but is also present in many pollutants.

Silicon

Silicon is normally derived from rock weathering and dissolves in water as orthosilicic acid [$Si(OH)_4$]. Thus, silicon comes largely from natural processes, and human discharges do not significantly contribute to its levels in aquatic environments. Some algae (primarily the diatoms and synurophytes) use silicon to synthesize cell coverings. Dense diatom growths in the spring may severely deplete silicon in the water and lead to nonsilicon-requiring algae replacing the diatoms.

Nitrogen and phosphorus enrichment of fresh and coastal waters may cause the diatoms to decline. Initially, such enrichment stimulates diatom growths, but as diatom cells sink, they remove silicon from the upper waters. Eventually, the dissolved silicon is insufficient to support the diatoms, and other phytoplankton dominate (Conley, Schelske, and Stoermer 1993). In cultures of marine phytoplankton, diatoms are dominant at high silicon to nitrogen ratios, while nonsilicon-requiring species are favored by low silicon to nitrogen ratios (Sommer 1994).

Flotation and Sinking

Most phytoplankton are denser than water and tend to sink. The loss of cells by sinking is probably significant in many populations (Alvarez Cobelas and García-Morato 1990). During diatom blooms, over half of the phytoplankton production may be lost in this fashion (Passow 1991).

Mixing of the water is important in maintaining cells in the photic zone as well as in returning cells to this zone. However, a pycnocline normally limits vertical mixing to the upper region of lakes or oceans. Once the phytoplankton sink into a pycnocline, they cannot be resuspended by mixing and are lost from the active population.

Adaptations that help keep cells suspended are both physiologic and morphologic. Viscous or frictional drag results from the interaction of a cell's surface with the surrounding water. For small objects such as the phytoplankton, water viscosity creates a "syrupy" environment that sticks to cellular surfaces. Viscous drag significantly affects cell movement through water and slows a cell's rate of sinking. Size and shape determine a cell's viscous drag. Small size is beneficial since smaller cells have a greater surface-to-volume ratio than larger cells, and therefore, proportionally more surface is in contact with the water for drag. Elongate rather than spherical shapes also have greater surface areas (spheres have the minimum surface area for a given volume). Many phytoplankton have a needlelike form or bear spines as extensions of their walls (plate 3*b*). A large surface area also aids in nutrient uptake. Colonial arrangements may also increase drag as well as cause colonies to tumble in the water.

The phytoplankton may accumulate low-density material to increase their buoyancy. Density is reduced most with the gas vesicles of the cyanobacteria or bubbles within a colony of cells, which may produce a positive buoyancy (fig. 7.12). However, floating on the surface, exposed to high irradiance and with

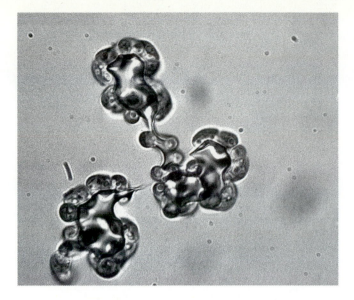

Figure 7.12
An unusual freshwater green alga *Helicodictyon*, with a bubble in the center of its colony.

limited access to nutrients, may not be beneficial for cells. Other low-density materials are more common than gases. Lipids have a density less than water and often replace carbohydrates as the principal reserves in many planktonic species. Other species show ionic regulation by selectively accumulating lighter ions in their vacuoles. Mucilage secretion may reduce a cell's density, as well as aid in nutrient uptake and deter grazers.

Another adaptation of phytoplankton is to swim with flagella. Swimming allows cells to move vertically in the water in response to the direction of light and thus show a positive **phototaxis** (swim toward light). This requires energy expenditure, however, so many flagellated algae or phytoflagellates also have form modifications for increased drag.

Sinking varies with the physiologic condition of cells and stage of population growth. During rapid growth of a young population, loss by sedimentation may be less than in a stable or senescent population. As growth continues, cells collide and stick to each other and other suspended particles, forming aggregations that sink faster than solitary cells (Kiorboe, Andersen, and Dam 1990). At times, sinking may be beneficial. When cells deplete nutrients in their immediate vicinity, slow sinking may expose cells to fresh nutrient supplies. After a period of active growth, some algae form dormant cysts or zygotes that sink to the bottom.

The following equation (modified Stokes equation) summarizes the rate of sinking of a passive cell:

$$v = \frac{2gr^2(\rho' - \rho)}{9\eta\phi}$$

where v = sinking rate, g = gravitational acceleration, r = radius of cell, ρ = density of medium (water), ρ' = density of cell, η = coefficient of viscosity of the medium, and ϕ = coefficient of form resistance (cell shape). Note that r, ρ', and ϕ are properties of the cell, while ρ and η are properties of the surrounding water. Both water viscosity and density are inversely related to temperature, and therefore, a cell's sinking rate will increase with rising temperature.

Water is almost always in motion, and these movements help keep cells suspended. Surface waves and currents are the more obvious forms of water movement, but small-scale, gentle movements often are more important to the phytoplankton.

Figure 7.13
Zooplankton. (*a*) Freshwater
copepods and cladoceran
(*upper left*). (*b*) Feeding
structure of *Euphasia* (krill)
composed of delicate setae.
(*b* from Quetin and Ross
1991, courtesy *American
Zoologist.*)

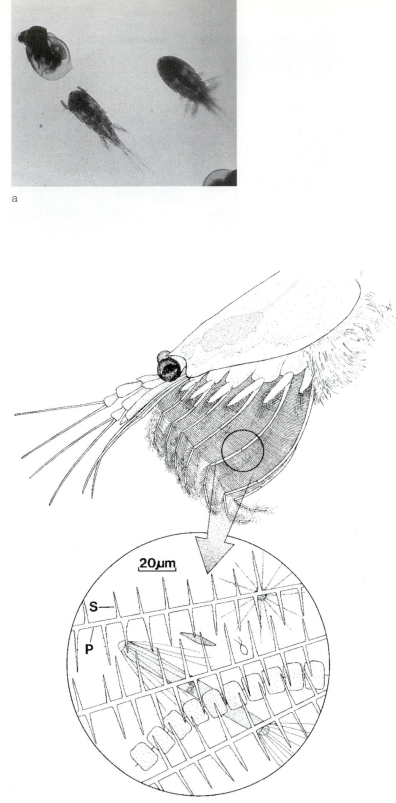

a

b

Chapter 7

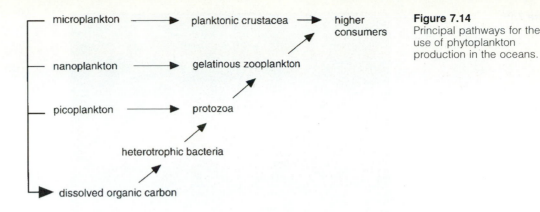

Figure 7.14
Principal pathways for the
use of phytoplankton
production in the oceans.

These movements include mixing by the winds and the convection caused when water in contact with the air cools and sinks.

Grazing

Grazing by herbivorous animals removes cells from phytoplankton populations. In the oceans and freshwaters, both herbivorous and carnivorous species comprise the zooplankton. Important herbivores include protozoa, rotifers, and crustaceans. Copepods are common crustaceans in both lakes and oceans, while cladocerans are important herbivores in lakes, as are euphausiids (krill) in the oceans (fig. 7.13*a*). Near shore, benthic, suspension-feeding animals, such as barnacles and mussels, consume the phytoplankton, and many benthic invertebrates release planktonic larvae that feed on the phytoplankton while temporary members of the zooplankton.

The zooplankton feed either by seizing individual algae or by filtering the water (**suspension-feeding**). Food is collected in filters either by allowing water to flow through a relatively passive sieve or by creating a flow to sweep the particles into a collecting structure. Suspension-feeders collect bacteria and particles of detritus, as well as phytoplankton. Most planktonic herbivores are suspension-feeders and select primarily for a specific size range of food particles; using chemical cues to sense the quality of food is of secondary importance. Herbivorous crustaceans create currents that collect phytoplankton and other particles on delicate hairs (setae) of their appendages (fig. 7.13*b*). They may adjust their filtering rates in response to the size and density of food particles. Typically, the rate is reduced both when food is scarce and when it is concentrated. Other suspension-feeders in the oceans feed on smaller phytoplankton and include planktonic tunicates (salps and appendicularians) and pteropods (planktonic snails called sea butterflies), which use sticky mucilage to collect particles. The protozoa may be the principal consumers of picoplankton (both photosynthetic and nonphotosynthetic) by phagocytosis of individual cells. Figure 7.14 summarizes the major pathways for the use of phytoplankton production.

Zooplankton grazing may reduce overall phytoplankton abundance, selectively remove a specific size or less-resistant phytoplankton species, and produce patchy distributions of the phytoplankton. Grazing may also be important in recycling nutrients released from cells broken during ingestion ("sloppy feeding") or excreted in wastes. Phytoplankton reproduction may increase in response to nutrients that zooplankton and benthic invertebrates release (Sterner 1986; Asmus and Asmus 1991). Thus, moderate grazing by zooplankton may be beneficial in recycling nutrients, such as ammonium and phosphate. In contrast, bacterial

Figure 7.15
Chytrids attached to the
diatom *Asterionella*.

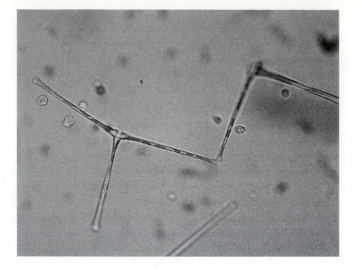

decomposition is slower and often occurs below the mixed layer, so that any nutrients released are not immediately available to actively growing phytoplankton.

The phytoplankton may respond to grazing pressure by increased rates of cell production and increased abundance of species that are less susceptible to herbivory because of their size or specific defenses. Mucilaginous sheaths, thick walls, and other external coverings may provide some protection against herbivores. However, spines are probably more important for flotation than for defense. Colonies may be less susceptible because they are too large for most zooplankton to handle. The common freshwater green alga *Scenedesmus* forms colonies in response to herbivory (Lampert et al. 1994). Large size and copious mucilage probably protect colonies of the marine haptophyte *Phaeocystis*. Only some cyanobacteria, dinoflagellates, and a few other species depend on chemical deterrents.

Planktivorous fishes that feed on zooplankton may affect the abundance and composition of the phytoplankton. By reducing the abundance of the herbivores, the fishes decrease algal loss to grazing zooplankton and may aid in nutrient recycling (Mazumder et al. 1990; Vanni and Findlay 1990).

Other Influences

In addition to the environmental factors already discussed, other factors may influence phytoplankton growth but probably not as frequently or to the same degree. In a dense population, mutual shading of cells and buildup of inhibitory chemicals in the water as by-products of normal metabolism may result in self-regulation of growth (new cell production). Chemical interactions are also possible between different algal populations. In one of the few instances that presence of such allelopathic substances was identified, winter growths of cyanobacteria in a small lake released an unidentified chemical that inhibited spring diatom growth (Keating 1977, 1978).

Potential algal parasites include some fungi, protists, and bacteria. Although parasites such as chytrids and bacteria sometimes attach to algal cells, the role of these organisms in natural communities is not always clear (fig. 7.15). Parasites may delay or prevent a bloom, reduce algal abundance, or cause a population decline (Kudoh and Takahashi 1990; van Donk 1989), but in other instances, parasites may infect cells in an already declining population. When healthy, many algae produce substances that inhibit parasites (Jones 1988; Patterson et al. 1993).

The abundance of viruses in seawater and the widespread viral infection of phytoplankton, especially cyanobacteria, suggest possible viral regulation of the phytoplankton (Proctor and Fuhrman 1990; Suttle, Chan, and Cottrell 1990). At times, a viral infection may control "brown tides" of *Aureococcus* in bays along the U.S. Northeast coast (Milligan and Cosper 1994).

Another potential means of cell loss from a population is physiologic death. Cells may die from various forms of stress, such as extreme temperatures, exposure to ultraviolet radiation, and starvation when nutrients are unavailable. In an oligotrophic lake, Crumpton and Wetzel (1982) found that physiologic death was unimportant compared to losses due to grazing and sinking.

Adaptive Strategies of the Phytoplankton

The planktonic algae show four general adaptive strategies (after Fogg and Thake 1987):

1. The picoplankton. The picoplankton are the phytoplankton with cells less than 2 micrometers in diameter; they may be flagellated or nonflagellated. Picoplankton are characterized by a large surface-to-volume ratio that allows them to use low nutrient concentrations and to create sufficient viscous drag so that sinking is negligible. The picoplankton are common in oligotrophic waters and may show brief periods of abundance in eutrophic waters, where they behave as opportunists. This category includes solitary cyanobacteria, small green algae (prasinophytes), and chrysophytes, such as *Aureococcus,* responsible for "brown tides."

2. The phytoflagellates. The phytoflagellates include solitary flagellated cells larger than 2 micrometers and colonies of flagellated cells. The phytoflagellates show a positive phototaxis and may migrate daily between the surface and deeper waters with higher nutrient concentrations. They are often common during periods of density stratification. Many algal groups, including the dinoflagellates and the haptophytes in the oceans, have phytoflagellate representatives.

3. Cells and colonies dependent on mixing of the water for suspension. Because their cells are denser than the water, these phytoplankton depend on mixing of the water for suspension and usually show form modifications to increase viscous drag. Cell movement caused by sinking and mixing may be beneficial in keeping cells exposed to fresh nutrient supplies. This category includes the planktonic diatoms and the chlorococcalean green algae.

4. Filaments and large colonies with gas-filled structures. The primary representatives of this category are the cyanobacteria with gas vesicles in their cells. Also, in small ponds under calm conditions, some filamentous eukaryotic algae form mats of entangled filaments in which gas bubbles are trapped (discussed in chapter 8).

Primary Production by Phytoplankton

Primary production is the conversion of light energy and inorganic carbon into organic material by photosynthesis (or chemosynthesis). The phytoplankton are the primary producers in the oceans and many freshwater environments. Other producers are autotrophic bacteria (sulfur bacteria and chemosynthetic bacteria), plants (bryophytes and vascular plants), and benthic algae.

In figure 7.16, photosynthesis is represented as a box with inputs and outputs. The important inputs are photosynthetically active radiation (PAR) and inorganic

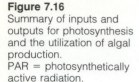

Figure 7.16
Summary of inputs and
outputs for photosynthesis
and the utilization of algal
production.
PAR = photosynthetically
active radiation.

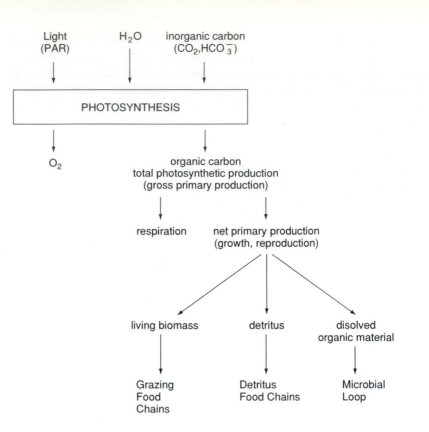

carbon (as either carbon dioxide or bicarbonate). Water only becomes critical in regulating photosynthetic activity in terrestrial environments. Outputs from the photosynthesis box are organic compounds and oxygen. Cells use some of the total photosynthetic production (or gross primary production) to maintain their structure and function (respiration). Any organic material in excess of these needs is net primary production, which is available for growth and reproduction.

Figure 7.16 indicates the two principal ways of measuring phytoplankton primary production. When photosynthetic activity is fairly high, measuring oxygen release is an easy way of estimating production. A more sensitive method, especially when photosynthetic activity is low, uses a radioactive isotope of carbon. Production is estimated by adding a known amount of inorganic carbon containing carbon-14 to a water sample containing phytoplankton and later measuring the amount of carbon-14 incorporated into organic material. Alternatively, with macroalgae, changes in biomass indicate net production.

The organic material that phytoplankton produce is available to consumers in three forms: living material composing cells, dissolved organic material that leaks from cells or is released when cells rupture, and nonliving, particulate, organic material from cell parts. Ecologists divide food chains on the basis of which of these three forms of food is consumed at the step after the producers. In grazing food chains, herbivores consume producers directly. In detrital food chains, detritivores eat detritus from producers. We shall see in chapter 9 that benthic algal communities provide a substantial amount of detritus to support detrital food chains in coastal waters. In the microbial loop, heterotrophic bacteria take up dissolved organic material. The protozoa, in turn, then feed on the heterotrophic bacteria.

With respect to the role of the phytoplankton (and other algae) in primary production, important questions to ask are how much production occurs on an annual basis, when does maximum annual production occur, and what environmental factors regulate the amount of production at different times. We shall see in the next sections that availability of specific inorganic nutrients, (phosphorus, nitrogen, silicon) often limits production, as well as growth.

Marine Phytoplankton

The oceans, consisting of the open oceans overlying the deep abyss, and the coastal oceans and marginal seas over the continental shelves, cover more than 70% of the earth's surface. In the open oceans, the major current systems circulate in a clockwise direction in the Northern Hemisphere and in a counterclockwise pattern in the Southern Hemisphere, and enclose regions called subtropical gyres. Major oceanic regions are (1) the subtropical gyres, which show little seasonal change; (2) the equatorial regions, where the trade winds drive equatorial currents from east to west; (3) temperate regions, which show seasonal variations in temperature that affect density stratification; (4) polar regions, where the annual formation and melting of sea ice is more important than temperature changes; and (5) coastal oceans over the continental shelves.

In the oceans, water layers with different densities (denser layers are below less dense layers) produce density stratification, which is determined by measuring vertical profiles of temperature and salinity (or conductivity). In the open oceans, except in polar regions, the stratification pattern consists of a permanent thermocline extending from approximately 200–300 meters to a depth of 1,000 meters. In the layer above the permanent thermocline, stratification is on a smaller scale. In temperate oceans, a distinct seasonal thermocline often forms in the spring and persists into the fall (see fig. 7.6). In polar regions, the surface waters remain cold (-2 to $4°$ C) throughout the year, without strong stratification except in regions of ice melt.

The nanoplankton and picoplankton are often more common in the open ocean than in coastal waters, where larger diatoms and dinoflagellates predominate. The picoplankton are widespread, especially the prokaryotes such as *Synechococcus* (cyanobacterium) and *Prochlorococcus* (prochlorophyte). Common eukaryotic groups with picoplanktonic or nanoplanktonic representatives include both flagellated and nonflagellated haptophytes, chrysophytes, prasinophytes, chlorophytes, and smaller diatoms.

Remote satellite sensing of the color of the ocean surface is used to estimate chlorophyll concentration in the surface layer as a measure of phytoplankton abundance. Satellites also can detect high coccolithophore concentrations by light reflected from their calcareous scales.

Dense phytoplankton growths may influence the weather in two major ways. First, absorption of radiant energy by cells may increase surface water temperature (Sathyendranath et al. 1991), and second, some phytoplankton may release dimethylsulfide (DMS), which seeds clouds (see fig. 5.4). DMS comes primarily from the haptophytes and dinoflagellates (and some benthic macroalgae).

Overall, 30–50% of photosynthetic production on earth occurs in the oceans as a result of algal photosynthesis. Nutrient availability in different parts of the oceans largely determines the differences observed in phytoplankton production (table 7.1) and in seasonal patterns (fig. 7.17). Many of the important changes in the oceans affecting the phytoplankton occur in the vertical direction: With depth, light and temperature decrease, density increases, and nutrients increase.

Figure 7.17
Seasonal patterns of
phytoplankton abundance in
(*a*) polar, (*b*) temperate, and
(*c*) tropical oceans.

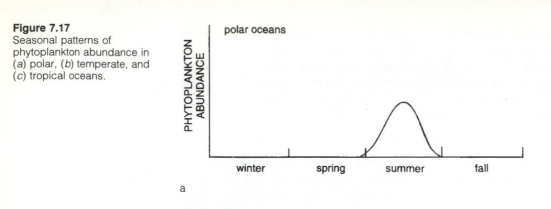

a

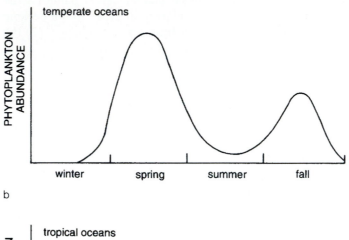

b

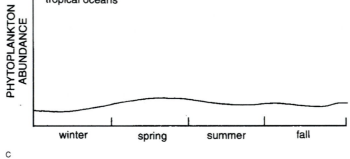

c

Table 7.1 *Annual Production of Phytoplankton in Different Ocean Regions**

Region	Production (g C $m^{-2}yr^{-1}$)
Open oceans	
Tropical/subtropical	40
Temperate	120
Arctic	<1
Southern (Antarctic)	100
Coastal oceans	
Non-upwelling	200
Upwelling	300

Source: After Sumich 1988.
*Annual production expressed as grams of carbon per square meter of ocean surface.

Temperate Oceans

In temperate oceans, phytoplankton abundance normally peaks in the spring and fall (fig. 7.17*b*). Temperature, as it affects density stratification, controls nutrient availability and is responsible for these two peaks. In winter, deep vertical mixing extends below the photic zone to 200–300 meters, where the permanent thermocline begins. With the photic zone occupying only the upper portion of the mixed layer, planktonic cells spend only a small part of their time there, and consequently, their production and growth are low, even though vertical mixing brings nutrients into the photic zone. As solar radiation warms the ocean surface in spring, a seasonal thermocline forms and restricts the depth of vertical mixing so that the phytoplankton spend all or most of their time in the photic zone. As a result, the phytoplankton, especially diatoms, become abundant.

In spring, the diatoms at first benefit from the availability of light and nutrients, but their continued growth depletes nutrients in the photic zone, especially nitrogen and silicon. Nutrients are lost from the mixed layer as the diatoms and other phytoplankton clump and sink into deeper water (Waite, Bienfang, and Harrison 1992; Waite et al. 1995). Also, because the diatoms are abundant as a food source, populations of herbivorous zooplankton increase. Both the reduced nutrient availability and grazing contribute to the decline of the spring diatom bloom and keep the phytoplankton at low levels through the summer. During this time, the dinoflagellates and other phytoflagellates are normally more common than the diatoms, but overall abundance remains low. Flagellated algae may migrate daily between the photic zone at the surface and deeper waters with higher nutrient concentrations (Lieberman, Shilo, and van Rijn 1994). The mixed layer may show a rough subdivision into an upper layer, where light is saturating for photosynthesis and nutrients are depleted, and a lower layer, where light limits phytoplankton but nutrients are available in higher concentration. The dinoflagellates may migrate into this lower layer at night and return to the surface layer during daylight periods. Also, the mixotrophic phytoplankton, including most dinoflagellates, may use organic material in the seawater to supplement their requirements.

Increased vertical mixing and resuspension of nutrients accompany breakdown of the seasonal thermocline as the water cools in the fall. Responding to this enrichment, the diatoms show a second peak of abundance, although not as great as in spring. As the water continues to cool and mixing is deeper, phytoplankton activity declines to a low winter level. During the winter, nutrients mix throughout the layer above the permanent thermocline, but algal activity remains low until the seasonal thermocline forms in the spring.

Tropical and Subtropical Oceans

The waters within the subtropical gyres bounded by the major current systems are marine deserts. Despite the availability of sunlight throughout the year, gyre surface waters remain low in nutrients, since no deep vertical mixing replenishes the surface waters. These conditions are similar to those found in temperate oceans in summer. Here, however, they result in low phytoplankton activity all year (fig. 7.17*c*).

The picoplankton and nanoplankton often predominate in the gyres. Their large surface areas relative to cell volumes are beneficial for uptake of nutrients at low concentrations. Also, nitrogen-fixing cyanobacteria may be important sources of nitrogen compounds. *Trichodesmium,* a filamentous cyanobacterium with gas vesicles (see fig. 2.11), is widespread in tropical and subtropical oceans, where it contributes to production and fixes nitrogen (Carpenter and Romans 1991;

Carpenter, Capone, and Reuter 1992). *Trichodesmium* forms large bundles of filaments that may be visible in the water as flakes of "sawdust" and is unusual in being able to fix nitrogen and photosynthesize without heterocysts. Other nitrogen-fixing cyanobacteria are symbionts in some diatoms or grow attached to floating mats of *Sargassum* found in a region of the North Atlantic gyre called the Sargasso Sea. Many phytoplankton in low-nutrient, tropical waters show vertical movement, tending to sink when cellular levels of nutrients are low (Villareal and Lipschultz 1995). Individual species of the phytoplankton may show seasonal patterns related to the availability of specific trace nutrients, but the peaks of overall phytoplankton activity seen at higher latitudes do not occur here, and annual production is lower than in temperate oceans.

Adjacent to the equator in the Atlantic, Pacific, and Indian Oceans, the trade winds drive the North and South Equatorial Currents. As these currents flow westward, they diverge. Such divergence of the surface waters brings deeper waters to the surface, as equatorial upwelling. The nutrients brought up in the rising water support greater phytoplankton activity than otherwise occurs in tropical and subtropical regions.

A proposal that iron availability regulates the phytoplankton when nitrogen and phosphorus are sufficient is controversial (Chisholm and Morel 1991). Evidence for possible iron limitation in the oceans, especially in polar regions and the equatorial Pacific, is based on low iron concentrations in the seawater and on bioassay studies in which the addition of iron stimulates algal growth. Airborne dust and upwelling of deep water may be important iron sources (Coale et al. 1996). Two large-scale tests of the iron limitation hypothesis involved enriching 64-square-kilometer areas of the equatorial Pacific near the Galápagos Islands. Both production and chlorophyll levels increased temporarily in the enriched areas (Martin et al. 1994; Monastersky 1995).

Polar Oceans

In both the Arctic and Southern (Antarctic) Oceans, annual changes in water temperature are small and have less effect on density stratification than at lower latitudes. Instead, the annual freezing of seawater and the melting of sea ice are of greater importance. As sea ice forms, the salts in seawater are excluded, increasing the salinity in the remaining unfrozen water and producing pockets of brine within the ice. Later, when the ice with its relatively low salt content melts, low-salinity, low-density water spreads at the surface, forming a weakly stratified condition. Even though their growing season is relatively short, the phytoplankton at high latitudes benefit from continuous daylight during summer. Diatoms usually dominate, but large colonies of the haptophyte *Phaeocystis* are sometimes abundant, and the dinoflagellates and other phytoflagellates also may be common.

The difference in production between the Arctic Ocean and the Southern Ocean is due primarily to nutrient availability. In the Arctic, a single summer peak of the diatoms is associated with the period of summer stratification, but because of the short growing season and lack of upwelling, the overall annual phytoplankton production is low (fig. 7.17*a*).

The Antarctic marine region between the Antarctic continent and the Antarctic Convergence (located between 50–60°S latitude) consists of three zones: an ice-free zone, a seasonal sea-ice zone, and a permanent sea-ice zone. Phytoplankton production is low in the ice-free zone because of the deep mixing and in the permanent ice zone because of the low light levels that penetrate the ice cover.

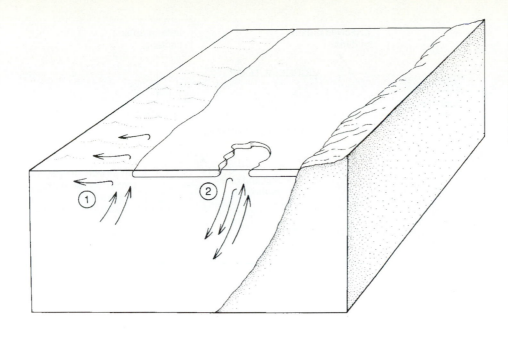

Figure 7.18
Upwelling in the Antarctic caused by (1) winds blowing surface water away from the edge of the ice, and (2) cooling and sinking of water in areas of open water, called polynyas, within the ice. In both cases, deeper water rises to replace the surface water. (Courtesy C. Feller.)

Production is relatively high, however, at the edge of the pack ice or marginal ice zone (MIZ). As the sea ice recedes in the spring, the MIZ moves south, and as the sea freezes in the fall, the MIZ moves north.

The MIZ is responsible for a relatively high overall phytoplankton production in the Southern Ocean. Even though water temperatures generally remain between −2 and 4° C and the growing season is only four months, the phytoplankton in the MIZ benefit from exposure to continuous daylight in summer and enrichment by upwelling. High production in the MIZ is associated with a shallow, well-illuminated, mixed layer formed by low-salinity water from ice melt. Polynyas—large areas of open water within the cover of permanent sea ice—also support high production.

Nutrient concentrations in the Southern Ocean are relatively high as a result of upwelling of deeper water. Thus, Antarctic phytoplankton do not experience the prolonged periods of nutrient limitation that occur at lower latitudes. Areas of upwelling are produced by winds blowing the surface water away from the edge of the ice, causing deeper water to rise to replace it, and by vertical circulation from water cooling and sinking when it comes in contact with the air (fig. 7.18). Upwelling also occurs farther from the edge of the ice at the Antarctic Divergence (boundary between the Circumpolar Current and the Coastal Current), where separation of the surface water draws up deeper water.

Antarctic phytoplankton represent a wide range of sizes. Larger diatoms and *Phaeocystis* produce blooms in the MIZ, while the nanoplankton and picoplankton usually dominate under nonbloom conditions and include smaller diatoms, other haptophytes, cryptophytes, green flagellates, and unicellular cyanobacteria. The phytoplankton are food for krill (euphausiids) and other zooplankton near the edge of the ice, and the krill, in turn, form the link with the rest of the rich Antarctic marine life, which includes penguins, seals, and whales. The phytoplankton and ice algae (described later in the chapter) also release into the seawater large quantities of dissolved organic material that support heterotrophic bacteria and protozoa in the microbial loop, and when they sink are food for a rich benthic community (fig. 7.19).

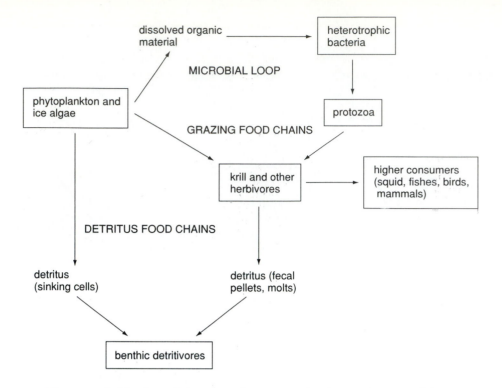

Figure 7.19
Food chains in the marginal ice zone of the Antarctic. (Based on Knox 1994.)

The annual thinning of the ozone layer over the Antarctic, caused by ozone-destroying pollutants, may adversely affect phytoplankton production. Since 1985, scientists have measured a "hole," or more correctly, a 50% reduction in the ozone layer in the stratosphere during the Antarctic spring (September through November). Less ozone means an increase in ultraviolet-B radiation (UVB, 280–320 nanometers) reaching the earth's surface. The overall affect on phytoplankton production is uncertain. Estimates range from no effect to as much as a 12% decrease, with perhaps 5% being a reasonable value (Smith et al. 1992; Schofield, Kroon, and Prézelin 1995). The shallow mixed layer exposed to sunlight and enriched with nutrients makes the MIZ favorable for production but also keeps the phytoplankton near the sea surface, where they may be exposed to harmful UVB levels. Ultraviolet radiation inhibits photosynthesis, damages DNA, and reduces nutrient uptake (Behrenfeld, Lean, and Lee 1995). However, algae vary in their sensitivity to ultraviolet radiation (Karentz, Cleaver, and Mitchell 1991; Davidson et al. 1994). To different degrees, they are able to repair UV damage, absorb the damaging energy with special molecules such as carotenoids and superoxide dismutase, or produce protective "sunscreens" (Vincent and Roy 1993). Migration into deeper water or under sea ice also may allow phytoplankton to avoid UV exposure.

During the Antarctic winter, a vast area of the ocean is covered with sea ice, creating a habitat for ice algae.

But in March, the Southern Hemisphere's first month of autumn, a phenomenon occurs that has been called the greatest seasonal event on earth. As the air temperature begins dropping to as low as −40° C, the ocean, which turns to ice at −1.8° C, starts freezing at the incredible average rate of 5.75 square kilometers (2.22 square miles) per minute. By the end of the Antarctic winter, in September, the ice pack measures 19 million square kilometers (7,334,000 square miles)—nearly twice the area of the United States—in a layer usually no more than one meter thick. (Stevens 1995).

In both the Arctic and Antarctic regions, **ice algae** make an important contribution to production. These algae grow in abundance on the bottom of sea ice, producing a greenish brown discoloration. They are also found in brine pockets in the ice and in melt water on its surface. Several hundred species of diatoms have been described from ice algal communities. The pennate diatoms are the most common group, but the prasinophytes, haptophytes, chrysophytes, and centric diatoms are also represented. Ice algae are adapted to low levels of irradiance filtered by the ice (Robinson et al. 1995), near-freezing temperatures, and widely varying salinity (0–100‰). Ice algae are food for protists, amphipods, copepods, krill, and other zooplankton, and release large quantities of organic material that support heterotrophic bacteria. Ice algae also may seed the annual growths of the phytoplankton. They begin to grow before the phytoplankton, and cells released into the water as the ice melts may initiate planktonic growths (Garrison, Buck, and Fryxell 1987). As the sea ice advances in the fall, ice algae may continue to grow after the phytoplankton decline.

The long eight-month winter in polar regions, when irradiance is insufficient to support photosynthesis, presents a particular survival problem. Some algae form cysts and become dormant, but others remain metabolically active by consuming reserves and possibly by assimilating organic material in the water (mixotrophy). Special lipids in cell membranes may represent an adaptation to very low temperatures (Kirst and Wiencke 1995).

Coastal Oceans

The coastal regions over the continental shelves are normally less than 200 meters deep. Although the photic zone is usually shallower than in offshore oceans because of the greater amount of suspended material, coastal regions are enriched by rivers and terrestrial runoff and by vertical mixing that reaches to the sea bottom during unstratified periods. Generally, levels of phytoplankton production are greater than in offshore waters but show similar seasonal patterns—spring and fall peaks in temperate coastal waters. On a local scale, bottom topography may cause upwelling of nutrient-rich bottom water and produce patches of high production. The next section discusses large-scale and persistent movements of water away from the shore, which cause more extensive regions of upwelling and high production.

The planktonic algae in coastal waters may survive as dormant stages on the bottom during unfavorable growth periods. Following a bloom in the photic zone, diatoms form dormant cysts that settle to the bottom and survive there under cold, dark conditions (Smetacek 1985). A reserve of bottom stages is important in seeding growths of the dinoflagellates responsible for red tides (see fig. 5.14).

Many benthic invertebrates form planktonic larvae that feed on the phytoplankton as the larvae disperse through the water. Dense phytoplankton growths release chemicals that stimulate spawning by mussels and sea urchins, thus assuring abundant food for their larvae (Starr, Himmelman, and Therriault 1990).

Recently, phytoplankton blooms in coastal waters, especially red tides caused by dinoflagellates and blooms of "exotic species," such as the chrysophyte *Aureococcus* and haptophytes such as *Chrysochromulina*, have increased worldwide. The primary cause of these blooms is the increased eutrophication of coastal waters, but a contributing factor to the widespread occurrence of nuisance species has been their transport worldwide in the ballast of ships or with shellfish stocks. Also, closer monitoring of coastal waters has led to a greater awareness of toxic blooms, especially when blooms may threaten fish farms. Information on the

Table 7.2 *Toxic Phytoplankton*

Effect on Humans	Principal Genus	Principal Toxin
Paralytic shellfish poisoning	*Alexandrium*	Saxitoxin
Neurotoxic shellfish poisoning (ichthyotoxicity)	*Gymnodinium*	Brevetoxin
Ciguatera fish poisoning	*Gambierdiscus, Prorocentrum*	Ciguatoxin and maitotoxin
Diarrhetic shellfish poisoning	*Dinophysis*	Okadaic acid
Amnesic shellfish poisoning	*Pseudonitzschia*	Domoic acid
Cyanobacterial neurotoxins	*Anabaena, Aphanizomenon*	Anatoxins
Cyanobacterial hepatotoxins	*Microcystis, Nodularia*	Microcystin, nodularin
Dermatitis	*Lyngbya*	Lyngbyatoxin, aplysiatoxin

Principal sources: Ikawa and Sasner 1990; Hallegraef 1993; Carmichael 1994.

worldwide occurrence of blooms is periodically exchanged at international conferences on toxic marine phytoplankton and through *Harmful Algal News*, published by the Intergovernmental Oceanographic Commission of UNESCO (United Nations Educational, Scientific, and Cultural Organization).

Dense phytoplankton growths may have a number of effects: (1) phytoplankton respiration at night and cell decomposition at the end of a bloom may severely deplete oxygen; (2) siliceous frustules of the diatoms may clog or damage gills of fishes; (3) less-desirable bloom species may outcompete species that are important as food for other marine organisms; and (4) toxins that some algae produce as defense against herbivores may accumulate in food chains (table 7.2). When these toxins concentrate in commercially important fishes or bivalves, human consumption of contaminated seafood may cause neuromuscular and gastrointestinal disorders, leading to death in extreme cases (see p. 141 for a discussion of toxic dinoflagellates). Also, marine mammals and birds may be poisoned (Anderson and White 1992). Fish and bivalve contamination by domoic acid from the diatom *Pseudonitzschia* (see p. 106) and diarrhetic shellfish poisoning caused by some dinoflagellates were reported recently for the first time in North America.

The increased incidence of blooms is linked to greater enrichment of coastal waters with nitrogen and phosphorus, especially from urban areas. Also decreased freshwater inflows, from damming rivers and diverting water for irrigation, may have reduced silicon inputs, producing a less favorable condition for diatom growth (Smayda 1990). Widespread toxic algal blooms may reflect a change to herbivore regulation of phytoplankton instead of regulation by nutrient availability. With increased grazing pressure acting as a selective factor on phytoplankton communities, species with special defenses such as toxins may be favored over less-resistant species.

Coastal Upwelling and Estuaries

In some areas, phytoplankton production is high where waters are naturally enriched by widespread coastal **upwelling** (see table 7.1). These regions of high, sustained phytoplankton production support rich marine communities, which

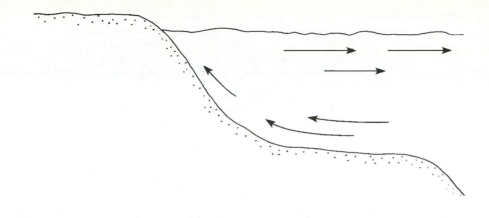

Figure 7.20
Coastal upwelling. Arrows indicate offshore movement of surface water and upwelling of deeper, nutrient-rich water.

often, in turn, are the basis for harvesting commercially valuable marine animals. Major areas of coastal upwelling occur along the west coasts of South America and Africa, and to a lesser extent along other coasts. Upwelling results from movement of surface water away from shore in response to prevailing offshore winds and the earth's rotation (Coriolis effect). As surface layers move offshore, deeper, nutrient-rich water is drawn to the surface as replacement (fig. 7.20).

One of the richest upwelling systems occurs in the South Pacific. Upwelling along the South American coast is part of a larger pattern of current circulation and winds in the South Pacific called the Southern Oscillation. Normally, the trade winds blow from east to west, driving the South Equatorial Current. As a result, warm water accumulates in the western Pacific, and cold water from the Antarctic flows northward along the coasts of Chile and Peru. This creates a very shallow thermocline and mixed layer (approximately 50 meters deep), and upwelling brings nutrient-rich water from below the thermocline into the mixed layer. In normal years, warming of the surface waters in summer (late December to April) temporarily reduces the upwelling. In El Niño years, however, the warming is more intense and prolonged from weakening of the trade winds and a decrease in the westward flow of water in the Equatorial Current. The mixed layer deepens along the South American coast, so that upwelling simply circulates low-nutrient water within the mixed layer.

In the South American system, enrichment from upwelling supports sustained growths of diatoms that are food for zooplankton and fishes, especially the commercially important anchoveta (Peruvian anchovy). The rich marine life supports seabirds (pelicans, boobies, cormorants), which produce valuable guano deposits on offshore islands. During an El Niño event, however, the fishes, birds, and seals decline, and red tides may occur. El Niño conditions occur approximately every three to five years and disrupt global weather patterns. One of the severest El Niño events occurred in 1982–1983, and scientists are concerned that the frequency of El Niño events may be increasing.

Estuaries are partially enclosed basins where seawater mixes with freshwater and include coastal plain estuaries that are extensions of rivers, fjords formed by glaciers, and lagoons behind coastal barrier islands (see fig. 9.9). In some estuaries, freshwater and salt water mix, and stratification is absent, but in other estuaries, freshwater forms a layer over seawater. Many coastal plain estuaries are stratified because of the density difference between freshwater and seawater, and the failure of these two water types to mix rapidly (fig. 7.21). River water flowing into an estuary is usually high in suspended material and nutrients for the phytoplankton.

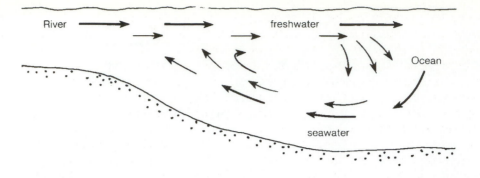

Figure 7.21
Circulation in a stratified estuary. Heavy arrows indicate surface flow of freshwater and bottom flow of seawater. Lighter arrows indicate nutrient recycling in counterflow.

River   freshwater   Ocean   seawater

In general, estuaries vary more in environmental conditions, such as temperature and salinity, than ocean water. Turbidity is often great, and light may limit photosynthetic activity.

The counterflow in stratified estuaries, with freshwater from rivers flowing over denser seawater, influences many biological activities. The two-directional flow helps to keep nutrients in an estuary. When cells die and sink from the surface layer into the deeper layer of seawater, they are carried back up the estuary. Cell decomposition releases nutrients, which later mix back into the surface layer. Thus, estuaries effectively recycle nutrients but may trap toxic pollutants. The counterflow may also help maintain the phytoplankton within the system. For example, the two-directional flow in the Chesapeake Bay controls the seasonality of the common dinoflagellate *Prorocentrum* (Tyler and Seliger 1978). In spring, bottom water carries *Prorocentrum* up the bay to shallow areas at the bay's north end. There, cells are brought into the photic zone near the surface, and rapid growth occurs. *Prorocentrum* spreads as it is carried in surface flow toward the mouth of the bay.

The high nutrient levels in estuaries may support large phytoplankton populations, which, in turn, are the basis of food chains. However, pollutants from urban areas and runoff from agricultural lands are causing eutrophication of many estuaries. The decline in estuarine water quality is shown by decreased water transparency because of phytoplankton blooms, increased amounts of suspended material, extensive anoxia in bottom waters as a result of decomposition, and a decline in submerged aquatic vegetation (SAV). The SAV is especially important in stabilizing bottom sediments, providing food for a variety of animals, and creating habitats for many aquatic species.

Chesapeake Bay, located in the middle Atlantic region, is one of the largest estuaries in North America. It is a coastal plain estuary that formed as sea level rose, flooding the lower part of the Susquehanna River. The photosynthetic activity of benthic communities (salt marshes and SAV) in shallow waters and the phytoplankton in open waters make the bay naturally productive. As a result, its waters have been an important source of commercially important oysters, crabs, and fishes, whose populations have been severely depressed by overharvesting and pollution (Horton and Eichbaum 1991). Increasingly, Chesapeake Bay is showing symptoms of environmental deterioration as a result of human activities. Enrichment of its waters has stimulated the phytoplankton at the expense of benthic producers. Such dense phytoplankton blooms, which herbivores do not consume, sink to the bottom and decompose, causing severe oxygen depletion during the summer and adversely affecting benthic animals. Photoplankton blooms also reduce light penetration, inhibiting SAV growth and thus causing a loss of habitat and food for many benthic animals.

Freshwater environments include standing waters, such as lakes and ponds, and flowing waters. In a typical pond, light penetrates to the bottom so that all of the water is in the photic zone, and water temperature is relatively uniform and does not produce density stratification. Most lakes have extensive regions below the photic zone and are stratified during at least part of the year. Normally, temperature determines the density of lake water, while salinity is unimportant. A typical temperate lake has a uniform temperature (isothermal condition) of 4° C from top to bottom during the periods of vertical mixing in the spring and fall, or the periods of overturn (see fig. 7.3). During the summer, the lake is thermally stratified, with the thermocline dividing the lake water into an upper epilimnion and a lower hypolimnion. In winter, the lake is inversely stratified and often covered with a layer of ice.

One system of lake classification is based on the seasonal pattern of density stratification. A lake that has periods of vertical mixing in the spring and fall and is stratified during the summer is **dimictic.** Many temperate lakes fall into this category. Other lakes have only one period of mixing and are **monomictic.** Very deep lakes and tropical lakes may remain above 4° C for most of the year and only mix in the winter. Cold monomictic lakes at high latitudes remain below 4° C for most of the year and warm sufficiently to mix only in summer. Other categories include amictic lakes without a period of mixing, polymictic lakes that mix frequently, and oligomictic lakes that rarely circulate. Many shallow lakes are polymictic, mixing frequently in response to storms. In contrast, oligomictic lakes include many deep tropical lakes. Finally, meromictic lakes never circulate completely because of a permanent halocline that a saline input produces. The water below the halocline never mixes with the water above it.

Nutrient content is also used to classify lakes. Oligotrophic lakes are low in nutrients. They are often relatively young lakes with deep basins that support only limited phytoplankton growth. As materials accumulate from biological activity, inputs from tributaries, and runoff from the surrounding land, a lake becomes progressively enriched. Nutrient-rich lakes, or eutrophic lakes, support extensive algal growths throughout the summer, and decomposition of organic material that sinks from the productive epilimnion may cause oxygen deletion in the hypolimnion. Sometimes, the term *mesotrophy* is used to describe an intermediate condition between oligotrophy and eutrophy. Table 7.3 compares production levels in different lake types.

The major groups of algae in the freshwater phytoplankton are the cyanobacteria, green algae, diatoms, synurophytes, chrysophytes, cryptomonads, and sometimes, the dinoflagellates.

Temperate Lakes

In oligotrophic lakes, the phytoplankton often consist of relatively small diatoms and phytoflagellates. Because of their cells' large surface areas, they are efficient at utilizing low nutrient concentrations and may show relatively high growth rates. Dimictic lakes often have spring and fall peaks of diatoms, stimulated by the formation of the thermocline in the spring and increased nutrient resuspension at the start of the fall overturn (fig. 7.22a).

Some oligotrophic lakes are poorly buffered, making them susceptible to acidification from atmospheric inputs (oxides of nitrogen and sulfur), acid-mine drainage, or naturally occurring organic acids. Below pH 6, the phytoplankton composition changes, as less tolerant species disappear. Schindler, Mills, and Malley (1985) described the results of artificially acidifying a small lake in north-

Figure 7.22
Seasonal changes in phytoplankton in typical temperature (*a*) oligotrophic and (*b*) eutrophic lakes. Spring and fall growths of diatoms and phytoflagellates (light line) occur in both lake types. During the summer, algal abundance is low in oligotrophic lakes, but green algae and cyanobacteria bloom in eutrophic lakes (heavy line in *b*).

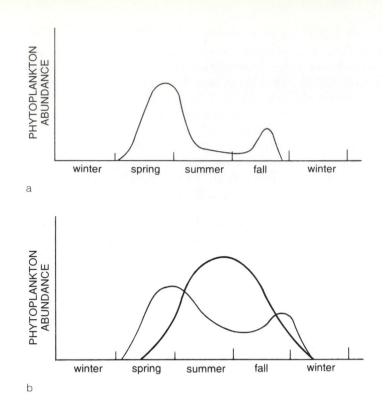

Table 7.3 *Annual Production of Phytoplankton in Different Types of Lakes*

Lake	Production (g C $m^{-2}yr^{-1}$)*
Oligotrophic	15–50
Mesotrophic	50–150
Eutrophic	150–500
Dystrophic	10–100

Source: After Whittaker (1975).
* Annual production expressed as grams of carbon per square meter of lake surface.

west Ontario by lowering its pH from 6.8 to 5 with sulfuric acid. Prior to acidification, chrysophytes were dominant. After acidification, the chrysophytes declined, while dinoflagellates, cyanobacteria, and chlorophytes increased. In the benthic community, the filamentous green alga *Mougeotia* became abundant. However, acidification did not change the lake's overall photosynthetic production.

As a lake becomes eutrophic, nutrient depletion by spring diatoms is less severe, and phytoplankton growth is greater during the summer. Green algae, particularly members of the Chlorococcales, and cyanobacteria often produce dense summer blooms, in addition to the spring and fall peaks of diatoms (fig. 7.22*b*). Phytoflagellates may also be common, occasionally blooming under a covering of ice. Compared to oligotrophic phytoplankton, the planktonic algae in eutrophic lakes are often larger and less efficient at nutrient uptake. Picoplankton make less of a contribution than in oligotrophic lakes (Takamura and Nojiri 1994). Large phytoplankton grow more slowly, and grazing impacts their abundance less.

Nutrients are important in controlling changes in the phytoplankton composition during the growing season. The diatoms often dominate when silicon concentrations are high relative to phosphorus (Tilman et al. 1986; Tilman and Kiesling 1984). Different diatom species are favored at different concentrations of silicon (Kilham 1971). The overall period of diatom dominance is often determined by the availability of silicon, which is depleted during the spring bloom. As the ratio of silicon to phosphorus decreases, either by diatoms removing silicon or by addition of phosphorus in pollutants, nonsilicon-requiring species replace the diatoms.

The effects of nutrients on the interactions between species of phytoplankton have been studied in cultures. In mixed cultures of diatoms, *Asterionella* is favored over *Cyclotella* when phosphorus is low, but when silicon is low, *Cyclotella* outcompetes *Asterionella* (Tilman 1977). Similarly, when *Asterionella* is grown with the cyanobacterium *Microcystis*, the latter dominates when silicon is limiting, but *Asterionella* is favored when phosphorus is low (Holm and Armstrong 1981).

The nitrogen-to-phosphorus ratio may determine whether green algae or cyanobacteria dominate in summer (Schindler 1977; Smith 1983). Average cellular requirements are approximately fifteen atoms of nitrogen for every atom of phosphorus. When the nitrogen to phosphorus ratio is high (greater than 15:1), nitrogen is in excess relative to the cellular requirement for phosphorus, and the green algae are favored because of their higher growth rates. When the nitrogen to phosphorus ratio is low, cyanobacteria capable of fixing nitrogen have an advantage.

The progressive enrichment of a lake, called **eutrophication,** is a natural process associated with biological activity and sediment accumulation in a lake basin. However, human activities around a lake and in its watershed may accelerate the process. Fertilizer and animal waste in runoff from agricultural lands and wastes discharged from urban areas may support massive blooms and lead to anoxia in the hypolimnion as cells sink and decompose. Air pollution is also a potential source of nutrients, such as oxides of nitrogen.

Nutrient inputs can be reduced to prevent unwanted algal growths. Wastes from urban areas normally are treated before discharge, but removal of such nutrients as nitrogen and phosphorus is expensive on a large scale. An alternative to waste treatment is to reduce use of materials likely to stimulate algal growths. This objective can be achieved by applying fertilizers in a way that lessens runoff and reducing use of products with high nutrient contents. Banning the sale of detergents with phosphates has helped reduce algal booms in some lakes (Makarewicz and Bertram 1991).

Tropical and Polar Lakes

As in the oceans, annual solar-heating patterns in tropical and polar regions affect the pattern of density stratification of lakes and lead to many of the differences from temperate lakes. In the tropics, relatively high temperatures with little annual variation favor prolonged periods of density stratification and nutrient depletion in the epilimnion. Winds associated with severe storms may be important in breaking up stratification and mixing the waters of relatively shallow lakes. The accompanying nutrient resuspension may temporarily increase algal production. Since such storms occur irregularly and vary greatly in intensity, annual patterns of phytoplankton activity are less regular than in temperate lakes. Lakes in arid regions may

Figure 7.23
Succession of phytoplankton
in a tropical lake. (After Lewis
1978.)

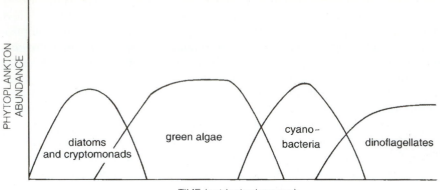

TIME (nutrients decrease)

experience substantial evaporation and become saline. Such soda lakes in the African Rift Valley support rich communities of cyanobacteria (such as *Spirulina*), which are, in turn, food for the zooplankton and flamingos. Blooms of the unicellular green alga *Dunaliella* are common in saline lakes.

Lake Lanao, in the Philippines, is a monomictic lake that mixes during winter. Following these periods of nutrient enrichment, diatoms and cryptomonads are initially common (fig. 7.23). As nutrients decrease in the water, the diatoms and cryptomonads are succeeded by green algae, which then give way to the cyanobacteria, and finally, dinoflagellates (Lewis 1978).

In the Arctic and Antarctic, lakes occur in regions that are free of snow during part of the year. Shallow lakes may freeze completely in winter but thaw in summer when daylight is continuous. Deeper lakes may have permanent ice covers. Snow on the ice cover and bubbles in the ice may greatly reduce light penetration into the underlying water. Because of the slow release of nutrients from the soil, many polar lakes are oligotrophic. Thus, algae in polar lakes tend to be adapted to low levels of irradiance and low nutrient concentrations.

In Arctic lakes, diatoms and small flagellates (cryptomonads and chrysophytes) may be common. Thick glacial ice covers most of the Antarctic continent. However, lakes are found in "dry valleys" that receive little snow. These lakes may show inverse stratification with relatively weak vertical mixing (Lizotte and Priscu 1992). Filamentous cyanobacteria, green algae, chrysophytes, diatoms, and cryptomonads may be common either in the plankton or on the bottoms of these lakes (Heywood 1984).

Rivers

A true phytoplankton community develops only in slowly flowing stretches of rivers. The phytoplankton collected at any time in a river consist of a mixture of true planktonic forms able to grow while being carried along in the water, temporary members of the phytoplankton washed out of the benthos but able to grow in the open water, and benthic algae unable to grow when floating. For a population to be successful in a river, its growth rate must exceed the loss of cells downstream. Phytoplankton growths are most likely to develop during periods of summer drought, when flow is reduced.

Slowly moving water benefits the phytoplankton. It aids in suspension (rivers normally do not thermally stratify), keeps cells supplied with nutrients, and reduces herbivory. Nutrient availability and grazing on the phytoplankton are less important in regulating river communities than in lakes.

Summary

1. Planktonic algae grow free-floating in freshwater and the oceans. They normally grow near the water surface in the mixed layer (or epilimnion of a lake). Density stratification of the water from seasonal temperature changes and sometimes salinity determines the depth of the mixed layer. The photic zone, which may occupy part or all of the mixed layer, has enough light for net photosynthesis. Phytoplankton growth by new cell production depends on spending sufficient time in the photic zone and on nutrient availability. Besides requiring inorganic nutrients, many algae need organic growth factors, and some algae are mixotropic using organic compounds as energy sources. The ability of an alga to utilize a specific nutrient is expressed in terms of models derived from enzyme kinetics. Phosphorus, nitrogen, and silicon often limit phytoplankton growths.

2. Cells may be lost from a planktonic community through grazing or by sinking below the mixed layer. Sinking depends on the mixing of the water, cells' ability to move vertically by means of flagella, cells' density difference with the water, and viscous drag between the cell surface and the water. The phytoplankton show four general adaptive strategies: the picoplankton, which are very small; the phytoflagellates, which have flagella and positive phototaxis; larger cells and colonies, which depend on mixing for suspension; and large colonies and filaments with gas-filled structures.

3. Phytoplankton production in the oceans is related to nutrient availability. In temperate oceans, phytoplankton growth is greatest in the spring and fall. Formation of the seasonal thermocline reduces vertical mixing, which stimulates the spring peak of the diatoms. Low abundance in summer is due to grazing and low nutrient levels. Enrichment of the surface waters in the fall produces a second diatom peak. In polar waters, sea-ice formation and melting significantly affect annual growth, characterized by a summer peak of ice algae and the phytoplankton. Phytoplankton production is greater in the Antarctic than the Arctic because of upwelling. In general, tropical and subtropical oceans support low levels of phytoplankton throughout the year because of a permanent thermocline that prevents nutrient resuspension in the mixed layer.

4. Phytoplankton growth in coastal waters is greater because of enrichment from rivers, land runoff, and localized upwelling of water. Especially high production is associated with areas of extensive and prolonged coastal upwelling and with estuaries, where water counterflow helps retain nutrients. Phytoplankton blooms, especially of toxic species, are increasing in coastal waters worldwide, possibly reflecting greater nutrient inputs from human activities.

5. Many temperate lakes show periods of vertical mixing in the spring and fall. Diatoms and phytoflagellates normally peak in the spring and fall in oligotrophic lakes. In eutrophic lakes, greater nutrient availability supports summer growths of green algae and cyanobacteria. Periods of mixing that resuspend nutrients are often irregular in tropical lakes, resulting in annual phytoplankton patterns that are more variable than in temperate lakes. Phytoplankton communities may develop in the more slowly flowing regions of rivers, where moderate flow may be beneficial in resupplying nutrients and keeping cells suspended.

Further Reading

Carpenter, E. J., D. G. Capone, and J. G. Rueter. 1992. *Marine Pelagic Cyanobacteria*: Trichodesmium *and other Diazotrophs*. Kluwer Academic Publishers.

Cosper, E. M., V. M. Bricelj, and E. J. Carpenter, eds. 1989. *Novel Phytoplankton Blooms*. Springer-Verlag.

El-Sayed, S. Z., and G. A. Fryxell. 1993. Phytoplankton. In E. I. Friedmann, ed., *Antarctic Microbiology*, 65–122. Wiley-Liss.

Hallegraeff, G. M. 1993. A review of harmful algal blooms and their apparent global increase. *Phycologia* 32: 79–99.

Harris, G. P. 1986. *Phytoplankton Ecology*. Chapman and Hall.

Knox, G. A. 1994. *The Biology of the Southern Ocean*. Cambridge University Press.

Raymont, J. E. G. 1980. *Plankton and Productivity in the Oceans*. Vol. 1, *Phytoplankton*. Pergamon Press.

Reynolds, C. S. 1984. *The Ecology of Freshwater Phytoplankton*. Cambridge University Press.

Sandgren, C. D., ed. 1988. *Growth and Reproductive Strategies of Freshwater Phytoplankton*. Cambridge University Press.

Smayda, T. J., and Y. Shimizu, eds. 1993. *Toxic Phytoplankton Blooms in the Sea*. Elsevier.

8

Freshwater Benthic and Terrestrial Algae

Algae grow on a variety of surfaces. In this chapter, we will consider benthic algae (phytobenthos) growing on submerged surfaces in freshwater environments and terrestrial algae growing on moist surfaces exposed to the air, including symbiotic algae in these habitats. Chapter 9 discusses benthic algae in marine environments.

Freshwater habitats include flowing water in rivers and streams and standing water in lakes, ponds, and bogs. Each of these habitats can contain a variety of substrates for algal growth, including rocks, loose sediments of sand and mud, and the surfaces of other organisms. A few algae grow as symbionts inside other organisms.

Freshwater Benthic Algae

Benthic algae range in form from nonflagellated cells to filaments, and occasionally, to simple parenchymatous thalli. Solitary cells and small colonies may loosely associate with a substrate or attach firmly by mucilage. Larger algae may grow appressed to the substrate as crusts or securely attached by a basal holdfast. Sometimes, the term *periphyton* is used to describe the assemblage of microorganisms, including algae, growing on a submerged surface. Major algal groups represented in the freshwater benthos are the pennate diatoms, green algae, and cyanobacteria, but other algae also may be present.

In addition to algae, vascular plants and some bryophytes grow in benthic communities (and are commonly referred to as macrophytes, but the term sometimes includes larger algae). Some aquatic vascular plants grow partially submerged, with their leaves and stems extending above the water surface, while others are completely underwater. A few, such as duckweed, float unattached on the surface. Besides being photosynthetic producers, these plants help stabilize loose sediments of sand or mud with their root systems, and their submerged stems and leaves provide extensive surfaces for algae to attach. However, plants also compete with algae for light and nutrients.

Light and nutrients may affect the seasonality, distributions, and abundance of benthic algae in much the same way they affect phytoplankton. Algal growth is limited to the photic zone, where light is sufficient for net production. The photic zone may not reach to the bottom in deeper parts of lakes and rivers. Benthic algae, however, are less able than many of the phytoplankton to adjust their vertical position relative to available light. After an initial dispersal phase, benthic algae remain in one place on the substrate—either attached firmly or capable of only limited movement. The nutritional requirements of benthic algae are similar to other algae.

Table 8.1 *Terms That Describe Benthic Algae According to Substrate*

Epilithic: Growing on rocks

Endolithic: Growing within rocks

Epiphytic: Growing on other algae or plants

Endophytic: Growing within other algae or plants

Epizoic: Growing on animals

Epilignic: Growing on wood of trees

Epipelic: Growing on mud

Endopelic: Growing within mud

Epipsammic: Growing on sand

Endopsammic: Growing among sand grains

Grazing, water movement, and substrate type affect benthic growths. Important grazers on benthic algae include snails, crustaceans, insect larvae, and fishes. These aquatic herbivores show a range of different feeding strategies (Cummins and Klug 1979). Browsers, such as fishes and crustaceans, feed by removing parts of algae. Snails use a chitinous tongue or radula to scrape off films of microalgae. Some insects have specialized mouthparts to pierce and suck the protoplasm from cells. Sometimes, herbivores graze preferentially on epiphytes and thus benefit host plants by removing coverings of smaller algae (Underwood, Thomas, and Baker 1992).

Water movement affects substrate stability and produces drag on benthic algae. Drag may break the thallus or completely detach it. Water disturbance of a sandy or muddy bottom may prevent successful algal attachment or bury algae after attachment. Even stable surfaces may lack algal growths if water-transported sand particles scour them. On the other hand, flowing water supplies algae with nutrients and inhibits herbivores and competitors.

Benthic algae grow on firm or consolidated substrates, such as rocks or the surfaces of larger plants, and on loose or nonconsolidated surfaces of mud or sand. Firm substrates may be abiotic (nonliving) or biotic (living). Biotic surfaces show seasonal variation, as in macrophytes' annual production and loss of leaves. Algae that grow on other algae or submerged vascular plants are **epiphytes.** Table 8.1 summarizes terms used to describe benthic algae according to substrate. Sometimes, filamentous algae detach from the bottom and continue to grow as floating mats of entangled filaments in ponds and ditches, often buoyed by gas bubbles trapped among the filaments.

Epiphytism is well developed among benthic algae. Smaller algae often grow on the surfaces of coarser filamentous algae and vascular plants. Normally, epiphytes securely attach to their hosts (fig. 8.1). Larger algae that lack mucilaginous sheaths, such as *Cladophora* or *Oedogonium,* are more susceptible to epiphyte colonization than filaments with extensive mucilage, such as *Spirogyra* and other zygnemataceans. Some epiphytes are confined to specific hosts, while others grow on a wide range of host species. The relationship between epiphyte and host may vary:

1. The host provides a surface, but other interaction between host and eiphyte is limited.
2. Dense growths of epiphytes may interfere with the host's growth by competing for nutrients, reducing light, or secreting inhibitory substances.

a   b

Figure 8.1
Epiphytic diatoms.
(*a*) Filaments of *Cladophora*
with stalked diatoms
(*Rhoicosphenia*).
(*b*) Scanning electron
micrograph of mixed
diatoms. (*b* from Luttenton
and Rada 1986, courtesy
Journal of Phycology.)

3. A host may benefit from chemicals that epiphytes release or from epiphytes' greater susceptibility to grazing by herbivores, which may reduce grazing on the host.

4. The host may supply epiphytes with organic material.

Some algae attach to animals. The calcareous shells of mollusks and the exoskeletons of crustaceans provide firm surfaces. Crustaceans rid themselves of epizoic algae when they replace their exoskeleton during a molt. The filamentous green alga *Basicladia,* a relative of *Cladophora,* grows only on turtle shells.

While algae on living substrates normally attach permanently in one place, many smaller benthic algae on abiotic surfaces move considerably. When in contact with a solid surface, the pennate diatoms and desmids secrete mucilage to move. Some filamentous cyanobacteria, such as *Oscillatoria,* move to adjust the density of their mats in response to environmental conditions. Most benthic algae depend on flagellated stages for dispersal and colonization of new surfaces. Filaments also propagate by fragmentation.

Standing Water

Standing water, or **lentic,** environments include lakes, ponds, and bogs. In ponds, all the water is in the photic zone, and benthic algae may grow over the entire pond bottom. However, ponds often support substantial populations of aquatic vascular plants, which may or may not benefit algae (Wetzel 1983). The leaves of vascular plants may reduce light penetration into the water and shade the bottom sufficiently to reduce algal growth. Duckweed is an angiosperm that grows unattached, with its leaves floating on the water surface. Dense duckweed growth may significantly limit algal growth beneath it. On the other hand, epiphytic algae may colonize leaves of submerged vascular plants and benefit from a favorable position in the water column to receive light and nutrients.

Floating mats of filamentous green algae or cyanobacteria often develop in calm waters of small ponds (fig. 8.2). Common mat-forming algae are *Spirogyra,*

Figure 8.2
Oedogonium mats in shallow
water of a small lake.

Mougeotia, Oedogonium, Oscillatoria, Cladophora, and *Pithophora.* Some species
of *Spirogyra* grow best in the early spring, when the water is cool but irradiance is
high (Graham et al. 1995). Their filaments begin growth attached to the bottom.
As filaments lengthen, they reproduce by breaking (sometimes aided by replicate
end walls). The fragments produced may continue to grow while floating in the
water and eventually accumulate to form mats of entangled filaments. Oxygen
bubbles trapped among the filaments float some mats on the surface, where they
are exposed to high irradiance and sometimes are yellow because of reduced
chlorophyll levels. Healthy *Spirogyra* filaments secrete copious mucilage to keep
external cell surfaces free of epiphytes. Heavy rain may wash the mats out of a
pond.

   Because of their limited water volume, ponds may vary widely in temperature
and other conditions during a year. Periods of low precipitation may reduce water
level, and ponds may even dry for part of the year. Such fluctuating conditions
require that most algae survive unfavorable periods as dormant stages. For example, mat-forming green algae have a variety of methods for surviving from one
growing season to the next. Both *Spirogyra* and *Oedogonium* undergo sexual reproduction, leading to the formation of a thick-walled, dormant zygote. In contrast,
Pithophora forms akinetes directly from vegetative cells (fig. 8.3) (extensively
studied by O'Neal and Lembi 1995). The cyanobacteria tolerate desiccation and
high temperatures. Vegetative cells of filamentous cyanobacteria may survive
stressful conditions without forming special structures or may produce akinetes.
Resistant algal stages may be carried from one pond to another by the wind or in
mud stuck to the feet of birds.

   Lakes are deeper than ponds (although no absolute criteria separate these
bodies of water) and often stratify (described in chapter 7). Benthic algae in lakes
are confined to shallow areas within the photic zone. A young lake often has poorly
developed shores and little benthic growth, but as sediments accumulate and shore
areas become less steep, algae may grow on bottoms receiving sufficient light.
Benthic algae often proliferate on solid surfaces, such as rocks or jetties (fig. 8.4*a*).

Figure 8.3
Pithophora with akinetes.

Figure 8.4
(*a*) *Ulothrix* forming a fuzzy growth on rocks in a wooded stream in winter.
(*b*) *Cladophora* growing in flowing water, and snails grazing on epiphytes.

a b

Growth of benthic algae may be more extensive slightly below the surface, where wave disturbance and variation in water conditions are less.

Generally, algae are less successful on loose sediments that waves disturb. Rooted macrophytes (vascular plants) usually grow better than algae on sandy or muddy bottoms. An exception is the green alga *Chara* and its relative (Charales), which form extensive rhizoidal systems that allow them to attach to soft bottoms. On relatively undisturbed sediments, mats of cyanobacteria and films of diatoms may form. Pennate diatoms often have raphes and are motile, allowing them to escape burial by sediments. *Navicula* and *Nitzschia* are common representatives.

Along the shores of many temperate lakes in the northern United States and Canada, *Ulothrix zonata* is common in the spring. To explain the seasonality of *Ulothrix,* Graham and associates studied the effects of light and temperature on photosynthesis, respiration, and zoospore formation in culture (Graham, Graham, and Kranzfelder 1985; Graham, Kranzfelder, and Auer 1985). Growth was best at low temperatures (5° C) and high irradiance, while zoospore production was greatest at warmer temperatures (above 10° C). Thus, during early spring, conditions are favorable for growth, but as the water warms, asexual reproduction leads to zoospore release and filament disintegration. Short filaments and germlings from zoospores probably persist over the summer, and as the water cools, may produce a second period of growth in the fall.

Colonization of benthic algae onto substrates often depends on the formation of a coating of mucilage. Such coatings, which bacteria and diatoms produce, trap spores and aid in adhesion of cells (Roemer, Hoagland, and Rosowski 1984). Stages in algal colonization of submerged surfaces are: (1) an initial stage with an organic coating and bacteria, (2) an intermediate stage with low-profile diatoms growing close to the substrate, and (3) a third stage with stalked and large rosette diatoms and filamentous green algae growing upright and extending outward from the substrate (Hoagland, Roemer, and Rosowski 1982).

Sometimes, development of a benthic community is inversely related to phytoplankton growth (Jones 1990). In lakes, planktonic algae initially dominate, but as benthic growths become more extensive, the phytoplankton are less successful in competing for nutrients. Larger benthic algae and vascular plants are more efficient at removing nutrients from the water and may release chemicals inhibitory to the phytoplankton. The trend may reverse when enrichment stimulates phytoplankton, which sufficiently reduce light penetration to limit benthic growth.

Bogs may form as a lake fills with sediments and mosses and other plants overgrow the water surface. Spongy mats of the moss *Sphagnum* ("peat moss") are common in many bogs. Bog water that is brown and has a high organic content is described as **dystrophic.** Bacterial decomposition is very slow, and organic material accumulates as peat. The concentrations of dissolved inorganic material, including most algal nutrients, are often low, and the water is acidic. Bogs' low pH and low nutrient levels may limit the diversity of algal species. The desmids are one of the few groups that may be diverse in bogs. Other green algae and cyanobacteria also may be present. Typically, however, algal production is low in bogs and probably reflects the low levels of nitrogen and phosphorus and poor light penetration.

The Everglades are an extensive freshwater wetland in southern Florida. Algae are important producers there, growing epiphytically on vascular plants and forming extensive floating mats of filamentous cyanobacteria, pennate diatoms, filamentous green algae, and desmids (Browder, Gleason, and Swift 1994; Vymazal and Richardson 1995). These algal mats are an important food for many aquatic animals in the system, which, in turn, support the higher consumers, such as alligators and a variety of waterbirds. Under natural conditions in the oligotrophic Everglades, the green algae dominate in more acidic water with low calcium, while the cyanobacteria form mats in more basic water and precipitate calcium carbonate. At present, however, extended periods of drought, diversion of water for agriculture, and phosphorus enrichment from sugarcane field runoff are seriously disrupting the Everglades ecosystem.

Symbiotic Algae

Some algae grow as symbionts inside other organisms. Such relationships may be mutualistic or parasitic, depending, respectively, on whether the host benefits or is harmed. Hosts provide symbionts with protection, a constant environment, and nutrients. Parasitic algae living within their hosts often lack photosynthetic pigments and are obligate heterotrophs, obtaining organic nutrients from their hosts. In other associations, symbionts are photosynthetic, and the host benefits from nutrients that symbionts excrete.

In freshwater environments, unicellular green algae called **zoochlorellae** occur as endosymbionts in cells of *Hydra,* protozoa (amoebae, ciliates), sponges, and flatworms. *Chlorella* is common in species of *Hydra* and *Paramecium* (fig. 8.5). Zoochlorellae are probably ingested as food particles but persist in special vesicles

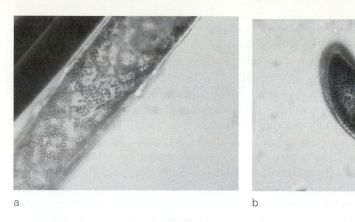

a   b

within host cells without being digested. Each perialgal vesicle contains one algal cell, and as cells divide, they separate in their own vesicles. In these mutualistic associations, the alga provides oxygen and organic material (simple sugars, alanine), while receiving carbon dioxide, phosphates, and nitrogenous compounds from its host. For example, *Chlorella* actively excretes maltose and alanine for its host to use. The host may regulate release of organic material by changing the pH in the perialgal vesicles, with low pH favoring greater release.

Flowing Water

Flowing-water, or **lotic,** environments range from small streams to broad rivers. Their one-way water flow makes them much more open systems than lakes or ponds. Water sources may be springs, rain falling directly on the stream, surface runoff from precipitation, snowmelt, or groundwater that has passed through the soil. Streams are generally shallow and have rocky bottoms. Stream flow ranges from riffles of rapidly flowing water to pools. Seasonal flow may vary greatly, depending on precipitation and spring snowmelt. Some streams dry during part of the year, subjecting their inhabitants to periods of desiccation. Rivers are deeper, and water flows more slowly. Loose sediments of silt and mud often cover river bottoms. Suspended sediments may prevent light from penetrating to the river bottom, limiting plant and algal growths to shallower areas along the banks. Marshes and swamps often border rivers.

Important photosynthetic producers in rivers and streams are rooted vascular plants, bryophytes, and benthic algae, especially pennate diatoms, filamentous green algae, and cyanobacteria. Green algae may be abundant at higher light levels. The filamentous green alga *Cladophora* is common in well-illuminated, rapidly flowing water, where its growths in spring and fall may become dense enough to clog the water (see fig. 8.4*b*). Although most red algae are marine, a few species are widespread in flowing waters. Their pigments and life histories (*Batrachospermum* type, p. 175) may be modified compared to their marine relatives. Generally, phytoplankton are absent from rapidly flowing waters but may develop in more slowly flowing stretches of rivers (p. 210).

Water flow affects the distribution of benthic algae. In pools, algae may experience relatively little water movement, but where flow is greater, algae need a secure attachment to the substrate. Since the force of water is reduced in the boundary layer immediately above submerged surfaces, some benthic algae largely avoid water motion by growing closely appressed to the substrate. Larger algae, however, especially filaments, extend from the substrate into rapidly moving water. Typically, they are firmly attached by holdfasts and have a flexible form to orient with the direction of water flow and thus minimize drag. Nevertheless, an entire thallus or parts of a thallus may be torn away. To be successful, algae in flowing water must have growth and reproductive rates great enough to compensate for these losses.

Living in flowing water has benefits. Flowing water continually exposes algae to a fresh nutrient supply. In flowing water, nutrients derived from land runoff less commonly become limiting than they do in standing water. The composition of inorganic and organic material entering streams as suspended and dissolved material from the surrounding watershed depends on rock and soil erosion, inputs from neighboring terrestrial communities, and human activities, including runoff of fertilizer and wastes from agricultural areas, and discharges from urban wastes.

Flowing waters can be viewed as a continuum from small, headwater streams to medium-sized streams (or small rivers), and finally, to large rivers. Photosynthetic production in each of these regions depends on how much solar radiation reaches the water (Vannote et al. 1980). Many headwater streams flow through forests, which shade the water sufficiently to minimize photosynthetic activity by aquatic producers. However, the forest vegetation provides leaves and twigs as the major source of organic material for stream consumers. Thus, requirements of the biological community in small streams exceed actual stream production, and stream animals depend on this organic input from the surrounding land. In medium-sized streams, surrounding terrestrial vegetation has less effect, and production by algae and aquatic plants is a major food source for herbivorous animals. In large rivers, depth and water clarity often limit photosynthetic production on the bottom, and fine particles of suspended organic material may be an important food source for animals.

Where a stream flows through a forest, the leaf canopy may shade the water sufficiently to limit algal abundance and moderate the water temperature. The importance of shading by tree canopies was shown dramatically in Rhode Island streams when algal abundance increased in 1981 after gypsy moth larvae defoliated the trees (Sheath et al. 1986). In other streams, greater growth occurs in winter, when the tree canopy is absent.

Even though herbivores and epiphytes are often less abundant in streams than in standing waters, grazing by stream animals, including snails and aquatic insects, sometimes limits algal abundance. Grazing may be more important when irradiance and nutrients are high, and filamentous green algae are dominant rather than diatoms (Rosemond 1993; Rosemond and Brawley 1996). Bothwell and associates studied the effects of ultraviolet radiation on communities in artificial streams and found that herbivores were more sensitive than some stream algae (Bothwell et al. 1993; Bothwell, Sherbot, and Pollock 1994). Exposure to ultraviolet radiation initially reduced diatom growth, but after several weeks, stream composition changed to tolerant species, and overall abundance of the diatoms increased because of a decline in herbivorous insects.

Extreme conditions in some flowing waters may limit algal diversity. Thermal streams that originate from geothermally heated springs or from outflows of power-generating plants (fossil or nuclear) may have temperatures above ambient air temperatures. When water temperatures exceed 35° C, algal species are noticeably limited, and cyanobacteria usually dominate. Acidification of streams may occur naturally or may be the result of drainage from mines or atmospheric pollutants. In general, when pH is less than 6, algal diversity is noticeably reduced (Keithan, Lowe, and DeYoe 1988).

Researchers studied the recovery of stream communities after the eruption of Mount Saint Helens in May 1980 (Rushforth, Squires, and Cushing 1986; Steinman and Lamberti 1988). Volcanic material scoured and buried streams. Diatoms recolonized within a few months, and later, a mixed community with diatoms, green algae, and cyanobacteria developed.

Algae inhabited the land long before the first true land plants (the bryophytes, vascular plants) appeared 400 million years ago. Although plants have replaced algae as the major photosynthetic producers in terrestrial habitats, some algae grow exposed to the air on the surfaces of soil, rocks, trees, and snow, where they may form visible films when abundant (plate 5). Terrestrial or aerial algae normally require moist conditions for active growth and tolerate greater variation in temperature and irradiance than algae living in water. When environmental conditions become stressful, terrestrial algae survive by forming dormant stages. Some algae are able to live in the Antarctic dry valleys under some of the most extreme conditions on the earth. Other terrestrial algae live in symbiotic associations with other organisms or within rocks, where environmental conditions are less stressful and less variable. Except for algae in the spray zone along seashores (discussed in chapter 9), most terrestrial algae are related taxonomically to freshwater groups. The green algae and cyanobacteria are the most common groups, but the diatoms, tribophytes, and red algae are also represented.

Algae grow on or just below the soil surface, where light is sufficient for photosynthesis. In general, they are more abundant in fields than in forests (Hunt, Floyd, and Stout 1979), and the green algae are more common in acidic soils and the cyanobacteria in basic soils (King and Ward 1977). Irradiance, temperature, and moisture help control their abundance and production (Shimmel and Darley 1985). Many of the green algae in soil have spherical cells or grow in compact packets (fig. 8.6). Individual genera and species are difficult to recognize and usually require culturing under standard conditions for identification.

Soil algae may have beneficial effects. Their metabolic activities enrich the soil with organic materials, and nitrogen-fixing cyanobacteria release nitrogenous compounds. Many soil algae secrete mucilaginous sheaths, composed of mucopolysaccharides that help aggregate soil particles. Palmelloid stages of *Chlamydomonas mexicana* are especially important in this respect. In arid regions, biocrusts (or cryptogamic crusts) form on the ground surface and are composed of algae, fungi, lichens, mosses, and soil particles (Johansen 1993). As major components of these crusts, the cyanobacteria help bind soil particles and reduce erosion.

Algae grow on bark and leaves of trees. The "moss" on the trunks and branches of trees often is actually a film of *Apatococcus* (see fig. 3.28*d*) or another green algae. In tropical regions, members of the Trentepohliales (Chlorophyta) grow on leaves, sometimes forming conspicuous orange spots, and producing and shedding sporangia that wind or animals disperse (fig. 8.7).

When moisture and light are adequate, rocks, bricks, and plaster surfaces may also provide suitable substrates for algal growth (see plate 5*a*). In fact, one of the negative effects of aerial algae is their growth on works of art, such as cave paintings, Mayan stone carvings, and frescoes (Caiola, Forni, and Albertano 1987). The caves at Lascaux, France, have been closed to the public in an attempt to reduce algae's destructive effect on the paleolithic drawings.

In arid environments, such as in polar, alpine, and desert regions, algae may grow within rocks (R. A. Bell 1993). These **endolithic** algae are usually coccoid cyanobacteria, resembling *Gloeocapsa,* or compact packets of green algae. Endolithic algae in porous rocks grow a short distance beneath the surface, where environmental stresses are less. In this microenvironment, light may be reduced to more optimal levels, moisture (largely derived from dew in deserts) is conserved,

Figure 8.6
Packets of a green alga.

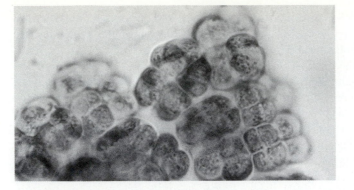

Figure 8.7
Trentepohlia with sporangia
(sp).

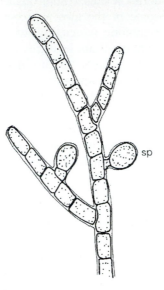

and temperatures fluctuate less. The role of endolithic algae in breaking down rocks is uncertain.

Algal growths sometimes produce conspicuous red or green discolorations on snowfields (see plate 5*b*). Chlorophytes are common **snow algae,** especially the green flagellates *Chloromonas* and *Chlamydomonas*. Many snow algae are red and, when abundant, discolor the snow surface to resemble the inside of a watermelon. The red color comes from large quantities of carotenoids, such as astaxanthin, which protect cells from exposure to high levels of irradiance (Bidigare et al. 1993). Snow algae must tolerate freezing and thawing. Their vegetative phases may be active for only a few days a year, when meltwater collects on the snow surface. During the remainder of the year, cells survive in a dormant condition, usually as zygotes, but germinate rapidly when the environment is favorable.

Major problems that terrestrial algae face relate to desiccation, exposure to high irradiance, and dispersal. Unlike vascular plants, algae lack waxy, waterproof coverings around their cells, but they may have thick walls and mucilage to retard water loss. They are usually confined, however, to relatively moist environments. Many terrestrial algae are spherical or form compact packets (see fig. 8.6) to minimize the surface area across which water is lost by evaporation.

Terrestrial algae may be exposed to levels of irradiance above the optimal range for photosynthesis. To prevent photoinhibition, carotenoids dispersed in their cells may act as light screens, often masking the photosynthetic pigments and giving cells a red color. A pigment in the sheath of cyanobacteria protects their

Figure 8.8
Lichens. (*a*) Crustose lichens.
(*b*) Fruticose lichen.

a   b

cells from high irradiance, especially ultraviolet radiation (Garcia-Pichel and Castenholz 1991).

Flagellated stages are the principal means of dispersal in a liquid medium but do not function out of water. Most terrestrial algae have a range of reproductive structures. They can form flagellated zoospores when flooded, but produce aplanospores in the absence of water. The wind may disperse aplanospores and zygotes. In some cases, animals may aid in dispersal, as in the transport of algae in mud stuck to the feed of birds. Dormant stages, such as thick-walled zygotes and cysts, allow survival during periods unfavorable for growth. Some algae may survive for many years in dry soil (Trainor and Gladych 1995; Sun 1996).

Some algae live successfully in terrestrial environments by forming symbiotic associations with other organisms. Some cyanobacteria enter into associations with bryophytes (mosses and liverworts), ferns, cycads, and one angiosperm *(Gunnera)* (see table 2.4). In many of these associations, *Nostoc*, a nitrogen-fixer, is the symbiont. A few algae associate with animals, growing on the exoskeletons of insects and spiders or on the shells of land snails. In neotropical rain forests, green algae grow on the hair of sloths, giving the sloths a degree of protective coloration.

Algae commonly associate with fungi to form lichens. A **lichen** is a fungus that contains algae in a stable association that produces a distinctive thallus (fig. 8.8). These associations may have evolved independently in several different groups of fungi (Gargas et al. 1995). The algal component of a lichen is called the **phycobiont,** while the fungus is the **mycobiont.** Growth of the fungal hyphae forms a pseudoparenchymatous thallus. Algal cells may be scattered throughout the thallus but more commonly are in a layer near the surface (fig. 8.9*a*). The algae are not inside the fungal cells, but fungal hyphae grow around the algae, adhering to their surfaces and sending extensions into their walls. Cyanobacteria and green algae are the algal components of lichens. Usually, only a single species of alga is present. In the majority of lichens found in temperate regions, the phycobiont is the unicellular green alga *Trebouxia*, which is rarely found outside of lichens (fig. 8.9*b*). However, over twenty other green algae, including *Coccomyxa*, *Myrmecia*, and *Trentepohlia*, are also found in lichens. Approximately 10% of lichens have a cyanobacterial phycobiont, usually *Nostoc*; these lichens are more common in tropical regions. Lichens reproduce by releasing fragments, called soredia, which consist of small clumps of hyphae with algal cells.

The mycobiont clearly benefits from the algal presence. Much of the phycobiont's photosynthetic production may transfer to the host as glucose, erythritol, or

Figure 8.9
(a) Cross section of a lichen showing algal cells in a layer below the surface.
(b) *Trebouxia*, a common green alga in lichens. (a from Hale, Mason E., *How to Know the Lichens.* © 1969 Wm. C. Brown Publishers, Dubuque, Iowa. All Rights Reserved. Reprinted by permission.)

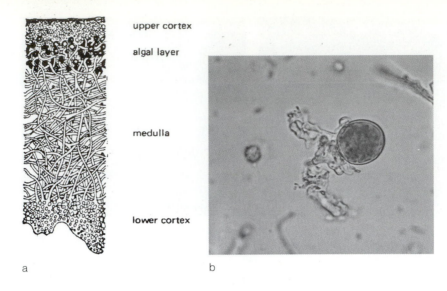

upper cortex

algal layer

medulla

lower cortex

a   b

sorbitol. The benefit to the algal component is less clear. Possibly, the fungus provides a degree of protection from physical conditions. Rather than a mutualistic relationship, however, lichens are often viewed as a form of slow parasitism, in which the alga grows fast enough and moderates infection by the fungus sufficiently not to be killed. In cultures, lichens form when nutrients are low. Fungal hyphae grow around algal cells, sometimes killing them.

Most lichens are terrestrial. The alga/fungus combination allows lichens to grow under conditions where neither component could live separately and leads to the production of organic chemicals unique to the association. Although lichens grow very slowly, they survive in such harsh environments as the arctic or alpine tundra, the coastal spray zone, and the Antarctic dry valleys. Lichens commonly colonize exposed rock surfaces. Lichen distribution is sometimes limited by their sensitivity to air pollution.

Summary

1. Diatoms, green algae, and cyanobacteria commonly grow on submerged surfaces in freshwater environments. Benthic algae grow better on rocks, or as epiphytes on larger algae and aquatic vascular plants, than on loose sediments of sand or mud. Substrate, light, temperature, current velocity, and herbivores help control distributions of benthic algae in lakes, ponds, and flowing waters. Floating mats of entangled filamentous algae commonly develop in small ponds.

2. Terrestrial algae grow exposed to the air in moist habitats. Green algae and cyanobacteria are sometimes common, and may discolor the surfaces of soil, snow, rocks, and trees. Terrestrial algae are adapted to minimize water loss from their cells and to reduce exposure to high irradiance. They often have means of dispersal other than flagellated stages.

3. Some algae enter into symbiotic associations with other organisms. Symbionts in protozoa and invertebrates are green algae called zoochlorellae. Lichens are fungi containing either cyanobacteria or green algae. In these associations, the algal cells provide organic nutrients to their hosts.

Further Reading

Hoffmann, L. 1989. Algae of terrestrial habitats. *Botanical Review* 55: 77–105.

Reisser, W., ed. 1992. *Algae and Symbiosis.* Biopress.

9

Benthic Marine Algae

Ocean shores support some of the most productive and diverse biological communities on the earth. Coastal systems include coral reefs, rocky shores, salt marshes, seagrass meadows, mangrove swamps, and unvegetated bottoms of sand and mud. In many of these environments, algae are the major organic producers. In general, the macroalgae, which require a firm substrate for attachment, are most diverse on rocky shores and coral reefs. Algae do not grow as well on nonconsolidated bottoms of sand or mud, except where protected from waves. Vascular plants are often more important on soft bottoms. Table 9.1 compares photosynthetic production in different coastal ecosystems.

Benthic marine algae show a greater range in size than do members of the freshwater benthos. The larger marine algae, commonly called seaweeds, are readily visible and include representatives of the red, brown, and green algae. Microscopic algae are also common in coastal environments and include cyanobacteria, diatoms, and smaller green algae. Microalgal growths are sometimes dense enough to noticeably discolor rocks, and diatoms are common epiphytes on larger algae.

Morphologic Types

Introduction to Marine Macroalgae

In ecologic studies, describing the groups of algae and species present is important. However, taxonomic groupings do not necessarily describe the roles of the species in a community. Among attempts to classify macroalgae into functional groups, one of the more useful, proposed by Dr. Mark M. Littler, divides algae into six

Table 9.1 *Photosynthetic Production in Different Benthic Communities*

Benthic Community	Photosynthetic Production (g C m^{-2}yr^{-1})*
Coral reefs	2,000–5,000
Rocky shores	
Kelp systems (subtidal)	1,000
Fucoid systems (intertidal)	100
Seagrass communities	300–1,000
Coastal phytoplankton	50–250

Source: Mann (1982).
* Annual production expressed as grams of carbon per square meter of ocean bottom.

Table 9.2 *Functional Forms of Littler*

Functional-Form Group	External Morphology	Internal Anatomy	Texture
Sheet group	Thin tubular or sheetlike (foliose)	Uncorticated, one to several cells thick	Soft
Filamentous group	Delicately branched (filamentous)	Uniseriate, multiseriate, or lightly corticated	Soft
Coarsely branched group	Coarsely branched, upright	Pseudoparenchymatous or parenchymatous	Fleshy to wiry
Thick-leathery group	Thick blades and branches	Differentiated parenchymatous or pseudoparenchymatous, thick-walled	Leathery to rubbery
Jointed-calcareous group	Articulated, calcareous, upright	Calcified segments, flexible joints	Stony
Crustose group	Prostrate, encrusting	Calcified or uncalcified, compact rows of cells	Stony, tough

morphologic types on the basis of their photosynthetic production and susceptibility to grazing (Littler 1980; Littler and Arnold 1982). As shown in table 9.2, sheetlike algae, with little cell specialization and filaments, in which all the cells are photosynthetic, have the greatest rates of photosynthesis (per unit biomass). In contrast, photosynthetic activity in algae with thick thalli is limited to the outer cortical layers, and production levels are much lower. Upright and encrusting algae with calcareous coverings have very low production rates.

These six functional forms also differ in their resistance to herbivores and thallus toughness (Littler, Littler, and Taylor 1983). The sheet and filamentous groups, with the highest photosynthetic productivities, show the least resistance to grazing and the least toughness. The thick-leathery, jointed-calcareous, and crustose groups, with low production, are more resistant to grazing and have tougher thalli. Thus, there is a trade-off between having a delicate thallus with a high photosynthetic capacity and having a tough thallus with lower photosynthetic rates. In the latter, more of an alga's resources are used for structural components.

Littler and Littler (1980) contrasted algal characteristics during different stages of succession (table 9.3). In early stages, opportunistic species are common. More persistent species replace the opportunistic species later in succession. Colonizing species, which are usually sheets and filaments, tend to have rapid growth rates, high photosynthetic levels, and relatively simple thallus forms. Generally, they do not have specific defenses against herbivores. Late successional forms tend to have slower growth rates, lower photosynthetic levels, and more complex thalli. They often have hard coverings or special chemical defenses against herbivores.

Table 9.2 *Continued*

Mean Photosynthetic Production (Net) (mg C g⁻¹h⁻¹)*		Palatability	Toughness	Representative Genera
Pacific	**Caribbean**			
5.16	5.06	High	Low	*Ulva, Enteromorpha, Porphyra*
2.47	5.65	High	Low	*Chaetomorpha, Cladophora, Ceramium*
1.30	1.09	Moderate	Low	*Gigartina, Chondrus, Agardhiella*
0.76	0.88	Low	High	*Fucus, Laminaria, Sargassum, Padina*
0.45	0.18	Low	Very high	*Halimeda, Corallina*
0.07	0.02	Very low	Very high	*Hildenbrandia, Ralfsia,* encrusting corallines

Sources: Littler (1980); Littler and Arnold (1982); Littler, Littler, and Taylor (1983).
*Expressed as milligrams of carbon per gram of algal tissue converted by photosynthesis in an hour.

Table 9.3 *Survival Strategies of Macroalgae*

	Opportunistic Forms	Late Successional Forms
Functional forms	Sheets and filaments	Jointed-calcareous and encrusting
Colonizing ability	Rapid	Slow
Growth	Rapid	Slow
Net primary production	High	Low
Thallus form	Simple	Differentiated
Area: Volume	High	Low
Resistance to physical stress	Low	High
Escape from herbivory	Temporal and spatial unpredictability; rapid growth	Reduced palatability, toughness, chemical defenses
Reproductive capacity	High	Low
Life cycle	Simple	Complex
Alternation	Isomorphic	Heteromorphic

Source: After Littler and Littler (1980).

Figure 9.1
Ecological cycle of a
macroalga with dispersal,
recruitment, vegetative
growth, and reproductive
phases. Important
environmental conditions
necessary for transition
between phases are
indicated.

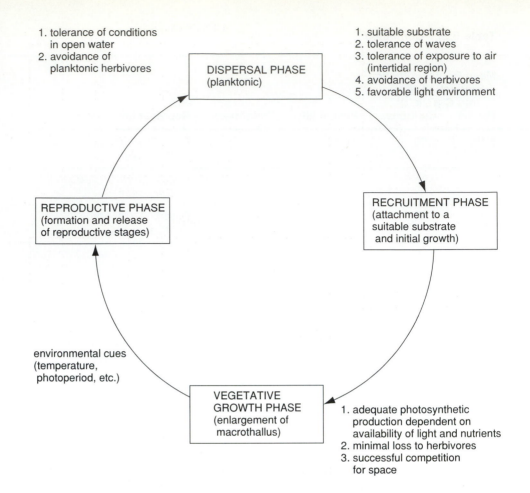

1. tolerance of conditions
   in open water
2. avoidance of
   planktonic herbivores

DISPERSAL PHASE
(planktonic)

1. suitable substrate
2. tolerance of waves
3. tolerance of exposure to air
   (intertidal region)
4. avoidance of herbivores
5. favorable light environment

REPRODUCTIVE PHASE
(formation and release
of reproductive stages)

RECRUITMENT PHASE
(attachment to a
suitable substrate
and initial growth)

environmental cues
(temperature,
 photoperiod, etc.)

**VEGETATIVE
GROWTH PHASE**
(enlargement of
macrothallus)

1. adequate photosynthetic
   production dependent on
   availability of light and nutrients
2. minimal loss to herbivores
3. successful competition
   for space

Life Cycles

Figure 9.1 shows a simplified macroalgal life cycle, with dispersal, recruitment, vegetative growth, and reproductive phases. Successful transition from one phase to the next depends on a number of environmental factors, some of the more important of which the figure indicates and are briefly discussed here (review articles cited). Specific environmental factors will be discussed in more detail in subsequent sections on specific coastal systems.

Most macroalgae depend on unicellular spores or zygotes as planktonic dispersal stages, although some species release larger, multicellular propagules (Clayton 1992). Recruitment involves passing through the boundary layer of water and settling on the substrate, followed by secretion of adhesive materials for initial attachment, germination, and development of the holdfast (Fletcher and Callow 1992). Successful transition from the dispersal phase to a benthic condition depends on substrate suitability, tolerance of the physical conditions encountered during initial settlement, and avoidance of herbivores during susceptible germling stages. Benthic algae generally attach to and grow better on rock substrates or on the surfaces of other organisms as epiphytes, than on sand and other loose sediments that waves disturb.

Physical conditions determining initial success often relate to vertical position on the shore. Algae growing in the upper part of the intertidal region must tolerate exposure to the air during the tidal cycle. Figure 9.2 shows the three types of tidal

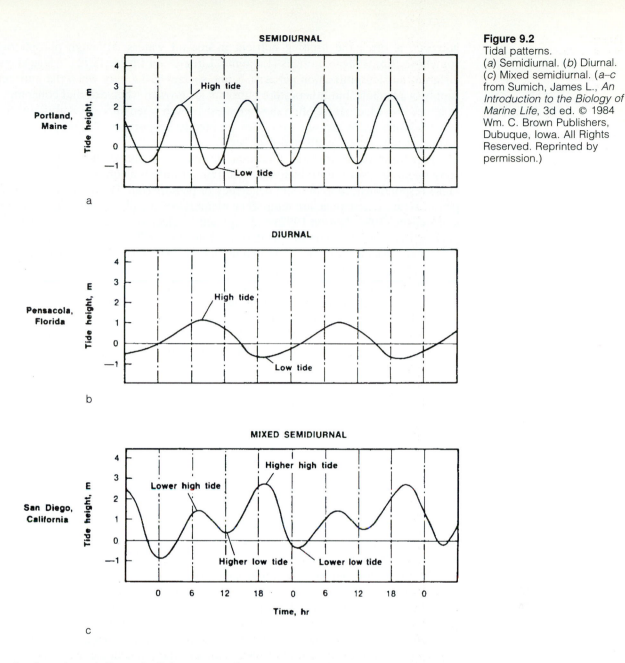

Figure 9.2
Tidal patterns.
(*a*) Semidiurnal. (*b*) Diurnal.
(*c*) Mixed semidiurnal. (*a–c*
from Sumich, James L., *An
Introduction to the Biology of
Marine Life*, 3d ed. © 1984
Wm. C. Brown Publishers,
Dubuque, Iowa. All Rights
Reserved. Reprinted by
permission.)

cycles based on daily periodicity. Depending on geographic location, tidal range may vary from less than a meter to several meters. In shallow water, waves also impact algae (Denny 1995), and light levels are high enough to damage algal cells. Deeper in the subtidal region, algae must adapt to a light environment with lower levels of irradiance and a reduced spectral composition, compared to conditions at the water's surface (Saffo 1987).

Some macroalgae are annuals, with their macrothalli appearing, growing, and disappearing in less than a year. Other macroalgae persist for several years but usually show seasonality in their growth. Growth and development of the macrothallus depends on receiving sufficient solar energy and an adequate nutrient supply, while at the same time competing successfully with other organisms for space on the substrate and minimizing losses to herbivores (Duffy and Hay 1990).

Nitrogen is often the limiting nutrient in coastal waters, although phosphorus sometimes limits algae on reefs (Lapointe, Littler, and Littler 1992). Algae show different nutrient utilization patterns. Some species grow only when the nutrient supply is adequate, but others store nutrients for growth when external concentrations are low. Macroalgae often show greatest growth when nitrogen is available for protein synthesis. When nitrogen is low, a greater proportion of algal production forms carbohydrates (walls and reserves).

Particular environmental conditions normally stimulate the formation and subsequent release of reproductive stages. Temperature is a common cue, but some algae respond instead to nutrient deficiency or photoperiod. During their dispersal phase, future macroalgae are temporary members of the plankton (Amsler, Reed, and Neushul 1992; Norton 1992) and experience selective forces similar to permanent members of the plankton, as described in chapter 7.

Some macroalgae have alternating generations (usually sporophytes and gametophytes) that differ in appearance. These distinct growth phases may represent adaptations to different environmental conditions encountered at different times of the year or in different habitats. For example, in the brown alga *Scytosiphon* (p. 118), the upright thallus grows in winter, and the crustose phase grows in summer, while in the red alga *Porphyra* (p. 171), the blade phase grows in the intertidal region, and the conchocelis phase grows in mollusk shells in the subtidal region.

Production and Food Chains

In coastal waters, the phytoplankton in the open water and a range of benthic organisms on submerged substrates, including autotrophic bacteria, microalgae, macroalgae, and vascular plants, are the photosynthetic producers. As discussed in chapter 7 (p. 195), algae use the products of photosynthesis to maintain their basic structure and function (respiration), and any excess or net production is available for the synthesis of new biomass that contributes to growth and reproduction. This net production may enter food chains as living biomass, detritus, or dissolved organic material (see fig. 7.16). Figure 9.3 is a generalized diagram showing pathways for the movement of energy and organic materials in coastal waters. Both the phytoplankton and benthic producers are important. In general, benthic herbivores' direct consumption of macroalgae is relatively minor, and macroalgae are more important as sources of detrital material as waves erode their thalli. These particles of suspended organic material (with associated bacteria) may be consumed by suspension-feeders, such as bivalves, barnacles, or zooplankton, or settle to the bottom for deposit-feeding animals to use. The importance of the microbial loop, in which dissolved organic material from algal cells supports heterotrophic bacteria, which, in turn, are the primary food for heterotrophic flagellates and ciliates, has only recently been recognized.

In the remainder of this chapter, we will consider first coral reefs and then soft-bottom communities, where marine vascular plants are the principal producers. Discussions of rocky shore communities on the North American Atlantic and Pacific coasts will conclude the chapter. Knowledge of macroalgal communities on rocky shores in the polar regions is limited and consists primarily of surveys of species present. Sea ice in the Arctic Ocean and around Antarctica limits algal growth in shallow water because of ice abrasion of the shore and poor light penetration through the layer of ice covering the water. Kelp forests occur in deeper

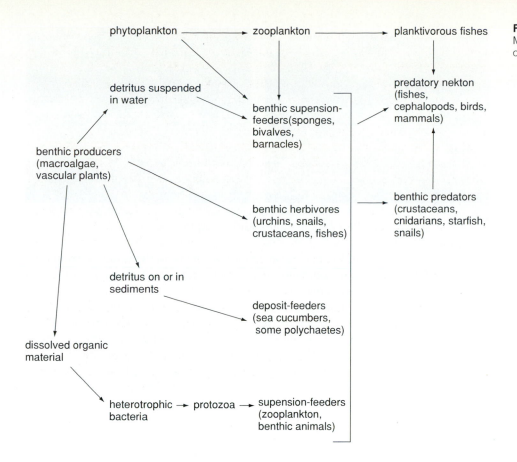

Figure 9.3
Major food chains in coastal oceans.

Arctic water, but in the Antarctic subtidal region, the brown algae in the order Desmarestiales replace the kelps. For reviews of Arctic macroalgae, see Wilce (1994), and for Antarctic algae, see Knox (1994) and Amsler et al. (1995).

Coral Reefs

Reefs are among the most productive ecosystems on the earth (see table 9.1), supporting a great diversity of organisms. Reef-forming corals are confined to tropical waters where temperatures remain above 22° C. Coral animals grow in large colonies in which each individual coral polyp secretes a cup of calcium carbonate. Reef structure is derived from accumulation of these carbonate structures over many generations of corals, with the living corals growing on the remains of earlier generations (fig. 9.4). Each coral polyp has a ring of tentacles for capturing zooplankton and other prey. Reef-forming corals also obtain nourishment from symbiotic dinoflagellates in their cells, called **zooxanthellae.**

Reefs form in shallow waters along the edges of continents and islands, or on tops of seamounts. Living corals normally grow at depths less than 25 meters, where their zooxanthellae receive adequate light to photosynthesize. Nutrient concentrations are low in the waters around reefs, and thus success of reef communities depends on effective nutrient retention and recycling. Reef algae rapidly take up nutrients released into the water as wastes or by death of organisms, and the nutrients are cycled among reef organisms. In addition, nitrogen fixation by cyanobacteria is an important source of organic nitrogen. In contrast, plankton

Figure 9.4
(*a, b*) Coral reefs.

a   b

communities with less ability to recycle materials are low in the waters surrounding reefs. Freshwater and high sedimentation, which decreases light penetration and smothers corals, normally reduce reef formation.

Corals vary in their dependence on the zooxanthellae in vesicles within their cells. Zooxanthellae supply much of the nutritional needs of some corals. These corals tend to have branched or platelike forms with small polyps (Porter 1976). The importance of zooxanthellae decreases with water depth (Muscatine, Porter, and Kaplan 1989). In the mutualistic association between corals and zooxanthellae, the algae receive protection and nutrients (carbon dioxide, nitrogen, and phosphorus) from their hosts. In turn, the algae supply the corals with organic compounds and oxygen. A large proportion of the photosynthetic production of the zooxanthellae may transfer to the corals as glycerol, but the algae may also supply amino acids, fatty acids, and sugars. The zooxanthellae affect corals' calcium carbonate deposition, but the exact role of the zooxanthellae in this process is not clear (Marshall 1996).

At times, corals expel their zooxanthellae, losing their normal color in an event called **coral bleaching.** Some form of stress, such as unusually high water temperatures, exposure to ultraviolet radiation, or prolonged shading, causes bleaching. Warming associated with an El Niño event in the Pacific (p. 205) causes bleaching of reefs in the eastern Pacific. Some scientists have suggested that widespread bleaching of reefs is the result of warmer temperatures associates with global warming (Brown and Ogden 1993).

Most reefs are one of three types (fig. 9.5). **Fringing reefs** border the land, while **barrier reefs** are separated from the land by a channel of relatively deep water. **Atolls** are island rings formed of corals. During the voyage of the HMS *Beagle* (1831–1836), Charles Darwin observed reefs in the Pacific and Indian Oceans. He later proposed that the reef around an oceanic island changes from a fringing reef to a barrier reef and finally into an atoll as the result of gradual sinking of the island. Today, Darwin's theory is generally accepted, with the recognition that changing sea level is also important. If reef growth keeps pace with a gradual rise in sea level, fringing reefs develop into barrier reefs and eventually into atolls. Rising sea level also is responsible for fringing reefs developing into barrier reefs along continental margins.

Figure 9.6 shows a typical tropical shoreline with a barrier reef. On the outer or seaward side, the forereef zone rises vertically from deep water to almost the water surface. There, the slope is more moderate, and incoming waves are channeled between coral buttresses. Waves break at the reef crest, where encrusting algae sometimes form a distinct ridge, and surge across the top of the reef or reef

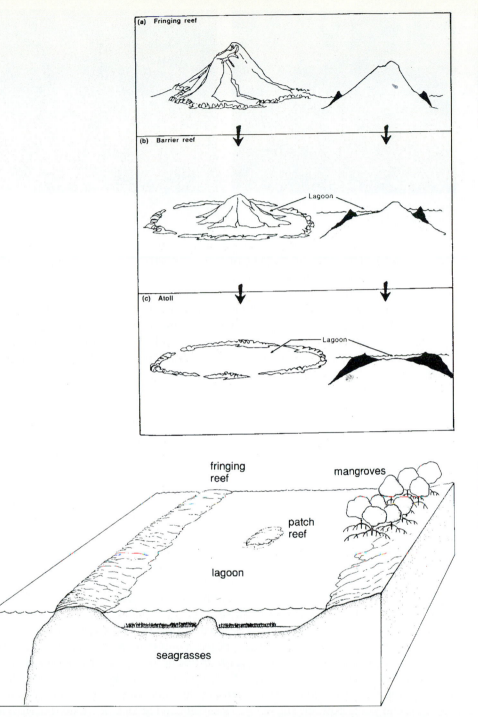

Figure 9.5
Types of reefs. (*a*) Fringing. (*b*) Barrier. (c) Atoll. (*a–c* from Sumich, James L., *An Introduction to the Biology of Marine Life*, 3d ed. © 1984 Wm. C. Brown Publishers, Dubuque, Iowa. All Rights Reserved. Reprinted by permission.)

Figure 9.6
Typical tropical shore with barrier reef and lagoon, and with seagrasses and mangroves growing in the lagoon. (Courtesy C. Feller.)

flat. In tropical waters with little tidal variation, reefs normally remain submerged, but parts of reef flats on some Indo-Pacific reefs may be exposed to the air during low tide. The back reef is largely protected from waves. Behind the reef is a sandy-bottomed lagoon. Here, the water is calm, and light usually penetrates to the bottom. Larger siphonous green algae and beds of seagrasses may be extensive. In contrast to the quartz sand on temperate shores, most of the sand in a tropical

Figure 9.7
(*a*) Turf algae covering dead
corals. (*b*) *Penicillus* growing
among *Thalassia* in lagoon.

a   b

lagoon is derived from the breakdown of calcareous structures of various algae and animals. Within the lagoon may be isolated reef outcrops or patch reefs. The shore area may be a sandy beach or a mangrove swamp on a muddy bottom.

Reef algae are divided into symbiotic algae, turf algae, crustose algae, and larger erect algae. Symbiotic algae include the zooxanthellae and cyanobacteria. The zooxanthellae are present in coral polyps, and also in other reef animals, including the tridacnid clams on Indo-Pacific reefs. Many sponges have zooxanthellae or symbiotic cyanobacteria. Endolithic algae are associated with the calcareous skeletons of corals and often include the green alga *Ostreobium* (Shashar and Stambler 1992).

Turf algae grow on coral skeletons, larger algae, and seagrasses (fig. 9.7*a*). They consist of filamentous and simple parenchymatous forms (sheet type), and belong to the brown, red, and green algae, and the cyanobacteria. Nitrogen fixation by the cyanobacteria in turf associations is an important nitrogen source for the reef community. Associated with the relatively simple structure of most turf algae are high rates of photosynthesis and growth. Rapid growth allows turf algae to replace quickly any losses to herbivores. Many turf species are heterotrichous, regrowing rapidly from their basal filaments after their erect systems are lost.

Encrusting algae, especially coralline reds, may cover much of the reef surface and are important in reinforcing the reef. Some reefs feature a thick algal ridge of encrusting algae exposed to breaking waves. Crusts are relatively resistant to waves and herbivores, but they are often poor competitors for space, and other algae may overgrow them in the absence of herbivores.

Larger fleshy and calcareous algae (coarsely branched, leathery, and jointed types of Littler) grow on parts of the reef protected from wave action. They may attach to the reef itself or grow on sandy bottoms in the lagoon. Larger green, red, and brown algae compose this group. Siphonous green algae (order Caulerpales), such as *Caulerpa* and *Udotea,* have extensive rhizoidal systems that allow them to grow well on sandy bottoms (fig. 9.7*b*). *Halimeda* and other calcareous macroalgae contribute to sediments when they die.

Herbivores, including fishes, sea urchins, crabs, and gastropods, control algal abundance and distribution on a reef. Important groups of herbivorous fishes are parrotfish, surgeonfish, and damselfish (Horn 1989; Hay 1991). Parrotfish teeth fuse to form a break, and a modified digestive system allows parrotfish to feed on calcareous algae. The calcareous particles that parrotfish release in their wastes contribute to sand. In addition to controlling overall abundance of macroalgae, herbivorous fish may graze selectively on particular species. Exclusion studies in which areas are kept free of fish grazing have shown that rapidly growing seaweeds have the potential to overgrow corals and encrusting algae in the absence of herbivores. Soft corals without calcareous exoskeletons or zooxanthellae are important consumers of the phytoplankton associated with reefs (Fabricius, Benayahu, and Genin 1995).

Algal adaptations to reduce susceptibility to grazing include: (1) avoidance of herbivores, (2) rapid growth, (3) hard coverings, and (4) chemical deterrents. Some algae live in places where herbivorous animals are less active, such as in surf on the reef flat. Damselfish maintain territories with lush growths of algal turf, driving off other herbivores. Some algae associate with other organisms and rely on the defenses of these other organisms for protection (Littler, Taylor, and Littler 1986; Littler, Littler, and Taylor 1987). Palatable algae may gain protection by growing near cnidarians with stinging cells (nematocysts) or other algae with chemical deterrents. Some algae grow too large for herbivores to handle.

While turf algae respond to losses by regrowing rapidly, larger algae with slower growth rates commonly have tough, leathery, outer layers or deposit calcium carbonate on the surfaces of their thalli as defenses against herbivores. However, calcium carbonate may not protect against the scraping radula of snails (Padilla 1989). Chemical defenses also are common among reef algae, such as caulerpin produced by *Caulerpa,* halogenated compounds of some red algae, and polyphenols in the brown algae. In *Halimeda,* defensive chemicals activate in response to physical damage from fish bites (Paul and Alystyne 1992). Chemical defenses may be more important in deterring fish-grazers than invertebrate herbivores (Duffy and Hay 1994). In a few cases, animals use algal chemicals as a defense against their predators. For example, one amphipod forms a portable cover from the brown alga *Dictyota* and relies on chemicals in the alga for predator protection (Hay et al. 1990).

Some snails puncture the cells of *Caulerpa* and its relatives and suck out the protoplasm containing chloroplasts. In chloroplast symbiosis, instead of being digested, the chloroplasts pass into the tissues of the animals and continue to photosynthesize for several days, providing organic material to their hosts (Clark 1992).

The long-spined black sea urchin *Diadema antillarum* is an important herbivore on Caribbean reefs. *Diadema* forages around patch reefs in the shallow waters of lagoons, where it grazes on a variety of algae, showing a preference for turf algae and germlings of larger algae. *Diadema* is important in limiting algal overgrowth of other sessile organisms. Artificial exclusion of *Diadema* leads to an increase in turf algae and larger upright macroalgae, often at the expense of encrusting algae (Morrison 1988). In 1983, a widespread die-off of *Diadema* occurred throughout the Caribbean (Lessios 1988). In some places, densities of the urchin declined as much as 95%, and algae increased greatly (Morrison 1988; Carpenter 1990). Subsequent recovery of *Diadema* has been slow on many reefs, allowing algae to persist at the expense of corals.

In addition, overfishing (including herbivorous fishes), direct damage from human activities, hurricane damage, and diseases are degrading Caribbean reefs (Hughes 1994). These stresses have probably all contributed to a decline in coral abundance and an increase in macroalgae. Since 1973, black-band disease, caused by the cyanobacterium *Phormidium,* has infected corals in the Caribbean and Indo-Pacific regions (Bruckner 1993). The disease is recognized by distinct dark bands between living coral polyps and the white skeletal remains of corals killed by the infection.

Intensity of grazing may determine the dominant algal group. From studies in the Caribbean, Hackney, Carpenter, and Adey (1989) proposed the model in table

Table 9.4 *Grazing Model*

	Intensity of Grazing		
	Low to Absent	**Moderate to High**	**High**
Predominant benthic algae	Erect macroalgae	Algal turfs	Crusts
Standing crop	Very high	Low	Very low
Primary productivity (per unit area)	Very high	Moderate	Low
Primary productivity (per unit biomass)	Low	Very high	Very low
Predominant types of adaptations	Maintain upright morphology	Growth	Resistance

Sources: From Hackney, Carpenter, and Adey (1989), courtesy *Phycologia*.

9.4. Macroalgae dominate at low levels of grazing, turf algae dominate at moderate to high levels of grazing, and crusts dominate when grazing levels are high.

Healthy coral reefs support an extremely high diversity of organisms, including many species of algae. Many explanations have been proposed for community structure and the high diversity (Connell 1978). In general, the explanations are divided into equilibrium and nonequilibrium models. In equilibrium models, interactions (competition, predation, mutualism) among different species are important, and the community reaches a stable state in species composition and relative abundance of component populations. In nonequilibrium models, the community is continually changing, with physical disturbance and chance events playing important roles. Several models of each type apply to reef diversity.

Equilibrium models address the problem of how intense competition for available resources is moderated to allow many species to live together. Specialization and predator cropping may reduce competition. According to the specialization (or niche diversification) hypothesis, organisms living under the constant conditions of tropical environments can specialize in the use of resources to reduce competition. The lack of seasonal variation and the stability of the reef environment over geologic time periods have allowed organisms to adapt to reef conditions and to specialize more than is possible in environments where food and living space vary seasonally. Except for differences seen in habitat preference, however, it is not obvious how reef algae have specialized, and cropping by herbivores also may be important in reducing competitive interactions. Herbivores may limit algal populations sufficiently that intense competition does not occur (Hixon and Brostoff 1996). Exclusion studies in which a few algal species become abundant in the absence of herbivores provide evidence for this.

Nonequilibrium models include the intermediate disturbance hypothesis and the variable recruitment hypothesis. Storm and wave disturbances may open up patches on a reef for recolonization. During recovery periods, when populations are increasing, competition is less intense than after full recovery. Thus, a moderate frequency and intensity of reef disturbance may help maintain patches at different stages of recovery, which support a greater diversity than fully recovered plots. The variable recruitment (or lottery) hypothesis also emphasizes patch dynamics and reef heterogeneity. This hypothesis postulates that the water has an excess of dispersal stages of different algal species and that which species happens to settle and grow on open reef patches is largely a matter of chance. Thus, different species may colonize neighboring patches. Sometimes, the initial colonist is able to persist on a patch by excluding other species, but in other cases, other species replace the initial colonist.

One simple overall explanation for reef diversity may not exist. The equilibrium and nonequilibrium models described provide partial explanations and are not necessarily mutually exclusive.

Soft-Bottom Communities

On sandy or muddy bottoms, waves disturb the loose sediments, making them poor substrates for macroalgae. Hence, algal abundance and diversity are generally lower on soft bottoms than on rocky shores and reefs, and marine vascular plants are the dominant producers. Most of the marine vascular plants are angiosperms, belonging to three general types: **seagrasses** that grow completely submerged in seawater, **mangroves** or woody plants with their roots in seawater, and salt-marsh plants that grow partially submerged. **Salt marshes** are a common feature of temperate shores, while mangrove swamps (mangals) are found in tropical and subtropical waters. Various species of seagrasses are found at different lattitudes.

The roots of marine vascular plants penetrate into the bottom, stabilizing loose substrates and aiding in sediment accumulation. A wide variety of marine organisms, including algae, grow on submerged plant surfaces and on the sediments around their roots. Marine vascular plants often show high rates of production but experience little direct grazing. Instead, much of their production enters food chains as detritus and may be exported to enrich other coastal environments.

Mangrove Swamps and Seagrass Meadows on Tropical Coasts

In the lagoon behind the reef of a typical tropical shoreline, marine vascular plants dominate two types of communities: seagrass meadows and mangrove swamps (see fig. 9.6).

Seagrasses grow completely submerged. Despite their appearance, seagrasses are not true grasses (in the taxonomic sense of belonging to the grass family). In tropical oceans, one of the most common seagrasses is *Thalassia* ("Turtle Grass"). Its narrow leaves provide surfaces for the growth of smaller marine organisms, including many species of epiphytic, encrusting and turf algae. Because of their abundance and rapid growth, these algae are important producers in seagrass meadows, and herbivores consume them more readily than the seagrasses themselves.

In the Caribbean, sandy bottoms show a succession from caulerpalean green algae to seagrasses. The rhizoidal systems of green algae, such as *Halimeda, Penicillus*, and *Udotea*, initially stabilize the sand. As these algae die and decompose, nutrients accumulate in the sediments, facilitating growth of *Halodule* ("Shoal Grass"), *Syringodium* ("Manatee Grass"), and ultimately, *Thalassia* (Williams 1990).

Mangrove trees and shrubs grow in shallow water. They have spreading root systems that extend into muddy bottoms. In the Atlantic and Caribbean, common mangroves are *Rhizophora* ("Red Mangrove") and *Avicennia* ("Black Mangrove"). *Rhizophora* forms prop roots that arch out from the lower part of the trunk into the water and create surfaces for macroalgal growth (fig. 9.8). Two algal associations are common. On sediments around roots are felty mats of filamentous algae, such as *Cladophora* and *Vaucheria*. A second association of "leafy" red algae grows on

Figure 9.8
(*a*) Red mangrove
(*Rhizophora*) with extensive
prop roots. (*b*) Bostrychian
community on prop roots.
(*b* courtesy Feller and
Rützler.)

a   b

Figure 9.9
Shore with barrier islands
and salt marshes.

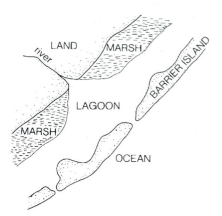

the roots and consists of *Caloglossa, Catenella,* and *Bostrychia.* This association is found on mangrove prop roots throughout the world. In addition to stabilizing sediments and creating surfaces for algal growth, mangroves provide shade from intense tropical sunlight and reduce desiccation during low tides.

Temperate Shorelines

Barrier islands border much of the Atlantic coast from Florida to Cape Cod and the Gulf of Mexico. These islands are large sandbars that shallow bays and lagoons separate from the mainland. A typical shoreline has beaches on the seaward side of the barrier islands (fig. 9.9). The shifting sand of exposed beaches prevents macroalgal attachment or buries macroalgae that successfully attach. Scouring by sand also may prevent algal growth on adjacent rocks and pilings. Occasionally, films of diatoms develop on the sand surface or bloom in the surf (Talbot, Bate, and Campbell 1990).

a

b

c

Figure 9.10
(*a*) Salt marsh with *Spartina alterniflora* dominant in the low marsh. (*b*) Filamentous algae growing on surface of mud. (*c*) Fucoids growing entangled with *Spartina*.

Behind barrier islands are lagoons, which are more protected from waves. In the shallow water of bays and lower parts of rivers, salt marshes develop. Marsh inhabitants live in an estuarine environment, where they must tolerate a mix of freshwater and ocean water. The freshwater input brings sediments and nutrients from terrestrial runoff. The dominant plants in marshes are grasses and other herbaceous plants that occupy the upper part of the intertidal region and above. The grass *Spartina alterniflora* ("Cordgrass") is common in the lower part of Atlantic marshes (fig. 9.10*a*). Higher in the marsh, *Spartina patens*, the rush *Juncus*, and other species are important.

Films of diatoms, cyanobacteria, green algae (such as *Enteromorpha* and *Rhizoclonium*), and the tribophyte *Vaucheria* may produce conspicuous growths on sediment surfaces (fig. 9.10*b*). Filamentous and foliose red algae may also be present in warmer waters. Distinctive forms of *Fucus vesiculosus* and *Ascophyllum nodosum* may grow entangled with the bases of grasses (fig. 9.10*c*) (Norton and Mathieson 1983). These growth forms are characterized by twisted or curled thalli, which branch profusely and lack holdfasts. They rarely produce reproductive structures (receptacles) but propagate vegetatively by breaking.

Seagrasses, such as *Zostera* ("Eel Grass"), often grow in deeper water of lagoons and estuaries. Like their tropical relatives, these seagrasses produce surfaces for epiphytic algae.

Atlantic Rocky Shores

North of Cape Cod on the Atlantic Coast of North America, the shoreline is rocky. During the last ice age, glaciers advanced over Canada and New England, scraping sediments from the land surface. When the Pleistocene glaciers retreated 10,000 years ago, bedrock was exposed.

Communities of macroalgae are well developed on rocky shores, where the substrate is firm for attachment, even under intense wave action. The brown algae, especially the fucoids and kelps, are more important on temperate rocky shores than in the tropics.

A typical rocky shore divides into three vertical regions. The **intertidal** (littoral) **region** is exposed to the atmosphere during low tide and submerged at high tide. Two intertidal habitats are **emergent surfaces,** which drain during low tide and are exposed to the air, and **tidepools,** which retain water but may experience substantial environmental variation during low tide (fig. 9.11). Above the intertidal region is the **spray** (supralittoral) **zone,** inhabited by organisms that depend on wave spray for moisture. Below the low-tide level, in the **subtidal** (sublittoral)

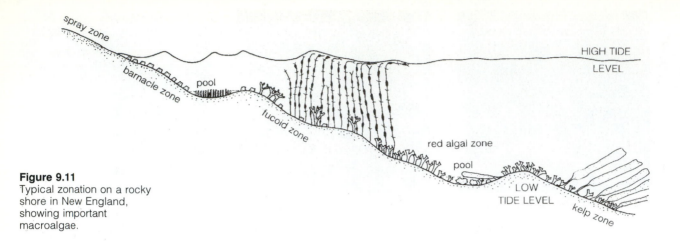

Figure 9.11
Typical zonation on a rocky shore in New England, showing important macroalgae.

region, organisms remain submerged and thus experience less variation in conditions.

On subtidal rocky substrates, light availability and herbivore activity are important in controlling algal distributions. In contrast, the intertidal region is well illuminated and supports fewer herbivores. During periods of exposure to the air, however, algae on emergent surfaces must withstand water loss by evaporation, greater temperature variation than in the water, and exposure to high irradiance. Duration of exposure to the atmosphere and intensity of wave action are important factors controlling the distributions of intertidal algae. The tidepool environment is intermediate between the subtidal region and emergent intertidal surfaces.

Semidiurnal tides (see fig. 9.2*a*) occur along much of the Atlantic coast. These tides have a periodicity of 12.5 hours; thus, two high tides and two low tides usually occur daily. The moon's gravitational attraction for the water and the inertial (centrifugal) force of the earth's rotation (about the common center of mass of the earth-moon system) cause semidiurnal tides. The sun's gravitational attraction is less than half that of the moon but still has an important effect. The sun reinforcing the moon's attraction produces the greatest tidal range, with the lowest low tides and the highest high tides. These tides are called spring tides (nothing to do with the season). The sun partially cancelling the moon's effect produces neap tides, with the minimum tidal range (highest low tides and lowest high tides). The spring/neap tide sequences have a fourteen-day periodicity. The variation of the moon above and below the earth's equatorial plane, natural oscillations of water within bays, and storm surges also influence tides. Thus, intertidal boundaries are not fixed elevations. Algae living above the average high-tide level may only be flooded during spring tides or storms, while algae lower on the shore may be exposed to the air only during unusually low tides.

Spray Zone

A spray-zone community is best developed where wave activity is great enough to dampen the rocks (fig. 9.12). Only a few species can tolerate the harsh conditions of the spray zone, but they may be abundant. Dense growths of some cyanobacteria, small green algae (*Ulothrix, Blidingia*), and sometimes bangean red algae (*Bangia, Porphyra*) may be conspicuous. The common cyanobacterium *Calothrix* is able to fix nitrogen. Guano from seabirds may also be an important nitrogen source. *Prasiola,* a foliose green alga resembling parsley flakes, has an interesting association with guano from seabird colonies.

a   b

Figure 9.12
Spray zone. (*a*) General view with blackening of rocks from cyanobacterial growth. (*b*) *Porphyra* dried during exposure to air.

Intertidal Emergent Surfaces

Emergent surfaces drain during low tide, exposing intertidal organisms to a variety of stresses. In particular, algae that live attached to rock surfaces experience desiccation, a wide temperature variation, and high irradiance during exposure to air, as well as physical disturbance from waves and material carried in the water. Despite these stresses, the benefits of living in the intertidal region have resulted in many algae successfully adapting to life there.

During air exposure, evaporative water loss from algal surfaces depends on the duration of exposure, the humidity of the air, and the surface area exposed. Desiccation disrupts cellular membranes, including the thylakoids reducing photosynthetic activity and the cell membrane causing organic material to leak from cells. Compared to subtidal algae, intertidal algae generally tolerate greater water loss before photosynthetic activity is severely reduced and recover better when resubmerged after a period of desiccation (Dring and Brown 1982; Brown 1987; E. C. Bell 1993). An extreme example is *Porphyra*, which lives high in the intertidal region and spray zone, where it sometimes dries and becomes brittle (like a potato chip) but is able to revive when submerged (fig. 9.12*b*). Some intertidal algae continue to photosynthesize at relatively high rates during at least part of the time they are out of the water, benefiting from the availability of carbon dioxide in air. *Fucus spiralis* growing high on the shore continues to show significant photosynthetic activity even when its water content has been reduced 70%, and it recovers in approximately 30 minutes when resubmerged (Madsen and Maberly 1990). Other algae survive in moist habitats under a canopy of larger algae or in cracks. Algae lack the waterproof, waxy coverings of vascular plants, but their thick cell walls and mucilage retard water loss to some degree. The saccate thalli of some intertidal algae, such as *Leathesia* (see fig. 4.32*a*) and *Colpomenia*, have a low surface-to-volume ratio and retain water in a hollow interior. Although these algae have low rates of photosynthetic activity, their water loss is relatively low, which allows them to maintain photosynthetic activity while out of the water (Oates 1985, 1986).

Intertidal algae are exposed to high irradiance, including ultraviolet radiation, which can inhibit photosynthesis and disrupt other cellular activities. Many intertidal algae have high concentrations of carotenoids in their cells, which function in photoprotection rather than as accessory photosynthetic pigments (p. 6). These carotenoids may mask the photosynthetic pigments, giving intertidal algae a brown color.

Intertidal algae directly experience the energy released when waves break, as well as the drag produced by the surge and backflow of water. In addition, material the water carries, such as sand, rocks, logs, or ice, may scrape against the shore. Some algae adapt to flow by firmly attaching to the substrate and having a flexible

Figure 9.13
Invertebrates of the intertidal region. (*a*) Barnacles (*Semibalanus*). (*b*) Mussels (*Mytilus*) with *Chondrus*.

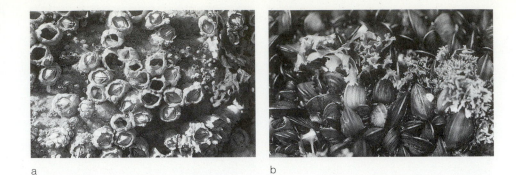

thallus that orients with water flow. Other algae grow as crusts appressed to the substrate in the boundary layer, where flow is reduced, or they avoid the full force of the water by living in cracks or beneath larger algae, which dissipate the wave force. Water motion also has beneficial effects. It may enhance production by preventing prolonged shading by overlapping fronds, maintaining a fresh nutrient supply, and hindering foraging by herbivores.

Spores' initial attachment to the substrate is critical. A newly settled spore secretes adhesive mucilage (Apple and Harlin 1995) and, as its holdfast develops, sends extensions into irregularities in the substrate. Once an alga is securely attached, it produces an upright system (except crusts). Holdfasts are also important as perennating structures and maintain a place in the community when the thallus is not actively growing.

The dominant intertidal organisms on a rocky shore often form distinct horizontal bands, with sharp boundaries between zones (see fig. 9.11; plate 6). In general, an algal population's ability to withstand exposure to the atmosphere (especially desiccation and heating) determines its upper limit in a zone, while biological interactions determine the lower boundary (Norton 1986; but see Hawkins and Hartnoll 1985). An important interaction is competition for space. Macroalgae and attached animals compete for the limited space available on rock surfaces. Given sufficient time, and unless limited by herbivores, predators, or wave action, better competitors overgrow other species.

Menge (1976) and Lubchenco and Menge (1978) described the pattern of intertidal communities on the New England coast. On shores exposed to intense wave activity, barnacles and mussels outcompete macroalgae for space on rocks (fig. 9.13), but where wave action is less intense, predation on these attached invertebrates reduces their coverage sufficiently for macroalgae to grow in abundance. A typical zonation pattern in these areas has a band of barnacles at the top of the intertidal region, an extensive midzone of fucoids, and a lower zone of red algae (see fig. 9.11; plate 6*a*). Several species of *Fucus* and *Ascophyllum nodosum* are common in the midzone (fig. 9.14*a*) (Chapman 1995 reviews fucoid ecology). *Ascophyllum* and *Fucus vesiculosus* have air bladders with sufficient buoyancy to support the algae vertically in the water (see figs. 4.42*a*, 4.44*b*). The dominant fucoids provide surfaces for smaller epiphytic algae (fig. 9.14*b*). Beneath the canopy of fucoids is an understory of smaller red algae, including *Chondrus*, *Mastocarpus*, and encrusting algae. A dense stand of fucoids reduces the force of breaking waves and creates a moist habitat where other organisms shelter during low tide. Snails are important herbivores in the fucoid zone, especially *Littorina littorea* ("Common Periwinkle") (fig. 9.14*c*; plate 8*c*). *Littorina*, which has been present on the East Coast only since the late nineteenth century, does not feed directly on the fucoids but grazes on more delicate forms (Lubchenco 1978).

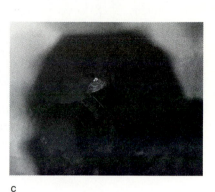

a

b

c

The boundary between the fucoids and the red algal zone is often sharp (plate 6*b*), produced by the red algae's competitive advantage over the fucoids in the lower intertidal region (Lubchenco 1980). The reds, however, tolerate less exposure to air than the fucoids. *Chondrus* and *Mastocarpus* are often the major species in the red algal zone, and *Littorina* is the principal herbivore. At high densities of *Littorina,* only encrusting algae, with tough thalli and chemical deterrents, may be able to grow (Bertness, Yund, and Brown 1983).

Like the fucoids, *Chondrus* and *Mastocarpus* are perennial species that are successful in holding space on the rocky substrate against potential macroalgal competitors. Both of these reds have tough thalli and are relatively resistant to herbivores. Part of the success of *Mastocarpus* is its ability to tolerate winter conditions (Dudgeon, Davison, and Vadas 1989). During exposure to subfreezing air temperatures, *Mastocarpus* experiences less disruption to its cellular structure and is able to continue photosynthetic activity better than *Chondrus*.

Physical disturbances are important in opening parts of the intertidal region for colonization. The movement of ice, logs, or rocks during storms may scrape off organisms attached to rock surfaces. Recolonization is often rapid, but complete recovery of an intertidal community is sometimes slow. A film of bacteria and diatoms first covers a cleared surface and may aid in the initial settlement and attachment of macroalgal spores. Germlings of macroalgae start growth within this film. An intermediate stage with ulvalean algae, such as *Enteromorpha,* usually develops before the fucoids, and later, the red algae become dominant again low on the shore. *Fucus vesiculosus* recruits well onto disturbed surfaces (fig. 9.15) (McCook and Chapman 1993). *Ascophyllum,* however, recolonizes very slowly because the settlement of its zygotes is sensitive to water movement (Vadas, Wright, and Miller 1990). Thus, full recovery of the fucoid zone may take a number of years. Red algae, such as *Chondrus* and *Mastocarpus,* develop initially below the *Fucus* canopy. After several years, when the thalli from the initial settlement of fucoids die, the red algal turf is uncovered and hinders further fucoid settlement on the lower shore. The frequency of disturbance and then slow recovery suggest that the intertidal region is a mosaic of plots at different stages of recovery.

Figure 9.14
Fucoid zone in the mid-intertidal region.
(*a*) *Ascophyllum nodosum* dominates, with a band of *Fucus spiralis* above.
(*b*) *Polysiphonia lanosa* is a common epiphyte on *Ascophyllum.* (*c*) *Littorina littorea* is an important intertidal herbivore (shown on *Laminaria*).

Figure 9.15
Recolonization following
disturbance of the intertidal
region in New England.
(Unpublished study by Sze
and Boden.)

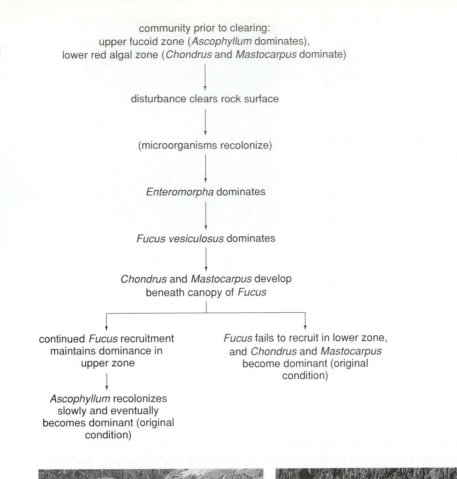

community prior to clearing:
upper fucoid zone (*Ascophyllum* dominates),
lower red algal zone (*Chondrus* and *Mastocarpus* dominate)

↓

disturbance clears rock surface

↓

(microorganisms recolonize)

↓

Enteromorpha dominates

↓

Fucus vesiculosus dominates

↓

Chondrus and *Mastocarpus* develop
beneath canopy of *Fucus*

continued *Fucus* recruitment
maintains dominance in
upper zone

↓

Ascophyllum recolonizes
slowly and eventually
becomes dominant (original
condition)

Fucus fails to recruit in lower zone,
and *Chondrus* and *Mastocarpus*
become dominant (original
condition)

Figure 9.16
Tidepools. (*a*) Pool in a rock
depression that retains water.
(*b*) Large pool with kelps.

a

b

Tidepools

Tidepools retain water during low tide and are flushed out during high tide
(fig. 9.16; plate 8*b*). Thus, tidepool inhabitants are rarely, if ever, exposed to the
air. Conditions in larger pools low in the intertidal region remain close to those of
the surrounding seawater. In small pools and pools high on shore, however, temperature, salinity, and pH may vary sufficiently between periods of flooding to limit the
species present. On sunny days, tidepool water may be much warmer than seawater.
Water salinity may increase as a result of evaporation or be diluted by rain. Photosynthetic activity by algae in a small tidepool may increase pH and reduce the
availability of carbon (p. 187).

Figure 9.17
High tidepool dominated by
Enteromorpha intestinalis.

Tidepool algae may experience a wide range of salinities. In general, algae respond to fluctuating salinities (or osmoregulate) by adjusting the concentration of dissolved materials in their cells in one of two ways. They can change the internal concentrations of ions, such as potassium, sodium, and chloride, by actively transporting them into or out of their cells. Alternatively, cells can convert polysaccharides to monosaccharides to increase their solute content as salinity rises. Low-weight molecules, such as mannitol in the brown algae (Reed et al. 1985), and a variety of molecules in other algae are important in osmoregulation.

In the North Atlantic, green algae (*Enteromorpha, Cladophora*) often dominate in high tidepools, while red and brown algae are abundant in lower tidepools (Wolfe and Harlin 1988). *Littorina littorea* is an important tidepool herbivore. Lubchenco (1978) showed the importance of *Littorina* in regulating algal communities of high tidepools. In the absence of *Littorina, Enteromorpha* grows in abundance, preventing the invasion of other algal species (fig. 9.17). Since *Enteromorpha* is a preferred food of *Littorina,* however, the presence of *Littorina* in a tidepool leads to a more diverse algal community (plate 8*a*). Wave action may reduce both the competitive advantage of *Enteromorpha* and the impact of *Littorina* on tidepool communities (Sze 1980).

Tidepools lower on shore usually have more diverse communities, sometimes with subtidal kelps (see fig. 9.16*b*). Although wave exposure influences the composition of tidepool communities, biological interactions such as competition and grazing also are important (Lubchenco 1982; Sze 1982). In addition to *Littorina*, other snails and amphipods are common herbivores in lower tidepools.

Subtidal Region

On the New England coast, the upper limit of kelp growth indicates the approximate boundary between the intertidal and subtidal regions. *Laminaria* is often dominant in the kelp zone (see fig. 1.13*d*; plate 7*a*) but *Alaria* replaces it on more wave-exposed shores (fig. 9.18*a*). A turf of *Chondrus* and other red algae of the low intertidal region continues underneath the kelp canopy. In deeper water, kelps such as *Agarum,* foliose red algae, and encrusting red algae become common (fig. 9.18*b*; plate 7*b*) (Vadas and Steneck 1988). In progressively deeper algal communities, the green algae are fewer, and perennial species are more important than annual species (Mathieson 1979).

Light is important in controlling the distributions of subtidal algae. With depth, the amount of light decreases, and its spectral composition is modified. As

Figure 9.18
(*a*) *Alaria* growing in the low intertidal region on a wave-exposed shore.
(*b*) Encrusting coralline on a bottle from the subtidal region.

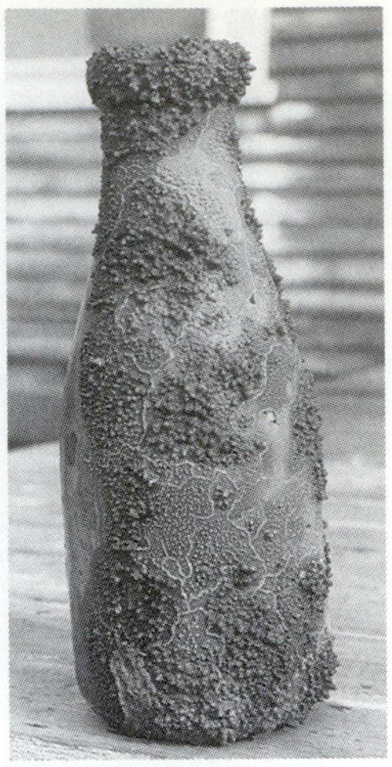

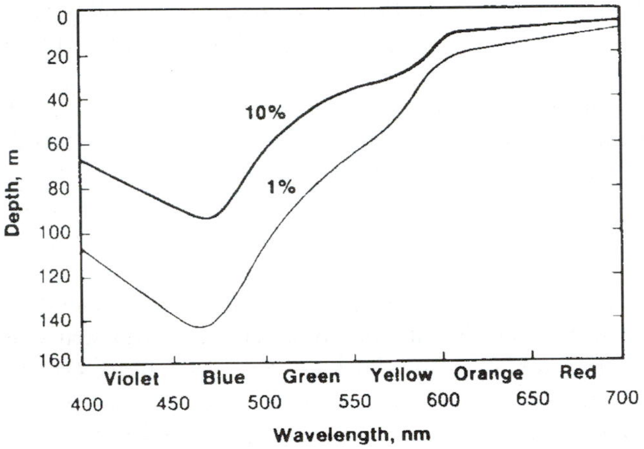

a

b

Figure 9.19
Depth at which surface radiation is reduced to 10% and 1% for different wavelengths of light. (From Sumich, James L., *An Introduction to the Biology of Marine Life*, 3d ed. © 1984 Wm. C. Brown Publishers, Dubuque, Iowa. All Rights Reserved. Reprinted by permission.)

described in chapter 7, light decreases exponentially in water (see fig. 7.2*a*). In more northern waters, this attenuation is usually more rapid than in tropical oceans because of the greater amount of plankton and other suspended material in the water. At the same time, the wavelengths of photosynthetically active radiation are unequally absorbed. Violet and red light at the ends of the visible spectrum are more rapidly absorbed than blue-green light, and thus, the spectral range of available light narrows (fig. 9.19). Since chlorophyll absorbs primarily in the red and violet regions of the spectrum, it is relatively ineffective in a light environment consisting primarily of blue-green light.

Subtidal benthic algae show two general adaptations to their light environment. In some, an accessory pigment becomes the principal light-collecting system (fig. 9.20). The red algae increase their phycoerythrin to collect light more effectively in the blue-green region of the spectrum (Sagert and Schubert 1995), fucoxanthin increases in some benthic microalgae (Robinson et al. 1995), and siphonaxanthin in some ulvophycean green algae may be an adaptation to deep water (Yokohama et al. 1977). However, other studies suggest that chromatic adaptation is not as important in macroalgae as previously thought (Dring 1981; Ramus 1983; Saffo 1987). Instead, many algae increase their overall pigment content in deep water to become more efficient at trapping whatever light is available.

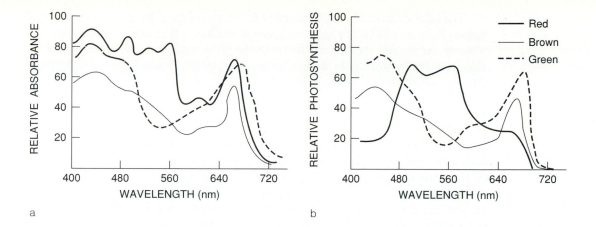

a

b

Figure 9.20
(*a*) Absorption spectra for red, brown, and green algae. (*b*) Photosynthesis at different wavelengths of light. (*a, b* from Saffo 1987, courtesy *Bioscience*, © 1987 American Institute of Biological Sciences.)

The deepest algae are crustose corallines. In general, availability of light determines the lower limit of macroalgal growth, which is where irradiance is approximately 0.02% of the surface level (Vadas and Steneck 1988). The greatest depth at which macroalgae have been found growing is 268 meters near the Bahamas (Littler et al. 1985).

Animals may graze more on macroalgae in the subtidal region than in the intertidal region. Although herbivorous fishes are less important at higher latitudes than in tropical oceans, sea urchins can have a major impact on algal communities. On the Atlantic coast, *Strongylocentrotus* ("Green Sea Urchin") is a relatively nonselective grazer. When abundant, this urchin can clear the bottom, leaving only crustose coralline algae. Such areas of largely bare bottom are called "urchin barrens" (plate 7*c*). Causes for periodic increases in urchins are poorly understood. In the 1970s, high urchin densities caused widespread destruction of the kelps, but kelp beds recovered in the 1980s as the urchins declined, possibly due to the spread of an amoeboid pathogen (Miller 1985; Scheibling 1986).

Kelp forests in the shallow subtidal region are among the most productive communities on earth. In Nova Scotia, the estimated annual production of laminarian beds is 1,750 g C m^{-2}yr^{-1} (grams of carbon per square meter of ocean bottom), but an average of 1,000 g C m^{-2}yr^{-1} is probably more typical of kelp beds in general (Mann 1982). A kelp such as *Laminaria* grows by production of new tissue at the base of its blade, and older tissue wears away at the end. In the laminarian beds in Nova Scotia, researchers punched holes in kelp blades, and used hole movement up a blade to estimate new tissue production (Mann 1973). High production represents blade turnover several times per year. Growth is greatest during winter, even though photosynthetic activity peaks in summer. Greater nitrogen availability in the seawater in winter determines this seasonal pattern (Chapman and Craigie 1977; Sjøtun and Gunnarsson 1995). Kelp growth slows when the spring phytoplankton deplete nutrients in the water. In estuaries and other environments where nitrogen is available throughout the year, kelp growth does not show a winter peak (Anderson, Cardinal, and Larochelle 1981; Gagné, Mann, and Chapman 1982). In the absence of dense sea urchin aggregations, grazing directly on the kelps is probably relatively minor, and the kelps instead release large quantities of detritus as their blades wear away. Suspended particles from the erosion of the kelps and other algae are an important food source for suspension-feeding animals, such as mussels and barnacles (Duggins, Simenstad, and Estes 1989). Algal detritus also settles to the bottom or sometimes accumulates on shore where other animals (detritivores) consume it or it rots.

The kelps (*Laminaria*) and mussels (*Modiolus*) compete for space in the subtidal region (Witman 1987). The kelps dominate in shallow water, where by attaching and growing on mussel shells, they increase the likelihood that the animals will be dislodged during storms. In deeper water, grazing by *Strongylocentrotus* helps to keep mussel shells free of the kelps.

Pacific Rocky Shores

Much of the Pacific coast of North America is a rocky shoreline with a narrow continental shelf and is enriched by upwelling. **Mixed tides,** in which successive tides differ greatly in their range, influence intertidal communities (see fig. 9.2*c*). Overall, the macroalgal flora is richer than on the Atlantic Coast.

Table 9.5 summarizes a typical pattern of algal distribution. In the intertidal region, *Pelvetia* grows high on shore, *Mastocarpus* and *Gigartina* in the midzone, and a variety of foliose red algae lower on shore. The sea grass *Phyllospadix* ("Surf Grass") is also common in the intertidal region.

Low in the intertidal region, in areas receiving intense surf, the distinctive kelp *Postelsia* ("Sea Palm"), with a tough stipe and flexible blades (see fig. 4.39*c*), grows in dense stands. *Postelsia* competes with mussels, barnacles, and other algae for space. Its success in high surf results from spore settlement in the vicinity of parent thalli. Instead of dispersing in the water, the spores slide down grooves in the parent's blades and attach nearby. The germlings may overgrow competitors or attach to the shells of mussels and barnacles. Waves more readily rip away shells with algae, resulting in open patches on the shore for further spore settlement (Dayton 1973).

Sousa (1979) described intertidal succession on cobble beaches. Disturbance, usually by winter storms, clears rock surfaces for diatoms and then *Ulva* to colonize. As a result of grazing and physical stress, red algae later replace *Ulva*. A final stage dominated by *Gigartina canaliculata* sometimes follows these mixed reds. Thus, the middle successional stage is the most diverse, and disturbance is important in maintaining diversity.

The subtidal region along the Pacific coast has a greater diversity of the kelps than that on the East Coast (table 9.5). Common kelps include *Egregia* (see fig. 4.39*a*) in shallow water, *Macrocystis* (see fig. 4.41), and *Nereocystis* (see fig. 4.39*b*), as well as *Laminaria*. The giant kelp *Macrocystis* creates extensive submarine forests at depths of 5–20 meters, which are habitats for a rich assortment of understory algae and animals. In *Voyage of the Beagle,* Charles Darwin wrote of the *Macrocystis* beds in the South Pacific:

> There is one marine production, which from its importance is worthy of a particular history. It is the kelp, or *Macrocystis pyrifera.* . . . I know few things more surprising than to see this plant growing and flourishing amidst those great breakers of the western ocean, which no mass of rock, let it be ever so hard, can resist. . . . The beds of this sea-weed, even when of not great breadth, make excellent natural floating breakwaters. It is quite curious to see, in an exposed harbour, how soon the waves from the open sea, as they travel through the straggling stems, sink in height, and pass into smooth water.
>
> The number of living creatures of all Orders, whose existence intimately depends on the kelp, is wonderful. A great volume might be written, describing the inhabitants of one of these beds of sea-weed. . . . I can only compare these great forests of the southern hemisphere, with the terrestrial ones in the inter-tropical regions. Yet if in any country a forest was destroyed, I do not believe nearly so many species of animals would perish as would here, from the destruction of the kelp. (Darwin 1962)

Table 9.5   *Common Macroalgae in Central California*

Upper Intertidal
  Pelvetia fastigiata (P)
  Endocladia muricata (R)

Mid-intertidal
  Mastocarpus papillatus (R)
  Gigartina canaliculata (R)
  Crustose brown and red algae (on movable rocks)

Low Intertidal
  Phyllospadix (seagrass)
  Prionitis lanceolata (R)
  Iridaea cordata (R)
  Rhodoglossum affine (R)
  Mastocarpus papillatus (R)

Shallow Subtidal
  Egregia menziesii (P)
  Cystoseira osmundacea (P)

Macrocystis pyrifera zone (5–20 meters)
  Cystoseira osmundacea (P)
  Nereocystis luetkeana (P)
  Pterygophora californica (P)
  Laminaria farlowii (P)
  Understory of foliose and crustose red algae

Deep Subtidal
  Agarum fimbriatum (P)
  Laminaria farlowii (P)
  Foliose and crustose red algae

Sources: After Foster and Schiel (1985); Murray and Horn (1989).
P = brown alga; R = red alga.

Along the California coast, the kelps are harvested for their phycocolloid alginic acid. At times, herbivorous sea urchins have substantially damaged commercially valuable beds, and various methods to control urchins, such as the application of quicklime (calcium oxide), have been employed. Sea otters forage in kelp beds for bottom invertebrates, including urchins. When otters were much more abundant, they may have controlled urchins numbers and prevented significant kelp damage. In the nineteenth century, however, otters were hunted for their fur almost to extinction, and now they survive in abundance only in Alaska. The relationship among otters, urchins, and kelp abundance has been studied in the Aleutian Islands and Gulf of Alaska (Duggins 1980; Estes and Duggins 1995). On islands where otters are common, kelp beds of *Laminaria* are well developed, and urchins are scarce. In the absence of otters at other sites, algae are less abundant, and urchins are common. Because of the otter's importance in regulating these communities, ecologists describe it as a keystone predator. Related studies have shown the importance of kelp detritus in supporting the associated community of animals (Duggins, Simenstad, and Estes 1989). Thus, healthy kelp beds are the basis for a rich subtidal community, which is lacking when urchins are abundant.

Summary

1. Benthic algae growing in shallow water along coasts include both microalgae and macroalgae. The marine macroalgae can be divided into six functional-morphological types. The morphologically less complex filaments and sheets have high photosynthetic rates, while the more complex types show greater toughness and resistance to grazing. The firm substrates on coral reefs and rocky shores support some of the richest algal communities.

2. Reef-forming corals are found in tropical oceans. Even though the seawater is low in nutrients, algal diversity and production are high. Reef algae include symbionts associated with corals and other invertebrates, turf algae, encrusting algae, and larger erect algae. Herbivorous fishes and invertebrates are common on reefs. Some algae are restricted to parts of the reef where herbivory is low or compensate for grazing losses with rapid growth. Other algae depend on tough coverings or chemicals for protection against herbivores.

3. Soft bottoms of sand or mud are poor substrates for algae, especially where wave disturbance is common. Vascular plants, such as seagrasses, mangroves, and salt-marsh plants, usually dominate. They may benefit algal growth by stabilizing the substrate with their roots and providing submerged surfaces for the growth of epiphytic algae.

4. Rocky shores are divided into an upper spray zone, an intertidal region, and a subtidal region. Inhabitants of the spray zone and intertidal surfaces that drain during low tide must survive desiccation and high solar irradiance during exposure to air, and be firmly attached to the substrate to withstand the wave action. Many species are further restricted by competitive interactions with other attached organisms and by herbivore activities. Many macroalgae have adapted successfully to living on emergent surfaces.

5. Although tidepools retain water during low tide, their environmental conditions, such as salinity, temperature, and pH, fluctuate considerably. Biological interactions also regulate tidepool algae.

6. In the subtidal regions of rocky shores, light availability and grazing are important in controlling algal distributions. Kelp beds in shallow water often have high productivity. In deeper water, algae may increase their pigment levels or have special pigments to absorb the available wavelengths. On many shores, sea urchins are important subtidal herbivores.

Further Reading

Chapman, A. R. O. 1995. Functional ecology of fucoid algae: Twenty-three years of progress. *Phycologia* 34: 1–32.

Chapman, A. R. O., and C. R. Johnson. 1990. Disturbance and organization of macroalgal assemblages in the northwest Atlantic. *Hydrobiologia* 192: 77–121.

Duffy, J. E., and M. E. Hay. 1990. Seaweed adaptations to herbivory. *BioScience* 40: 368–75.

Lobban, C. S., and P. J. Harrison. 1994. *Seaweed Ecology and Physiology.* Cambridge University Press.

Lüning, K. 1990. *Seaweeds: Their Environment, Biogeography, Ecophysiology.* Wiley.

Mathieson, A. C., and P. H. Nienhuis, eds. 1991. *Intertidal and Littoral Ecosystems.* Elsevier.

Saffo, M. B. 1987. New light on seaweeds. *BioScience* 37: 654–64.

Santelices, B. 1990. Patterns of reproduction, dispersal, and recruitment in seaweeds. *Oceanography and Marine Biology, Annual Review* 28: 177–276.

A P P E N D I X

Useful References for Studies with Algae

Methods

Gantt, E., ed. 1980. *Handbook of Phycological Methods: Developmental and Cytological Methods*. Cambridge University Press.

Hellebust, J. A., and J. S. Craigie, eds. 1978. *Handbook of Phycological Methods: Physiological and Biochemical Methods*. Cambridge University Press.

Littler, M. M., and D. S. Littler, eds. 1985. *Handbook of Phycological Methods: Ecological Field Methods: Macroalgae*. Cambridge University Press.

Lobban, C. S., D. J. Chapman, and B. P. Kremer, eds. 1988. *Experimental Phycology*. Cambridge University Press.

Stein, J. R., ed. 1973. *Handbook of Phycological Methods: Culture Methods and Growth Measurements*. Cambridge University Press.

Identification of Marine Macroalgae

Abbott, I. A., and E. Y. Dawson. 1978. *How to Know the Seaweeds*. Wm. C. Brown Publishers.

Abbott, I. A., and G. J. Hollenberg. 1976. *Marine Algae of California*. Stanford University Press.

Bird, C. J., and J. L. McLachlan. 1992. *Seaweed Flora of the Maritimes 1. Rhodophyta—The Red Algae*. Biopress.

Burrows, E. M. 1991. *Seaweeds of the British Isles*. Vol. 2, *Chlorophyta*. Natural History Museum, London.

Christensen, T. 1987. *Seaweeds of the British Isles*. Vol. 4, *Tribophyceae (Xanthophyceae)*. Natural History Museum, London.

Dawes, C. J. 1974. *Marine Algae of the West Coast of Florida*. University of Miami Press.

Dixon, P. S., and L. M. Irvine. 1977. *Seaweeds of the British Isles*. Vol. 1, *Rhodophyta*: part 1, *Introduction, Nemaliales, Gigartinales*. Natural History Museum, London.

Edwards, P. 1976. *Illustrated Guide to the Seaweeds and Sea Grasses in the Vicinity of Port Aransas, Texas*. University of Texas Press.

Fletcher, R. L. 1987. *Seaweeds of the British Isles*. Vol. 3, *Fucophyceae (Phaeophyceae)*, part 1. Natural History Museum, London.

Gabrielson, P. W., R. F. Scagel, and T. B. Widdowson. 1989. *Keys to the Benthic Marine Algae and Seagrasses of British Columbia, Southeast Alaska, Washington and Oregon*. Phycological contribution no. 4. Department of Botany, University of British Columbia.

Hiscock, S. 1986. *A Field Key to the British Red Seaweeds*. Field Studies Council (England).

Humm, H. J. 1979. *The Marine Algae of Virginia*. University Press of Virginia.

Humm, H. J., and S. R. Wicks. 1980. *Introduction and Guide to the Marine Bluegreen Algae*. Wiley.

Irvine, L. M. 1983. *Seaweeds of the British Isles.* Vol. 1, *Rhodophyta;* part 2A, *Cryptonemiales (sensu stricto), Palmariales, Rhodymeniales.* Natural History Museum, London.

Irvine, L. M., and Y. M. Chamberlain. 1994. *Seaweeds of the British Isles.* Vol. 1, *Rhodophyta;* part 2B, *Corallinales, Hildenbrandiales.* Natural History Museum, London.

Kapraun, D. F. 1980. *An Illustrated Guide to the Benthic Marine Algae of Coastal North Carolina. I. Rhodophyta.* University of North Carolina Press.

Kapraun, D. F. 1984. *An Illustrated Guide to the Benthic Marine Algae of Coastal North Carolina. II. Chlorophyta and Phaeophyta.* J. Cramer.

Littler, D. S., M. M. Littler, K. E. Bucher, and J. N. Norris. 1989. *Marine Plants of the Caribbean.* Smithsonian Institution Press.

Maggs, C. A., and M. H. Hommersand. 1993. *Seaweeds of the British Isles.* Vol. 1, *Rhodophyta;* part 3A, *Ceramiales.* Natural History Museum, London.

Scagel, R. F., P. W. Gabrielson, D. J. Garbary, L. Golden, M. W. Hawkes, S. C. Lindstrom, J. C. Oliveira, and T. B. Widdowson. 1989. *A Synopsis of the Benthic Marine Algae of British Columbia, Southeast Alaska, Washington and Oregon.* Phycological contribution no. 3. Department of Botany, University of British Columbia.

Schneider, C. W., and R. B. Searles. 1991. *Seaweeds of the Southeastern United States.* Duke University Press.

South, G. R. 1984. A checklist of marine algae of eastern Canada, second revision. *Canadian Journal of Botany* 62:680–704.

South, G. R., and R. G. Hooper. 1980. *A Catalogue and Atlas of the Benthic Marine Algae of the Island of Newfoundland.* Occasional papers in biology, no. 3. Memorial University of Newfoundland (Canada).

Villalard-Bohnsack, M. 1995. *Illustrated Key to the Seaweeds of New England.* The Rhode Island Natural History Survey.

Wynne, M. J. 1986. A checklist of benthic marine algae of the tropical and subtropical western Atlantic. *Canadian Journal of Botany* 64: 2239–81.

Identification of Marine Phytoplankton

Dodge, J. D. 1982. *Marine Dinoflagellates of the British Isles.* Her Majesty's Stationery Office (England).

Hendey, N. I. 1964. *An Introductory Account of the Smaller Algae of British Coastal Waters.* Otto Koeltz.

Sykes, J. B. 1981. *An Illustrated Guide to the Diatoms of British Coastal Plankton.* Field Studies Council (England).

Tomas, C. R., ed. 1993. *Marine Phytoplankton–A Guide to Naked Flagellates and Coccolithophorids.* Academic Press.

Wood, E. J.F. 1968. *Dinoflagellates of the Caribbean Sea and Adjacent Areas.* University of Miami Press.

Identification of Freshwater Algae

Dillard, G.E. 1989–1991. *Freshwater Algae of the Southeastern United States.* Parts 1–5. J. Cramer.

Knutson, K. M., and J. C. Sterk. 1996. *Computer-Assisted Algal Recognition.* Wm. C. Brown Publishers. (computer software)

Patrick, R., and C. W. Reimer. 1966, 1975. *The Diatoms of the United States.* Vols. 1, 2. Academy of Natural Sciences, Philadelphia.

Prescott, G. W. 1962. *Algae of the Western Great Lakes Region.* Wm. C. Brown Publishers.

Prescott, G. W. 1978. *How to Know the Freshwater Algae.* Wm. C. Brown Publishers.

Prescott, G. W., H. T. Croasdale, W. C. Vinyard, and C. E. de M. Bicudo. 1975–1983. *A Synopsis of North American Desmids.* University of Nebraska Press.

Sheath, R.G., and M. M. Harlin, eds. 1988. Freshwater and Marine Plants of Rhode Island. Kendall/Hunt.

Whitford, L. A., and G.J. Schumacher. 1973. *A Manual of Fresh-Water Algae.* Sparks Press.

accessory pigment photosynthetic pigment that absorbs light and transfers the energy to a reaction center of chlorophyll *a*.

aerial growing exposed to the air.

agar mucopolysaccharide in the walls of some red algae; polymer of galactose.

aggregation group of cells, often surrounded by mucilage, in which the cells function as individuals.

akinete resistant cell formed when a vegetative cell transforms into a thick-walled dormant cell.

alga a prokaryotic or eukaryotic organism that is photosynthetic, produces oxygen, and lacks the complex reproductive structures with sterile tissues that bryophytes and vascular plants form.

alginic acid (=algin) mucopoly-saccharide in the walls of brown algae; polymer of mannuronic and guluronic acids.

allelochemical chemical that members of one species produce to inhibit growth of other species.

alternation of generations life cycle involving an alternation between a haploid, gamete-producing vegetative phase (gametophyte) and a diploid, spore-producing vegetative phase (sporophyte).

androspore in the Oedogoniales, spore that produces a reduced male filament.

anisogamy sexual fusion between flagellated gametes of distinctly different sizes.

anoxia condition where oxygen is absent.

antheridium reproductive structure in which sperm are produced.

antherozoid male gamete.

aphotic zone region in a body of water where light is insufficient for photosynthesis.

aplanospore nonflagellated spore (asexual reproduction).

asexual reproduction formation of reproductive cells that develop directly into new individuals (without sexual fusion).

assimilator filament composed of photosynthetic cells.

atoll ring-shaped reef surrounding a lagoon.

autocolony newly produced coenobium with the same form as the parent colony.

autospore nonflagellated spore that is similar in form but a miniature of its parent cell.

auxiliary cell in floridean red algae, cell that is not in the carpogonial branch but receives the zygote nucleus and bears a carposporophyte.

auxospore in diatoms, cell produced by enlargement of a zygote (or, occasionally, a vegetative cell).

auxotrophy requirement for one or more organic growth factors or vitamins.

barrier reef reef separated from land by a deep channel or lagoon.

benthos community of organisms associated with submerged substrates; sometimes used in a more restricted sense to refer to organisms living on the bottom of a body of water.

bloom dense growth of microscopic algae that noticeably discolors the water.

brevetoxin neurotoxin that some dinoflagellates produce and that causes neurotoxic shellfish poisoning and fish-kills.

C₄ metabolism uptake of carbon dioxide or bicarbonate to produce four-carbon acids.

carotenoid yellow, brown, or red pigment with various functions (accessory photosynthetic pigment, photoprotection) (see fig. 1.3).

carpogonium cell in the red algae that contains the egg (equivalent to an oogonium).

carpospore a spore formed by the carposporophyte stage of red algae.

carposporophyte carpospore-producing phase of some red algae that forms on the female gametophyte; composed of gonimoblast filaments.

carrageenan mucopolysaccharide in the walls of some red algae; sulfated polymer of galactose.

cell plate flattened vesicle in which new wall material is deposited during cell division; normally initiated at the center of a cell and enlarged to the periphery by addition of vesicles from Golgi bodies.

chantransia stage juvenile filamentous phase in some floridean red algae.

chemocline see *halocline*.

chlorophyll green photosynthetic pigment (see fig. 1.2).

chloroplast endoplasmic reticulum (CER) endoplasmic reticulum surrounding the chloroplast; usually continuous with the nuclear envelope.

chrysolaminarin reserve polysaccharide composed of β-1,3-linked polymers of glucose (see fig. 1.5*b*).

ciguatera poisoning from eating contaminated fishes; caused by ciguatoxin and other toxins that some tropical dinoflagellates produce.

ciguatoxin a toxin produced by some dinoflagellates; responsible for ciguatera.

clone population of genetically identical cells produced by asexual reproduction.

closed spindle mitotic spindle that the nuclear envelope encloses during mitosis.

coccoid spherical.

coccolith scale with calcium carbonate; characteristic of coccolithophores (haptophytes).

codiolum stage unicellular diploid stage in some green algae formed by enlargement of a zygote, which remains photosynthetic, often elongate, or club-shaped.

coenobium colony in which the number of cells is fixed at the time of formation and no further cells are added; usually, the cells are also in a distinctive arrangement.

coenocyte multinucleate cell.

colony group of cells that function as a unit.

commensalism relationship between two species in which one species benefits and the other is not affected.

compensation depth depth in a body of water where photosynthetic production equals respiration for a species.

conceptacle cavity that contains reproductive structures.

conchocelis phase microscopic, filamentous phase in the life cycles of some bangean red algae.

conjugation fusion of nonflagellated, amoeboid gametes; characteristic of members of the order Zygnematales.

connecting fiber part of the cytoskeleton of a flagellated cell; proteinaceous fiber connecting basal bodies.

coral bleaching color loss by corals resulting from loss of their zooxanthellae or loss of zooxanthellae pigmentation.

cortex outer layer of a multicellular thallus; usually composed of photosynthetic cells.

cortication outer covering of small cells; usually produced by special corticating branches growing over the axial cells of a branch.

cruciate root system arrangement of microtubules in four bundles radiating from the basal bodies.

cryptostomate (=cryptostoma) sterile cavity that some brown algae form.

cyanelle endosymbiotic cyanobacterium functioning as a chloroplast in a eukaryotic cell.

cyst thick-walled cell.

cystocarp compact carposporophyte that gametophytic tissue (pericarp) surrounds in floridean red algae.

desmid unicellular green alga in the order Zygnematales.

desmokont type of dinoflagellate with two anterior flagella.

detritivore   an animal that feeds on detritus.

detritus   nonliving particles of organic material.

diatomaceous earth (=diatomite)   deposit of diatom frustules.

dichotomous   splitting into two equal parts (as a "Y").

dimictic lake   lake with two periods of vertical mixing—in the spring and fall.

dinokaryotic   nuclear condition in the dinoflagellates with condensed chromosomes that lack histone proteins.

dinokont   dinoflagellate with an encircling flagellum and a posteriorly directed flagellum.

diplobiontic   having two vegetative phases (gametophyte and sporophyte) in an alternation of generations.

diplohaplont   life cycle with an alternation of haploid and diploid generations and sporic meiosis.

diplont   life cycle with a diploid vegetative phase and gametic meiosis.

diurnal tides   tidal pattern with one high and one low tide each tidal day (24 hours, 50 minutes).

domoic acid   neurotoxin that diatoms in the genus *Pseudonitzschia* produce and that causes amnesic shellfish poisoning.

dystrophy   condition in which water has a high content of organic material (humus) and a brown color.

edaphic   referring to soil.

ejectosome   in the cryptomonads, a coiled ribbon capable of discharge from its vesicle.

embryo   young plant (sporophyte) that develops within a protective covering the parent plant produces; formed by the bryophytes and vascular plants but not by algae.

emergent surface   surface that is exposed to the air, especially with reference to an intertidal surface that is submerged during high tide and exposed to the air during low tide.

encrusting   growing appressed to the substrate without upright structures.

endolithic   living within rocks.

endophyte   alga living inside another organism.

endosymbiont   symbiont living within the cells of its host.

epibiont   organism that lives on the surface of another organism.

epicone   anterior part of a dinoflagellate cell.

epilimnion   upper layer of warm water in a stratified lake.

epiphyte   alga growing on another alga or plant.

epithallium   layer of cells produced above or external to the meristem in the thallus of coralline red algae.

epitheca   epivalve and associated girdle bands of a diatom frustule.

epivalve   larger or outer valve of a diatom frustule.

estuary   partially enclosed basin in which freshwater and seawater mix.

eukaryote   organism with cells containing a nuclear region (defined by a nuclear envelope) and organelles, such as chloroplasts, mitochondria, Golgi bodies, and endoplasmic reticulum.

euphotic zone   see *photic zone*.

eutrophication   progressive enrichment of a body of water, either from natural or artificial causes.

eutrophy   nutrient-rich condition.

false branching   divergence of two parts of a trichome adjacent to a break in some cyanobacteria.

fibrous root   part of cytoskeleton that is composed of protein and that normally extends from basal bodies into cell.

filament   cells in a linear arrangement; filaments may be branched or unbranched; in the cyanobacteria, a filament is the trichome and its surrounding sheath.

flagellar apparatus   flagella and associated structures, which include basal bodies and cytoskeletal elements radiating from the basal bodies (see table 1.5).

fringing reef   reef bordering land.

frustule   siliceous wall of diatoms.

furrow   invaginating cell membrane during cell division; usually occurs at the cell equator.

fusiform   spindle-shaped, tapering at both ends.

gametangium   structure that produces gametes.

gametophyte   gamete-producing phase in an alternation of generations; usually haploid.

gas vesicle   vesicle containing gas in some cyanobacteria.

girdle band (=cingulum)   siliceous band between the valves of a diatom frustule.

girdle lamella thylakoid encircling the periphery of a chloroplast.

glaucophyte eukaryotic cell containing endosymbiotic cyanobacteria (cyanelle).

globule male reproductive structure of charalean algae.

gonidium asexual reproductive cell in *Volvox* that produces a daughter colony.

gonimoblast filament filament composing the carposporophyte of the red algae.

hair elongate, nonpigmented cell or series of cells that often taper to a point.

half-saturation constant concentration of a nutrient at half the maximum growth (or uptake) rate for a species.

halocline region of rapid salinity change with increasing depth in a body of water.

haplobiontic having only one vegetative phase (either haploid or diploid).

haplont life cycle with a haploid vegetative phase and zygotic meiosis.

hapteron fingerlike outgrowth (multicellular) composing a holdfast; characteristic of laminarian brown algae.

haptonema threadlike appendage of haptophytes that functions in food capture and, possibly, attachment; superficially resembles a flagellum.

heterocyst specialized cell the cyanobacteria form for nitrogen fixation.

heterokaryon cell containing genetically different nuclei.

heterokont cell with flagella of different lengths and usually of different types (one smooth flagellum, one flagellum with mastigonemes).

heterokontous flagellation condition in which a cell bears a smooth flagellum and a flagellum with mastigonemes; characteristic of the division Chromophyta.

heteromorphic generations vegetative phases in an alternation of generations that are distinctly different in form.

heteroplasty presence of different kinds of plastids (chloroplasts and leucoplasts) in a cell.

heterotrichy filamentous thallus showing differentiation into basal branches growing on the substrate and erect, usually more openly branching filaments (see fig. 3.26).

heterotrophy ability to use organic compounds in the dark as energy sources.

holdfast modified basal region for substrate attachment; may be multicellular or unicellular (see table 1.7).

homokaryon cell with genetically similar nuclei.

hormogonium short segment of cells that separates from the end of a trichome and initiates a new trichome in some cyanobacteria.

hydrocolloid see *phycocolloid*.

hypnospore thick-walled spore.

hypnozygote thick-walled, dormant zygote.

hypocone posterior part of a dinoflagellate cell.

hypolimnion lower layer of cold water in a stratified lake.

hypothallium basal layer in the thallus of coralline red algae.

hypotheca hypovalve and associated girdle bands of a diatom frustule.

hypovalve smaller or inner valve of a diatom frustule.

ice algae algae associated with sea ice.

intercalary located within the thallus (as opposed to terminal).

intertidal region part of a shore exposed to the air during low tide and submerged during high tide.

isogamy sexual fusion between flagellated gametes that are morphologically similar and usually the same size.

isokont cell with flagella of equal length and of the same type.

isomorphic generations vegetative phases in an alternation of generations that are similar in form.

laminarin reserve polysaccharide found in brown algae and composed of β-1,3-linked polymers of glucose (see fig. 1.5*b*).

lentic referring to standing water, as in a lake or pond.

leucosin see *chrysolaminarin*.

lichen fungus that contains symbiotic algae.

lorica cell covering that differs in form and composition from a wall and that is often larger than the cell it encloses.

lotic referring to flowing water, as in streams and rivers.

macrandry condition in the Oedogoniales in which antheridia are produced in normal-sized filaments.

macroalga an alga with a thallus that is large enough to be visible without a viewing aid.

macrophyte plant visible without a microscope; includes larger algae (seaweeds, macroalgae), vascular plants, and bryophytes.

macrothallus macroscopic vegetative phase in an algal life cycle.

mangrove woody plant (tree or shrub) that grows partially submerged in seawater.

mannan polymer of mannose.

mannitol sugar alcohol in the brown algae.

mastigoneme stiff, tubular hair on some flagella.

medulla inner layers of a multicellular thallus composed of nonpigmented cells that usually function in storage.

meristem (=meristematic region) specific region where cells divide.

meristoderm meristematic region on the thallus surface; characteristic of the brown algae in the Laminariales.

meromictic lake lake that never undergoes complete vertical mixing.

meroplankton organisms that spend only part of their existence as members of the plankton.

mesokaryotic see *dinokaryotic.*

microplankton members of the plankton that range from 20–200 micrometers in diameter.

microthallus microscopic vegetative phase in the life cycle of an alga.

microtubule part of the cytoskeleton and flagella of a cell; tubular rods composed of the protein tubulin.

mixed layer surface layer of the oceans above the thermocline.

mixed tides semidiurnal tidal pattern in which heights of successive high (and low) tides differ greatly.

mixotrophy ability to obtain energy by photosynthesis and by uptake of organic material (dissolved or particulate) from the environment.

monomictic lake lake with one annual period of vertical mixing.

monospore spore formed singly from the contents of a cell.

mucocyst (muciferous body) vesicle containing mucilage.

multilayered structure structure associated with the microtubular roots in some green algae.

multiseriate (=pluiseriate) filament filament with cells arranged in more than one series or row.

mutualism association between two species in which both benefit.

mycobiont fungus composing a lichen.

nannandry condition in some members of Oedogoniales in which reduced (dwarf) male filaments form.

nanoplankton members of the plankton that range from 2–20 micrometers in diameter and that normally can pass through the mesh of plankton nets.

neap tide tide with the least range between the high- and low-tide levels.

net plankton the larger component of the plankton that can be retained in plankton nets.

nitrogen fixation conversion of nitrogen (N_2) to ammonium (NH_4^+) by some cyanobacteria.

nucleomorph DNA-containing structure in the periplastidal space of the cryptomonads.

nucule female reproductive structure of charalean algae.

okadaic acid toxin that some dinoflagellates form and that produces diarrhetic shellfish poisoning.

oligotrophy low-nutrient condition in a body of water.

oogamy sexual fusion between a flagellated gamete (sperm, antherozoid) and a nonflagellated gamete (egg).

oogonium cell that produces one or more eggs.

palmelloid stage aggregation of nonflagellated cells in mucilage.

paraflagellar body swelling near the base of a flagellum.

paralytic shellfish poisoning illness produced from eating bivalves containing saxitoxin from certain dinoflagellates.

paramylon reserve polysaccharide in the euglenophytes; composed of β-1,3-linked units of glucose.

paraphysis sterile branch associated with a reproductive structure.

parasitism association between two species in which the parasite harms the host.

parenchymatous construction mass of cells produced by divisions in three dimensions to form a tube, blade, or cylinder.

pellicle in the euglenophytes, the outer covering of proteinaceous strips immediately beneath the cell membrane; in the dinoflagellates, a fibrous layer underlying the theca.

perialgal vesicle in a host cell, a vesicle containing a symbiotic alga.

pericarp layer of cells the female gametophyte forms around a carposporophyte of some floridean red algae.

pericentral cell one of a ring of cells cut off from a central axial cell.

periplast outer covering of the cryptomonads composed of organic plates associated with the cell membrane.

periplastidal space region around the chloroplast that chloroplast endoplasmic reticulum encloses.

perithallium layer of cells produced below or inward by the meristem in the thallus of coralline red algae.

phagotrophy ingestion of particles by a cell.

pheromone chemical that one member of a species secretes that influences other members of the same species; includes sexual attractants.

photic zone upper layer of a body of water where light is sufficient for net photosynthesis (photosynthesis exceeds respiration).

photoheterotrophy ability to take up organic materials in the light for use as energy sources.

photorespiration light-dependent uptake of oxygen and formation of glycolate.

photosynthetically active radiation (PAR) part of the electromagnetic spectrum used in photosynthesis (400–700 nanometers).

phototaxis movement in response to the direction of light (a positive phototaxis is movement toward the light).

phragmoplast structure composed of microtubules oriented perpendicular to the plane of cell division.

phycobilin water-soluble pigment found in the cyanobacteria, red algae, and cryptomonads; covalently linked to a protein to form a phycobiliprotein; phycocyanobilin (blue) and phycoerythrobilin (red) (see fig. 1.4, table 2.2).

phycobiliprotein see *phycobilins.*

phycobilisome granular structure composed of phycobiliproteins on the outer surface of thylakoids of the cyanobacteria and red algae.

phycobiont algal component of a lichen.

phycocolloid mucilaginous material extracted from the walls of some algae, especially some brown and red algae, and used as thickeners and gels in some commercial products.

phycology study of algae.

phycoplast assemblage of microtubules parallel to the plane of cell division.

physode vesicle in the brown algae that contains polyphenolic compounds.

phytobenthos algal component of a benthic community.

phytoflagellate flagellate that contains chloroplasts; a flagellated alga.

phytoplankton community of floating algae.

picoplankton members of the plankton that are less than 2 micrometers in diameter.

plakea multicellular developmental stage in volvocine algae that inverts to bring its cells into proper orientation.

plankton assemblage of organisms that float in water, unassociated with a substrate.

planozygote flagellated zygote.

plethysmothallus juvenile filamentous stage in the sporophytic generation of some brown algae.

plurilocular sporangium or **gametangium** multicellular reproductive structure that the brown algae form.

pluriseriate see *multiseriate.*

pneumatocyst inflated region of thallus containing gas.

polysiphonous filamentous construction with regular tiers of cells.

procarp condition in the red algae in which the carpogonial branch and auxiliary cell branch arise from the same supporting cell.

prochlorophyte a prokaryotic alga with chlorophylls *a* and *b* as its photosynthetic pigments (lacks phycobiliproteins).

prokaryote organism with cells that lack a nuclear region surrounded by a nuclear envelope, and that also do not have complex organelles, such as mitochondria, chloroplasts, endoplasmic reticulum, and Golgi bodies.

propagule multicellular reproductive structure formed for vegetative propagation.

pseudoflagellum (=pseudocilium) rigid extension from the cells of some tetrasporalean algae; superficially resembles a flagellum but does not provide motility.

pseudoparenchymatous construction filamentous mass in which individual filaments are difficult to distinguish.

pseudoraphe clear area that extends the length of the valve face of some pennate diatoms.

pusule system of vesicles in dinoflagellate cells that is involved with osmoregulation, secretion, and molecular uptake.

pycnocline region of rapid density change with depth in a body of water.

pyrenoid distinct structure in some chloroplasts that is composed of ribulose bisphosphate carboxylase and other proteins; may be embedded in the chloroplast or protrude from it; starch may accumulate on its surface.

pyriform pear-shaped.

raphe slit that extends the length of the valve of some pennate diatoms and assists with gliding movement.

receptacle in fucoid brown algae, specialized region of a thallus that produces reproductive structures.

red tide dense growth of microscopic algae, usually dinoflagellates, that discolors water red or brown.

replicate end wall fold in the end wall of some zygnematalean algae that aids in cell separation.

reticulate arranged in a net or having the appearance of a net.

rhizoid branch composed of nonpigmented cells; usually functions in attachment.

rhizoplast fibrous root extending from the basal bodies of a cell as part of the cytoskeleton.

salt marsh a tidal marsh that is partially flooded with seawater.

saturation constant see *half-saturation constant.*

saxitoxin toxin that some dinoflagellates produce and that causes paralytic shellfish poisoning.

scale platelike organic structure on the surface of some cells.

seagrass plant (not an alga) that lives completely submerged in seawater (herbaceous monocot but not a true grass).

seaweed macrophytic marine alga; loosely used to describe any alga visible without magnification.

semicell half of a cell of a desmid; a narrow isthmus joins two semicells.

semidiurnal tides tidal pattern of normally two high and two low tides each day (12.5-hour periodicity).

seta rigid extension from a cell.

sexual reproduction reproductive process involving gametic fusion to produce a zygote, which may develop into a new alga or form asexual reproductive cells.

sieve cell specialized cell found in some brown algae for the transport of organic material.

silica deposition vesicle vesicle in which the polymerization of silica forms a siliceous structure.

siphonous tubular, or without division into cellular units by cross walls (coenocytic).

snow algae algae that grow on the surface of snow.

somatic meiosis meiosis during vegetative growth of the thallus and not associated with formation of a reproductive stage.

sorus cluster of reproductive structures.

spermatium male gamete (nonflagellated) of the red algae.

spine thin extension of a cell wall.

sporangium structure in which spores form.

sporophyll specialized blade that bears sporangia.

sporophyte spore-producing phase in an alternation of generations; usually diploid.

spray zone part of the shore above the high-tide level that ocean spray wets but that rarely, if ever, is submerged.

spring tide tide with the greatest range between high- and low-tide levels.

starch polysaccharide composed of α-1,4-linked units of glucose; reserve material in some algal groups (see fig. 1.5a).

statocyst (=statospore) resistant stage that members of the Chrysophyceae and Synurophyceae form.

stephanokontous condition in which a cell has a ring of flagella.

stipe stemlike region of a thallus.

stratification (density) layers of water with different densities; usually, a pycnocline (thermocline) separates warmer, low-density water from colder, higher density water.

stromatolite columnar structure formed of interwoven filaments of cyanobacteria, other bacteria, and sedimentary material.

subtidal region part of the shore that remains submerged during low tides.

sulcus longitudinal depression in the hypocone of the dinoflagellates.

suspension-feeding a method of feeding by the collection of suspended particles in the water (also called filter-feeding).

terete composed of cylindrical branches.

tetraspore member of a cluster of four spores produced after meiosis.

tetrasporophyte tetraspore-producing phase in the life cycle of the red algae.

thallus body of an alga lacking true roots, stems, and leaves.

theca in the Chlorophyta, a continuous cell covering of fused scales; in the dinoflagellates, the outer covering of a layer of platelike vesicles.

thermocline region of rapid temperature change with depth in a body of water.

thylakoid flattened vesicle with photosynthetic pigments in its membrane; found in the cytoplasm of prokaryotic algae and in the chloroplasts of eukaryotic algae.

tidepool in the intertidal region, a depression in the substrate that retains water during low tide and is flushed with seawater during high tide.

trichoblast series of nonpigmented cells that form a tapering branch or hair in the red algae.

trichocyst vesicle in dinoflagellate cells that contains a crystalline rod that can be discharged.

trichogyne carpogonium extension to which spermatia attach in the red algae.

trichome in the cyanobacteria, the series of cells that compose a filament.

trichothallic meristem meristematic region at the base of a hair.

turf algae filamentous and other simple macroalgae, often appearing as a fuzzy growth on submerged surfaces.

unilocular sporangium unicellular sporangium in which meiosis occurs in the brown algae.

uniseriate filament filament with cells arranged in a single row or series.

upwelling vertical movement of deeper, often nutrient-rich water to the surface.

utricle enlarged branch tip containing chloroplasts in some caulerpalean algae.

valve one or two overlapping units composing the frustule of the diatoms.

vegetative cell division cell division that adds new cells to a thallus (rather than forming a reproductive cell).

vegetative growth increase in the size of an alga by new cell production (vegetative divisions) and cell enlargement.

vegetative propagation separation of part of a parent alga to form a new alga; often occurs by breakage without the formation of special reproductive structures.

viscous drag resistance to movement resulting from the interaction (friction) between an object's surface and the surrounding medium.

xylan polymer of xylose.

zoochlorella symbiotic green alga found in the cells of some protozoa and invertebrates.

zoospore flagellated spore (asexual reproduction).

zooxanthella symbiotic dinoflagellate found in the cells of some cnidarians and other invertebrates.

zygospore thick-walled zygote.

REFERENCES

Abbott, I. A. 1988. Food and food products from seaweeds. In C. A. Lembi and J. R. Waaland, eds., *Algae and Human Affairs*, 135–47. Cambridge University Press.

Abrahams, M. V., and L. D. Townsend. 1993. Bioluminescence in dinoflagellates: A test of the burglar alarm hypothesis. *Ecology* 74: 258–60.

Alvarez Cobelas, M., and C. R. García-Morato. 1990. Population dynamics of *Nitzschia gracilis* (Bacillariaceae) in a hypertrophic lake. *British Phycological Journal* 25: 263–73.

Amsler, C. D, D. C. Reed, and M. Neushul. 1992. The microclimate inhabited by macroalgal propagules. *British Phycological Journal* 27: 253–70.

Amsler, C. D., R. J. Rowley, D. R. Laur, L. B. Quetin, and R. M. Ross. 1995. Vertical distribution of Antarctic peninsular macroalgae: Cover, biomass, and species composition. *Phycologia* 34: 424–30.

Andersen, R. A. 1987. Synurophyceae *classis nov.*, a new class of algae. *American Journal of Botany* 74: 337–53.

Anderson, D. M., and P. S. Lobel. 1987. The continuing enigma of ciguatera. *Biological Bulletin* 172: 89–107.

Anderson, D. M., and A. W. White. 1992. Marine biotoxins at the top of the food chain. *Oceanus* 35 (3): 55–61.

Anderson, D. M., A. W. White, and D. G. Baden, eds. 1985. *Toxic Dinoflagellates*. Elsevier.

Anderson, M. R., A. Cardinal, and J. Larochelle. 1981. An alternate growth pattern for *Laminaria longicruris*. *Journal of Phycology* 17: 405–11.

Antia, N. J., P. J. Harrison, and L. Oliveira. 1991. The role of dissolved organic nitrogen in phytoplankton nutrition, cell biology, and ecology. *Phycologia* 30: 1–89.

Apple, M. E., and M. M. Harlin. 1995. Inhibition of tetraspore adhesion in *Champia parvula* (Rhodophyta). *Phycologia* 34: 417–23.

Asmus, R. M., and H. Asmus. 1991. Mussel beds: Limiting or promoting phytoplankton? *Journal of Experimental Marine Biology and Ecology* 148: 215–32.

Avron, M., and A. Ben-Amotz, eds. 1992. *Dunaliella: Physiology, Biochemistry, and Biotechnology*. CRC Press.

Ayers, G. P., J. P. Ivey, and R. W. Gillett. 1991. Coherence between seasonal cycles of dimethyl sulphide, methanesulphonate, and sulphate in marine air. *Nature* 349: 404–6.

Bacon, L. C., and R. L. Vadas. 1991. A model for gamete release in *Ascophyllum nodosum* (Phaeophyta). *Journal of Phycology* 27: 166–73.

Badger, M. R., and G. D. Price. 1994. The role of carbonic anhydrase in photosynthesis. *Annual Review of Plant Physiology and Plant Molecular Biology* 45: 369–92.

Balch, W. M., P. M. Holligan, S. G. Ackleson, and K. J. Voss. 1991. Biological and optical properties of mesoscale coccolithophore blooms in the Gulf of Maine. *Limnology and Oceanography* 36: 629–43.

Baldauf, S. L., J. R. Manhart, and J. D. Palmer. 1990. Different fates of the chloroplast *tuf*A gene following its transfer to the nucleus in green algae. *Proceedings of the National Academy of Sciences USA* 87: 5317–21.

Baldauf, S. L., and J. D. Palmer. 1990. Evolutionary transfer of the chloroplast *tuf*A gene to the nucleus. *Nature* 344: 262–65.

Beech, P. L., R. Wetherbee, and J. D. Pickett-Heaps. 1990. Secretion and deployment of bristles in *Mallomonas splendens* (Synurophyceae). *Journal of Phycology* 26: 112–22.

Behrenfeld, M. J., D. R. S. Lean, and H. Lee. 1995. Ultraviolet-B radiation effects on inorganic nitrogen uptake by natural assemblages of oceanic plankton. *Journal of Phycology* 31: 25–36.

Bell, E. C. 1993. Photosynthetic response to temperature and desiccation of the intertidal alga *Mastocarpus papillatus*. *Marine Biology* 117: 337–46.

Bell, R. A. 1993. Cryptoendolithic algae of hot semiarid lands and deserts. *Journal of Phycology* 29: 133–39.

Berdach, J. T. 1977. In situ preservation of the transverse flagellum of *Peridinium cinctum* (Dinophyceae) for scanning electron microscopy. *Journal of Phycology* 13: 243–51.

Bergman, B., and E. J. Carpenter. 1991. Nitrogenase confined to randomly distributed trichomes in the marine cyanobacterium *Trachodesmium thiebautii*. *Journal of Phycology* 27: 158–165.

Bertness, M. D., P. O. Yund, and A. F. Brown. 1983. Snail grazing and the abundance of algal crusts on a sheltered New England rocky beach. *Journal of Experimental Marine Biology and Ecology* 71: 147–64.

Bhattacharya, D., and L. Medlin. 1995. The phylogeny of plastids: A review based on comparisons of small-subunit ribosomal RNA coding regions. *Journal of Phycology* 31: 489–98.

Bhattacharya, D., L. Medlin, P. O. Wainwright, E. V. Ariztia, C. Bibeau, S. K. Stickel, and M. L. Sogin. 1992. Algae containing chlorophylls *a + c* are paraphyletic: Molecular evolutionary analysis of the Chromophyta. *Evolution* 46: 1801–17.

Bidigare, R. R., M. E. Ondrusek, M. C. Kennicutt, R. Iturriaga, H. R. Harvey, R. W. Hoham, and S. A. Macko. 1993. Evidence for a photoprotective function for secondary carotenoids of snow algae. *Journal of Phycology* 29: 427–34.

Bonneau, E. R. 1977. Polymorphic behavior of *Ulva lactuca* (Chlorophyta) in axenic culture. I. Occurrrence of *Enteromorpha*-like plants in haploid clones. *Journal of Phycology* 13: 133–41.

Bothwell, M. L., D. M. J. Sherbot, and C. M. Pollock. 1994. Ecosystem responses to solar ultraviolet-B radiation: Influence of trophic-level interactions. *Science* 265: 97–100.

Bothwell, M. L., D. Sherbot, A. C. Roberge, and R. J. Daley. 1993. Influence of natural ultraviolet radiation on lotic periphytic diatom community growth, biomass, accrual, species composition: Short-term versus long-term effects. *Journal of Phycology* 29: 24–35.

Boyle, R. H. 1996. Phantom. *Natural History* 105 (3): 16–19.

Broadwater, S. T., and J. Scott. 1982. Ultrastructure of early development in the female reproductive system of *Polysiphonia harveyi* Bailey (Ceramiales, Rhodophyta). *Journal of Phycology* 18: 427–41.

Bronk, D. A., P. M. Gilbert, and B. B. Ward. 1994. Nitrogen uptake, dissolved organic nitrogen release, and new production. *Science* 265: 1843–46.

Browder, J. A., P. J. Gleason, and D. R. Swift. 1994. Periphyton in the Everglades: Spatial variation, environmental correlates, and ecological implications. In S. M. Davis and J. C. Ogden, eds. *Everglades, the Ecosystem and Its Restoration* 379–418. St. Lucie Press.

Brown, B. E., and J. C. Ogden. 1993. Coral bleaching. *Scientific American* 268 (1): 64–70.

Brown, M. T. 1987. Effects of desiccation on photosynthesis of intertidal algae from a southern New Zealand shore. *Botanica Marina* 30: 121–27.

Bruckner, A. 1993. Outbreak of black band disease in Caribbean. *Harmful Algae News* 4: 3.

Bujak, J. P., and G. L. Williams. 1981. The evolution of dinoflagellates. *Canadian Journal of Botany* 59: 2077–87.

Burkholder, J. M., E. J. Noga, C. H. Hobbs, and H. B. Glasgow. 1992. New "phantom" dinoflagellate is the causative agent of major estuarine fishkills. *Nature* 358: 407–10.

Cain, J. R., and F. R. Trainor. 1976. Regulation of gametogenesis in *Scenedesmus obliquus* (Chlorophyceae). *Journal of Phycology* 12: 383–90.

Caiola, M. G., C. Forni, and P. Albertano. 1987. Characterization of the algal flora growing on ancient Roman frescoes. *Phycologia* 26: 387–90.

Canini, A., M. Grilli Caiola, P. Civitareale, and F. Galiazzo. 1992. Superoxide dismutase in symbiotic, free-living, and wild *Anabaena* and *Nostoc* (Nostocales, Cyanophyta). *Phycologia* 31: 225–30.

Carlton, J. T., and J. A. Scanlon. 1985. Progression and dispersal of an introduced alga: *Codium fragile* ssp. *tomentosoides* (Chlorophyta) on the Atlantic coast of North America. *Botanica Marina* 28: 155–65.

Carmichael, W. W. 1994. The toxins of cyanobacteria. *Scientific American* 270 (1): 78–86.

Carpenter, E. J., D. G. Capone, and J. G. Rueter. 1992. *Marine Pelagic Cyanobacteria:* Trichodesmium *and Other Diazotrophs.* Kluwer Academic Publishers.

Carpenter, E. J., and K. Romans. 1991. Major role of the cyanobacterium *Trichodesmium* in nutrient cycling in the North Atlantic Ocean. *Science* 254: 1356–58.

Carpenter, R. C. 1990. Mass mortality of *Diadema antillarum.* I. Long-term effects on sea urchin population-dynamics and coral reef algal communities. *Marine Biology* 104: 67–77.

Chapman, A. R. O. 1995. Functional ecology of fucoid algae: Twenty-three years of progress. *Phycologia* 34: 1–32.

Chapman, A. R. O., and J. S. Craigie. 1977. Seasonal growth in *Laminaria longicruris:* Relations with dissolved inorganic nutrients and internal reserves of nitrogen. *Marine Biology* 40: 197–205.

Chapman, R. L., and B. H. Good. 1983. Subaerial symbiotic green algae: Interactions with vascular plant hosts. In L. J. Goff, ed., *Algal Symbiosis,* 173–204. Cambridge University Press.

Charlson, R. J., J. E. Lovelock, M. O. Andreae, and S. G. Warren. 1987. Oceanic phytoplankton, atmospheric sulphur, cloud albedo, and climate. *Nature* 326: 655–61.

Chisholm, S. W., and F. M. M. Morel, eds. 1991. What controls phytoplankton production in nutrient-rich areas of the open sea? *Limnology and Oceanography* 36: 1507–1965.

Chow, W. S., A. Melis, and J. M. Anderson. 1990. Adjustments of photosystem stoichiometry in chloroplasts improve the quantum efficiency of photosynthesis. *Proceedings of the National Academy of Sciences USA* 87: 7502–6.

Clark, K. B. 1992. Plant-like animals and animal-like plants: Symbiotic coevolution of ascoglossan (sacoglossan) molluscs, their algal prey, and algal plastids. In W. Reisser, ed., *Algae and Symbioses* 515–30. Biopress.

Clayton, M. N. 1980. Sexual reproduction—A rare occurrence in the life history of the complanate form of *Scytosiphon* (Scytosiphonaceae, Phaeophyta) from southern Australia. *British Phycological Journal* 15: 105–18.

Clayton, M. N. 1981. Experimental analysis of the life history of the complanate form of *Scytosiphon* (Scytosiphonaceae, Phaeophyta) in southern Australia. *Phycologia* 20: 358–64.

Clayton, M. N. 1988. Evolution of life histories of brown algae. *Botanica Marina* 31: 379–87.

Clayton, M. N. 1992. Propagules of marine macroalgae: Structure and development. *British Phycological Journal* 27: 219–32.

Coale, K. H., S. E. Fitzwater, R. M. Gordon, K. S. Johnson, and R. T. Barber. 1996. Control of community growth and export production by upwelled iron in the equatorial Pacific Ocean. *Nature* 379: 621–24.

Coleman, A. W. 1985. Diversity of plastid DNA configuration among classes of eukaryote algae. *Journal of Phycology* 21: 1–16.

Colijn, F., and C. van den Hoek. 1971. The life history of *Sphacelaria furcigera* Kütz. (Phaeophyceae). II. The influence of day length and temperature on sexual and vegetative reproduction. *Nova Hedwigia* 21: 899–922.

Conley, D. J., C. L. Schelske, and E. F. Stoermer. 1993. Modification of the biogeochemistry of silica with eutrophication. *Marine Ecology Progress Series* 101: 179–92.

Connell, J. H. 1978. Diversity of tropical rain forests and coral reefs. *Science* 199: 1302–10.

Correa, J., I. Novaczek, and J. McLachlan. 1986. Effect of temperature and day length on morphogenesis of *Scytosiphon lomentaria* (Scytosiphonales, Phaeophyta) from eastern Canada. *Phycologia* 25: 469–75.

Cosper, E. M., V. M. Bricelj, and E. J. Carpenter, eds. 1989. *Novel Phytoplankton Blooms.* Springer-Verlag.

Cousens, R. 1985. Frond size distributions and the effects of the algal canopy on the behaviour of *Ascophyllum nodosum* (L.) Le Jolis. *Journal of Experimental Marine Biology and Ecology* 92: 231–349.

Crawford, R. M. 1973. The protoplasmic ultrastructure of the vegetative cell of *Melosira varians* C. A. Agardh. *Journal of Phycology* 9: 50–61.

Crawford, R. M., and F. E. Round. 1989. *Corethron* and *Mallomonas*—Some striking morphological similarities. In J. C. Green, B. S. C. Leadbeater, and W. L. Diver, eds., *The Chromophyte Algae,* 295–305. Clarendon Press.

Critchley, A. T., W. F. Farnham, and L. S. Morrell. 1983. A chronology of new European sites of attachment for the invasive brown alga, *Sargassum muticum,* 1973–1981. *Journal of the Marine Biological Association of the United Kingdom* 63: 799–811.

Crumpton, W. G., and R. G. Wetzel. 1982. Effects of differential growth and mortality in the seasonal succession of phytoplankton populations in Lawrence Lake, Michigan. *Ecology* 63: 1729–39.

Cumming, B. F., J. P. Smol, and H. J. B. Birks. 1992. Scaled chrysophytes (Chrysophyceae and Synurophyceae) from Adirondack drainage lakes and their relationship to environ-

mental variables. *Journal of Phycology* 28: 162–78.

Cummins, K. W., and M. J. Klug. 1979. Feeding ecology of stream invertebrates. *Annual Review of Ecology and Systematics* 10: 147–72.

Darwin, C. 1962. *Voyage of the Beagle.* Doubleday.

Davidson, A. T., D. Bramich, H. J. Marchant, and A. McMinn. 1994. Effects of UV-B irradiation on growth and survival of Antarctic marine diatoms. *Marine Biology* 119: 507–15.

Davidson, A. T., and H. J. Marchant. 1992. The biology and ecology of *Phaeocystis* (Prymnesiophyceae). In F. E. Round and D. J. Chapman, eds., *Progress in Phycological Research,* vol. 8, 1–45. Biopress.

Dawson, P. A. 1973. Observations on the structure of some forms of *Gomphonema parvulum* Kütz. II. The internal organization. *Journal of Phycology* 9: 165–75.

Dayton, P. K. 1973. Dispersion, dispersal, and persistence of the annual intertidal alga, *Postelsia palmaeformis* Ruprecht. *Ecology* 54: 433–38.

DeCew, T. C., and J. A. West. 1977. Culture studies on the marine red algae *Hildenbrandia occidentalis* and *H. prototypus.* *Bulletin of the Japanese Society of Phycology* 25 (supplement): 31–41.

de Jesus, M. D., F. Tabatabai, and D. J. Chapman. 1989. Taxonomic distribution of copper-zinc superoxide dismutase in green algae and its phylogenetic importance. *Journal of Phycology* 25: 767–72.

Delwiche, C. F., L. E. Graham, and N. Thomson. 1989. Lignin-like compounds and sporopollenin in *Coleochaete,* an algal model for land plant ancestry. *Science* 245: 399–401.

Demmig-Adams, B. 1990. Carotenoids and photoprotection in plants: A role for the xanthophyll zeaxanthin. *Biochimica et Biophysica Acta* 1020: 1–24.

Demmig-Adams, B., and W. W. Adams. 1992. Photoprotection and other responses of plants to high light stress. *Annual Review of Plant Physiology and Plant Molecular Biology* 43: 599–626.

Denny, M. 1995. Survival in the surf zone. *American Scientist* 83: 166–73.

Dodds, W. K., and D. A. Gudder. 1992. The ecology of *Cladophora.* *Journal of Phycology* 28: 415–27.

Dodds, W. K., D. A. Gudder, and D. Mollenhauer. 1995. The ecology of *Nostoc.* *Journal of Phycology* 31: 2–18.

Dodge, J. D. 1983. Dinoflagellates: Investigation and phylogenetic speculation. *British Phycological Journal* 18: 335–56.

Domozych, C. R., K. Plante, P. Blais, L. Paliulis, and D. S. Domozych. 1993. Mucilage processing and secretion in the green alga *Closterium*. I. Cytology and biochemistry. *Journal of Phycology* 29: 650–59.

Domozych, D. S., and T. T. Nimmons. 1992. The contractile vacuole as an endocytic organelle of the chlamydomonad flagellate *Gloeomonas kupfferi* (Volvocales, Chlorophyta). *Journal of Phycology* 28: 809–16.

Douglas, S. E. 1994. Chloroplast origins and evolution. In D. A. Bryant, ed., *The Molecular Biology of Cyanobacteria*, 91–118. Kluwer Academic Publishers.

Douglas, S. E., C. A. Murphy, D. F. Spencer, and M. W. Gray. 1991. Cryptomonad algae are evolutionary chimaeras of two phylogenetically distinct unicellular eukaryotes. *Nature* 350: 148–51.

Douglas, S. E., and S. Turner. 1991. Molecular evidence for the origin of plastids from a cyanobacterium-like ancestor. *Journal of Molecular Evolution* 33: 267–73.

Dring, M. J. 1981. Chromatic adaptation of photosynthesis in benthic marine algae: An examination of its ecological significance using a theoretical model. *Limnology and Oceanography* 26: 271–84.

Dring, M. J., and F. A. Brown. 1982. Photosynthesis of intertidal brown algae during and after periods of emersion: A renewed search for physiological causes of zonation. *Marine Ecology Progress Series* 8: 301–8.

Droop, M. R. 1973. Some thoughts on nutrient limitation in algae. *Journal of Phycology* 9: 264–72.

Dudgeon, S. R., I. R. Davison, and R. L. Vadas. 1989. Effect of freezing on photosynthesis of intertidal macroalgae: Relative tolerance of *Chondrus crispus* and *Mastocarpus stellatus* (Rhodophyta). *Marine Biology* 101: 107–14.

Duffy, J. E., and M. E. Hay. 1990. Seaweed adaptations to herbivory. *BioScience* 40: 368–75.

Duffy, J. E., and M. E. Hay. 1994. Herbivore resistance to seaweed chemical defense: The role of mobility and predation risk. *Ecology* 75: 1304–19.

Duggins, D. O. 1980. Kelp beds and sea otters: An experimental approach. *Ecology* 61: 447–53.

Duggins, D. O., C. A. Simenstad, and J. A. Estes. 1989. Magnification of secondary production by kelp detritus in coastal marine ecosystems. *Science* 245: 170–73.

Dyck, L. J., and R. E. DeWreede. 1995. Patterns of seasonal demographic change in the alternate isomorphic stages of *Mazzaella splendens* (Gigartinales, Rhodophyta). *Phycologia* 34: 390–95.

Eckhardt, R., R. Schnetter, and G. Seibold. 1986. Nuclear behaviour during the life cycle of *Derbesia* (Chlorophyceae). *British Phycological Journal* 21: 287–95.

Edgar, L. A., and J. D. Pickett-Heaps. 1984a. Diatom locomotion. In F. E. Round and D. J. Chapman, eds., *Progress in Phycological Research* 47–88. Biopress.

Edgar, L. A., and J. D. Pickett-Heaps. 1984b. Valve morphogenesis in the pennate diatom *Navicula cuspidata*. *Journal of Phycology* 20: 47–61.

Egeland, E. S., G. Johnsen, W. Eikrem, J. Throndsen, and S. Liaaen-Jensen. 1995. Pigments of *Bathycoccus prasinos* (Prasinophyceae): Methodological and chemosystematic implications. *Journal of Phycology* 31: 554–61.

Eikrem, W. 1996. *Chrysochromulina throndsenii* sp. nov. (Prymnesiophyceae). Description of a new haptophyte flagellate from Norwegian waters. *Phycologia* 35: 377–80.

Esaias, W. E., and H. C. Curl. 1972. Effect of dinoflagellate bioluminescence on copepod ingestion rates. *Limnology and Oceanography* 17: 901–6.

Estes, J. A., and D. O. Duggins. 1995. Sea otters and kelp forests in Alaska: Generality and variation in a community ecological paradigm. *Ecological Monographs* 65: 75–100.

Evans, L. V., J. A. Callow, and M. E. Callow. 1982. The biology and biochemistry of reproduction and early development in *Fucus*. In F. E. Round and D. J. Chapman, eds., *Progress in Phycological Research*. vol. 1, 67–110. Elsevier.

Fabricius, K. E., Y. Benayahu, and A. Genin. 1995. Herbivory in asymbiotic soft corals. *Science* 268: 90–92.

Farmer, M. A., and K. R. Roberts. 1990. Organelle loss in the endosymbiont of *Gymnodinium acidotum* (Dinophyceae). *Protoplasma* 153: 178–85.

Fletcher, R. L., and M. E. Callow. 1992. The settlement, attachment, and establishment of marine algal spores. *British Phycological Journal* 27: 303–29.

Floyd, G. L., K. D. Stewart, and K. R. Mattox. 1971. Cytokinesis and plasmodesmata in *Ulothrix*, *Journal of Phycology* 7: 306–9.

Floyd, G. L., K. D. Stewart, and K. R. Mattox. 1972. Comparative cytology of *Ulothrix* and *Stigeoclonium*. *Journal of Phycology* 8: 68–81.

Fogg, G. E., and B. Thake. 1987. *Algal Cultures and Phytoplankton Ecology*. 3d ed. University of Wisconsin Press.

Foster, M. S., and D. R. Schiel. 1985. *The Ecology of Giant Kelp Forests in California: A Community Profile*. U.S. Fish and Wildlife Service, Biological Report 85 (7.2).

Frank, H. A., A. Cua, V. Chynwat, A. Young, D. Gosztola, and M. R. Wasielewski. 1994. Photophysics of the carotenoids associated with the xanthophyll cycle in photosynthesis. *Photosynthetic Research* 41: 389–95.

French, F. W., and P. E. Hargraves. 1985. Spore formation in the life cycles of the diatoms *Chaetoceros diadema* and *Leptocylindrus danicus*. *Journal of Phycology* 21: 477–83.

Fries, L. 1988. *Ascophyllum nodosum* (Phaeophyta) in axenic culture and its response to the endophytic fungus *Mycosphaerella ascophylli* and epiphytic bacterie. *Journal of Phycology* 24: 333–37.

Fritsch, F. E. 1945. *The Structure and Reproduction of the Algae*. Vol. 2. Cambridge University Press.

Fujiwara, S., H. Iwahashi, J. Someya, S. Nishikawa, and N. Minaka. 1993. Structure and cotranscription of the plastic-encoded *rbc*L and *rbc*S genes of *Pleurochrysis carterae* (Prymnesiophyta). *Journal of Phycology* 29: 347–55.

Gagné, J. A., K. H. Mann, and A. R. O. Chapman. 1982. Seasonal patterns of growth and storage in *Laminaria longicruris* in relation to differing patterns of availability of nitrogen in the water. *Marine Biology* 69: 91–101.

Gantt, E. 1971. Micromorphology of the periplast of *Chroomonas* sp. (Cryptophyceae). *Journal of Phycology* 7: 177–84.

Gantt, E., K. Ohki, and Y. Fujita. 1984. *Trichodesmium thiebautii*: Structure of a nitrogen-fixing marine blue-green alga (Cyanophyta). *Protoplasma* 119: 188–96.

Garbary, D. J., and P. W. Gabrielson. 1990. Taxonomy and evolution. In K. M. Cole and R. G. Sheath, eds., *Biology of the Red Algae*, 477–98. Cambridge University Press.

Garbary, D. J., and K. A. MacDonald. 1995. The *Ascophyllum/Polysiphonia/Mycosphaerella* symbiosis. IV. Mutualism in the *Ascophyllum/Mycosphaerella* interaction. *Botanica Marina* 38: 221–25.

Garcia-Pichel, F., and R. W. Castenholz. 1991. Characterization and biological implications of scytonemin, a cyanobacterial sheath pigment. *Journal of Phycology* 27: 395–409.

Gargas, A., P. T. DePriest, M. Grube, and A. Tehler. 1995. Multiple origins of lichen symbioses in fungi suggested by SSU rDNA phylogeny. *Science* 268: 1492–95.

Garrison, D. L., K. R. Buck, and G. A. Fryxell. 1987. Algal assemblages in Antarctic pack ice and in ice-edge plankton. *Journal of Phycology* 23: 564–72.

Geider, R. J., and P. A. Gunter. 1989. Evidence for the presence of phycoerythrin in *Dinophysis norvegica*, a pink dinoflagellate. *British Phycological Journal* 24: 195–98.

Gibbs, S. 1990. The evolution of algal chloroplasts. In W. Wiessner, D. G. Robinson, and R. C. Starr, eds., *Experimental Phycology* 1: 145–57. Springer Verlag.

Gibson, M. T., and B. A. Whitton. 1987a. Hairs, phosphatase activity and environmental chemistry in *Stigeoclonium, Chaetophora*, and *Draparnaldia* (Chaetophorales). *British Phycological Journal* 22: 11–22.

Gibson, M. T., and B. A. Whitton. 1987b. Influence of phosphorus on morphology and physiology of freshwater *Chaetophora, Draparnaldia*, and *Stigeoclonium* (Chaetophorales, Chlorophyta). *Phycologia* 26: 59–69.

Gillott, M. A., and S. P. Gibbs. 1980. The cryptomonad nucleomorph: Its ultrastructure and evolutionary significance. *Journal of Phycology* 16: 558–68.

Goff, L. J., and A. W. Coleman. 1985. The role of secondary pit connections in red algal parasitism. *Journal of Phycology* 21: 483–508.

Goff, L. J., D. A. Moon, P. Nyvall, B. Stache, K. Mangin, and G. Zuccarello. 1996. The evolution of parasitism in the red algae: Molecular comparisons of adelphoparasites and their hosts. *Journal of Phycology* 32: 297–312.

Goodenough, U. W., E. V. Armbrust, A. M. Campbell, and P. J. Ferris. 1995. Molecular genetics of sexuality in *Chlamydomonas*. *Annual Review of Plant Physiology and Plant Molecular Biology* 46: 21–41.

Gorham, P. R., and W. W. Carmichael. 1988. Hazards of freshwater blue-green algae (cyanobacteria). In C. A. Lembi and J. R. Waaland, eds., *Algae and Human Affairs*, 403–31. Cambridge University Press.

Graham, J. M., L. E. Graham, and J. A. Kranzfelder. 1985. Light, temperature and photoperiod as factors controlling reproduction in *Ulothrix* zonata (Ulvophyceae). *Journal of Phycology* 21: 235–39.

Graham, J. M., J. A. Kranzfelder, and M. T. Auer. 1985. Light and temperature as factors regulating seasonal growth and distribution of *Ulothrix zonata* (Ulvophyceae). *Journal of Phycology* 21: 228–34.

Graham, J. M., C. A. Lembi, H. L. Adrian, and D. F. Spencer. 1995. Physiological responses to temperature and irradiance in *Spirogyra* (Zygnematales, Charophyceae). *Journal of Phycology* 31: 531–40.

Graham, L. E. 1984. *Coleochaete* and the origin of land plants. *American Journal of Botany* 71: 603–8.

Graham, L. E. 1993. *Origin of Land Plants*. Wiley.

Graham, L. E., and G. E. McBride. 1979. The occurrence and phylogenetic significance of a multilayered structure in *Coleochaete* spermazoids. *American Journal of Botany* 66: 887–94.

Granéli, E., B. Sundström, L. Edler, and D. M. Anderson. 1990. *Toxic Marine Phytoplankton*. Elsevier.

Greuel, B. T., and G. L. Floyd. 1985. Development of the flagellar apparatus and flagellar orientation in the colonial green alga *Gonium pectorale* (Volvocales). *Journal of Phycology* 21: 358–71.

Hackney, J. M., R. C. Carpenter, and W. H. Adey. 1989. Characteristic adaptations to grazing among algal turfs on a Caribbean coral reef. *Phycologia* 28: 109–19.

Häder, D. P., and E. Hoiczyk. 1992. Gliding motility. In M. Melkonian, ed., *Algal Cell Motility*, 1–38. Chapman and Hall.

Hagen, C., W. Braune, and L. O. Björn. 1994. Functional aspects of secondary carotenoids in *Haematococcus lacustris* (Volvocales). III. Action as a "sunshade." *Journal of Phycology* 30: 241–48.

Hale, M. F. 1969. *How to Know the Lichens*. Wm. C. Brown Publishers.

Hallegraeff, G. M. 1993. A review of harmful algal blooms and their apparent global increase. *Phycologia* 32: 79–99.

Hansen, P. J., T. G. Nielsen, and H. Kaas. 1995. Distribution and growth of protists and mesozooplankton during a bloom of *Chrysochromulina* spp. (Prymnesiophyceae, Prymnesiales). *Phycologia* 34: 409–16.

Harlin, M. M., and J. S. Craigie. 1975. The distribution of photosynthate in *Ascophyllum nodosum* as it relates to epiphytic *Polysiphonia Ianosa*. *Journal of Phycology* 11: 109–13.

Harris, E. H. 1989. *The Chlamydomonas Sourcebook*. Academic Press.

Haupt, W. 1983. Movement of chloroplasts under the control of light. In F. E. Round and D. J. Chapman, eds., *Progress in Phycological Research*, vol. 2, 227–81. Elsevier.

Hawkins, S. J., and R. G. Hartnoll. 1985. Factors determining the upper limits of intertidal canopy-forming algae. *Marine Ecology Progress Series* 20: 265–71.

Hay, M. E. 1991. Fish-seaweed interactions on coral reefs: Effects of herbivorous fishes and adaptations of their prey. In P. F. Sale, ed., *The Ecology of Fishes on Coral Reefs,* 96–119. Academic Press.

Hay, M. E., J. E. Duffy, and W. Fenical. 1990. Host-plant specialization decreases predation on a marine amphipod: An herbivore in plant's clothing. *Ecology* 71: 733–43.

Henriksen, P., F. Klipschildt, Ø. Moestrup, and H. A. Thomsen. 1993. Autecology, life history and toxicology of the silicoflagellate *Dictyocha speculum* (Silicoflagellata, Dictyochophyceae). *Phycologia* 32: 29–39.

Henry, E. C., and K. M. Cole. 1982. Ultrastructure of swarmers in the Laminariales (Phaeophyceae). I. Zoospores. *Journal of Phycology* 18: 550–69.

Herman, E. M., and B. M. Sweeney. 1976. *Cachonina illdefina* sp. nov. (Dinophyceae): Chloroplast tubules and degeneration of the pyrenoid. *Journal of Phycology* 12: 198–205.

Heywood, R. B. 1984. Antarctic inland waters. In R. M. Laws, ed., *Antarctic Ecology*, vol. 1, 279–344. Academic Press.

Hibberd, D. J. 1977. Ultrastructure of cyst formation in *Ochromonas tuberculata*

(Chrysophyceae). *Journal of Phycology* 13: 309–20.

Hibberd, D. J., and R. E. Norris. 1984. Cytology and ultrastructure of *Chlorarachnion reptans* (Chlorarachniophyta diviso nova, Chlorarachniophyceae classis nova). *Journal of Phycology* 20: 310–30.

Hixon, M. A., and W. N. Brostoff. 1996. Succession and herbivory: Effects of differential fish grazing on Hawaiian coral-reef algae. *Ecological Monographs* 66: 69–90.

Hoagland, K. D., S. C. Roemer, and J. R. Rosowski. 1982. Colonization and community structure of two periphyton assemblages, with emphasis on the diatoms (Bacillariophyceae). *American Journal of Botany* 69: 188–213.

Hoek, C. van den, D. G. Mann, and H. M. Jahns. 1995. *Algae*. Cambridge University Press.

Hoffman, L. R. 1973. Fertilization in *Oedogonium*. I. Plasmogamy. *Journal of Phycology* 9: 62–84.

Hoffman, L., L. Talarico, and A. Wilmotte. 1990. Presence of CU-phycoerythrin in the marine benthic blue-green alga *Oscillatoria* cf. *corallinae*. *Phycologia* 29: 19–26.

Holm, N. P., and D. E. Armstrong. 1981. Role of nutrient limitation and competition in controlling the populations of *Asterionella formosa* and *Mycrocystis aeruginosa* in semi-continuous culture. *Limnology and Oceanography* 26: 622–34.

Hoops, H. J., and G. L. Floyd. 1979. Ultrastructure of the centric diatom, *Cyclotella meneghiniana:* Vegetative cell and auxospore development. *Phycologia* 18: 424–35.

Horn, M. H. 1989. Biology of marine herbivorous fishes. *Oceanography and Marine Biology, Annual Review* 27: 167–272.

Horton, T., and W. M. Eichbaum. 1991. *Turning the Tide*. Island Press.

Hoshaw, R. W., C. V. Wells, and R. M. McCourt. 1987. A polyploid species complex in *Spirogyra maxima* (Chlorophyta, Zygnemataceae), a species with large chromosomes. *Journal of Phycology* 23: 267–73.

Howarth, R. W. 1988. Nutrient limitation of net primary production in marine ecosystems. *Annual Review of Ecology and Systematics* 19: 89–110.

Hughes, T. P. 1994. Catastrophes, phase shifts, and large-scale degradation of a Caribbean coral reef. *BioScience* 265: 1547–51.

Hunt, M. E., G. L. Floyd, and B. B. Stout. 1979. Soil algae in field and forest environments. *Ecology* 60: 362–75.

Hurd, C. L., R. S. Galvin, T. A. Norton, and M. J. Dring. 1993. Production of hyaline hairs by intertidal species of *Fucus* (Fucales) and their role in phosphate uptake. *Journal of Phycology* 29: 160–165.

Hutchinson, G. E. 1961. The paradox of the plankton. *American Naturalist* 95: 137–45.

Hutt, W., and G. Kochert. 1971. Effects of some protein and nucleic acid synthesis inhibitors on fertilization in *Volvox carteri*. *Journal of Phycology* 7: 316–20.

Ikawa, M., and J. J. Sasner. 1990. The chemistry and physiology of algal toxins. In I. Akatsuka, ed., *Introduction to Applied Phycology*, 27–65. SPB Academic Publishing.

Iverson, R. L., F. L. Nearhoof, and M. O. Andreae. 1989. Production of dimethylsulfonium propionate and dimethylsulfide by phytoplankton in estuarine coastal waters. *Limnology and Oceanography* 34: 53–67.

Janson, S., A. N. Rai, and B. Bergmann. 1995. Intracellular cyanobiont *Richelia intracellularis*. Ultrastructure and immuno-localisation of phycoerythrin, nitrogenase, Rubisco and glutamine synthetase. *Marine Biology* 124: 1–8.

Janson, S., P. J. A. Siddiqui, A. E. Walsby, K. M. Romans, E. J. Carpenter, and B. Bergman. 1995. Cytomorphological characterization of the planktonic diazotrophic cyanobacteria. *Trichodesmium* spp. from the Indian Ocean and Caribbean and Sargasso Seas. *Journal of Phycology* 31: 463–77.

Johansen, J. R. 1993. Cryptogamic crusts of semiarid and arid lands of North America. *Journal of Phycology* 29: 140–47.

Johnson, C. R., and K. M. Mann. 1986. The crustose coralline alga, *Phymatolithon Foslie*, inhibits the overgrowth of seaweeds without relying on herbivores. *Journal of Experimental Marine Biology and Ecology* 96: 127–46.

Jones, A. K. 1988. Algal extracellular products-antimicrobial substances. In L. J. Rogers and J. R. Gallon, eds., *Biochemistry of the Algae and Cyanobacteria*, 257–81. Clarendon Press.

Jones, R. C. 1990. The effect of submerged aquatic vegetation on phytoplankton and water quality in the tidal freshwater Potomac. *Journal of Freshwater Ecology* 5: 279–88.

Kantz, T. S., E. C. Theriot, E. A. Zimmer, and R. L. Chapman. 1990. The Pleurastrophyceae and Micromonadophyceae: A cladistic analysis of nuclear rRNA sequence data. *Journal of Phycology* 26: 711–21.

Karentz, D., J. E. Cleaver, and D. L. Mitchell. 1991. Cell survival characteristics and molecular responses of Antarctic phytoplankton to ultraviolet-B radiation. *Journal of Phycology* 27: 326–41.

Kawachi, M., and I. Inouye. 1995. Functional roles of the haptonema and the spine scales in the feeding process of *Chrysochromulina spinifera* (Fournier) Pienaar et Norris (Haptophyta=Prymnesiophyta). *Phycolocia* 34: 193–200.

Kawachi, M., I. Inouye, O. Maeda, and M. Chihara. 1991. The haptonema as a food-capturing device: Observations of *Chrysochromulina hirta* (Prymnesiophyceae). *Phycologia* 30: 563–73.

Kawai, H. 1988. A flavin-like autofluorescent substance in the posterior flagellum of golden and brown algae. *Journal of Phycology* 24: 114–17.

Keating, K. I. 1977. Allelopathic influences on blue-green bloom sequence in a eutrophic lake. *Science* 196: 885–87.

Keating, K. I. 1978. Blue-green algal inhibition of diatom growth: Transition from mesotrophic to eutrophic community structure. *Science* 199: 971–73.

Keithan, E. D., R. L. Lowe, and H. R. DeYoe. 1988. Benthic diatom distribution in a Pennsylvania stream: Role of pH and nutrients. *Journal of Phycology* 24: 581–85.

Khan, S., O. Arakawa, and Y. Onoue. 1996. A toxicological study of the marine phytoflagellate *Chattonella antiqua* (Raphidophyceae). *Phycologia* 35: 239–44.

Kilham, P. 1971. A hypothesis concerning silica and the freshwater planktonic diatoms. *Limnology and Oceanography* 16: 10–18.

King, J. M., and C. H. Ward. 1977. Distribution of edaphic algae as related to land usage. *Phycologia* 16: 23–30.

Kiorboe, T., K. P. Andersen, and H. G. Dam. 1990. Coagulation efficiency and aggregate formation in marine phytoplankton. *Marine Biology* 107: 235–45.

Kirst, G. O., and C. Wiencke. 1995. Ecophysiology of polar algae. *Journal of Phycology* 31: 181–99.

Kiss, J. Z., A. C. Vasconcelos, and R. E. Triemer. 1987. Structure of the euglenoid storage carbohydrate, paramylon. *American Journal of Botany* 74: 877–82.

Klaveness, D. 1972. *Coccolithus huxleyi* (Lohm.) Kamptn. II. The flagellate cell, aberrant cell types, vegetative propagation and life cycles. *British Phycological Journal* 7: 309–18.

Klemer, A. R., J. J. Cullen, M. T. Mageau, K. M. Hanson, and R. A. Sundell. 1996. Cyanobacterial buoyancy regulation: The paradoxical roles of carbon. *Journal of Phycology* 32: 47–53.

Klut, M. E., T. Bisalputra, and N. J. Antia. 1987. Some observations on the structure and function of the dinoflagellate pusule. *Canadian Journal of Botany* 65: 736–44.

Knox, G. A. 1994. *The Biology of the Southern Ocean*. Cambridge University Press.

Koeman, R. P. T., and A. M. Cortel-Breeman. 1976. Observations on the life history of *Elachista fucicola* (Vell.) Aresch. (Phaeophyceae) in culture. *Phycologia* 15: 107–17.

Kudoh, S., and M. Takahashi. 1990. Fungal control of population changes in the planktonic diatom *Asterionella formosa* in a shallow eutrophic lake. *Journal of Phycology* 26: 239–44.

Kugrens, P., R. E. Lee, and R. A. Andersen. 1986. Cell form and surface patterns in *Chroomonas* and *Cryptomonas* cells (Cryptophyta) as revealed by scanning electron microscopy. *Journal of Phycology* 22: 512–22.

Kugrens, P., and R. E. Lee. 1988. Ultrastructure of fertilization in a cryptomonad. *Journal of Phycology* 24: 385–93.

Kuhlenkamp, R., and R. G. Hooper. 1995. New observations on the Tilopteridaceae (Phaeophyceae). I. Field studies of *Haplospora* and *Phaeosiphoniella*, with implications for survival, perennation and dispersal. *Phycologia* 34: 229–39.

Kuhlenkamp, R., D. G. Müller, and A. Whittick. 1993. Genotypic variation and alternating DNA levels at constant chromosome numbers in the life history of the brown alga *Haplospora globosa* (Tilopteridales). *Journal of Phycology* 29: 377–80.

Kuhn, D. L., J. L. Plafkin, J. Cairns, and R. L. Lowe. 1981. Qualitative characterization of aquatic environments using diatom life-form strategies. *Transactions of the American Microscopical Society* 100: 165–182.

Kuhsel, M. G., R. Strickland, and J. D. Palmer. 1990. An ancient group I intron shared by eubacteria and chloroplasts. *Science* 250: 1570–73.

Laliberté, G., and J. A. Hellebust. 1991. The phylogenetic significance of the distribution of arginine deiminase and arginase in the Chlorophyta. *Phycologia* 30: 145–50.

Lampert, W., K. O. Rothhaupt, and E. von Elert. 1994. Chemical induction of colony formation in a green alga (*Scenedesmus acutus*) by grazers (*Daphnia*). *Limnology and Oceanography* 39: 1543–50.

Lapointe, B. E., M. M. Littler, and D. S. Littler. 1992. Nutrient availability to marine macroalgae in siliciclastic versus carbonate-rich coastal waters. *Estuaries* 15: 75–82.

Larkum, A. W. D., C. Scaramauzzi, G. C. Cox, R. G. Hiller, and A. G. Turner. 1994. The light-harvesting chlorophyll c-like pigment in *Prochloron*. *Proceedings of the National Academy of Sciences USA* 91: 679–83.

Leadbeater, B. S. C. 1990. Ultrastructure and assembly of the scale case in *Synura* (Synurophyceae Andersen). *British Phycological Journal* 25: 117–32.

Lebert, M., and D. P. Häder. 1996. How *Euglena* tells up and down. *Nature* 379: 590.

Lee, J. J. 1995. Living sands. *BioScience* 45: 252–61.

Leipe, D. D., P. O. Wainright, J. H. Gunderson, D. Porter, D. J. Patterson, F. Valois, S. Himmerich, and M. L. Sogin. 1994. The stramenopiles from a molecular perspective: 16S-like rRNA sequences from *Labyrinthuloides minuta* and *Cafeteria roenbergensis*. *Phycologia* 33: 369–77.

Lessios, H. A. 1988. Mass mortality of *Diadema antillarum* in the Caribbean: What have we learned? *Annual Review of Ecology and Systematics* 19: 371–93.

Lewin, R. A. 1975. A marine *Synechocystis* (Cyanophyta, Chroococcales) epizoic on ascidians. *Phycologia* 14: 153–60.

Lewin, R. A., P. A. Farnsworth, and G. Yamanaka. 1981. The algae of green polar bears. *Phycologia* 20: 303–14.

Lewis, J. G., N. F. Stanley, and G. G. Guist. 1988. Commercial production and applications of algal hydrocolloids. In C. A. Lembi and J. R. Waaland, eds., *Algae and Human Affairs*, 205–36. Cambridge University Press.

Lewis, W. M. 1978. Dynamics and succession of the phytoplankton in a tropical lake. *Journal of Ecology* 66: 849–80.

Lewis, W. M. 1984. The diatom sex clock and its evolutionary significance. *American Naturalist* 123: 73–80.

Lieberman, O. S., M. Shilo, and J. van Rijn. 1994. The physiological ecology of a freshwater dinoflagellate bloom population: Vertical migration, nitrogen limitation, and nutrient uptake kinetics. *Journal of Phycology* 30: 964–71.

Littler, M. M. 1980. Morphological form and photosynthetic performances of marine macroalgae: Tests of a function/form hypothesis. *Botanica Marina* 23: 161–65.

Littler, M. M., and K. E. Arnold. 1982. Primary productivity of marine macroalgal functional-form groups from southwestern North America. *Journal of Phycology* 18: 307–11.

Littler, M. M., and D. S. Littler. 1980. The evolution of thallus form and survival strategies in benthic macroalgae: Field and laboratory tests of a functional model. *American Naturalist* 116: 25–44.

Littler, M. M., and D. S. Littler. 1995. Impact of CLOD pathogen on Pacific coral reefs. *Science* 267: 1356–60.

Littler, M. M., D. S. Littler, S. M. Blair, and J. N. Norris. 1985. Deepest known plant life discovered on an uncharted seamount. *Science* 227: 57–59.

Littler, M. M., D. S. Littler, and P. R. Taylor. 1983. Evolutionary strategies in a tropical barrier reef system: Functional-form groups of marine macroalgae. *Journal of Phycology* 19: 229–37.

Littler, M. M., D. S. Littler, and P. R. Taylor. 1987. Animal-plant defense associations: Effects on the distribution and abundance of tropical reef macrophytes. *Journal of Experimental Marine Biology and Ecology* 105: 107–21.

Littler, M. M., D. S. Littler, and P. R. Taylor. 1995. Selective herbivore increases biomass of its prey: A chiton-coralline reef-building association. *Ecology* 76: 1666–81.

Littler, M. M., P. R. Taylor, and D. S. Littler. 1986. Plant defense associations in the marine environment. *Coral Reefs* 5: 63–71.

Livingstone, D., and B. A. Whitton. 1983. Influence of phosphorus on morphology of *Calothrix parietina* (Cyanophyta) in culture. *British Phycological Journal* 18: 29–38.

Lizotte, M. P., and J. C. Priscu. 1992. Photosynthesis-irradiance relationships in phytoplankton from the physically stable water column of a perennially ice-covered lake (Lake Bonney, Antarctica). *Journal of Phycology* 28: 179–85.

Lobban, C. S., and P. J. Harrison. 1994. *Seaweed Ecology and Physiology*. Cambridge University Press.

Löffelhardt, W., and H. J. Bohnert. 1994. Molecular biology of cyanelles. In D. A. Bryant, ed., *The Molecular Biology of Cyanobacteria*. 65–89. Kluwer Academic Publishers.

Lubchenco, J. 1978. Plant species diversity in a marine intertidal community: Importance of herbivore food preference and algal competitive abilities. *American Naturalist* 112: 23–39.

Lubchenco, J. 1980. Algal zonation in the New England rocky intertidal community: An experimental analysis. *Ecology* 61: 333–44.

Lubchenco, J. 1982. Effects of grazers and algal competitors on fucoid colonization in tidepools. *Journal of Phycology* 18: 544–50.

Lubchenco, J., and J. Cubit. 1980. Heteromorphic life histories of certain marine algae as adaptations to variations in herbivory. *Ecology* 61: 676–87.

Lubchenco, J., and B. A. Menge. 1978. Community development and persistence in a low rocky intertidal zone. *Ecological Monographs* 48: 67–94.

Lucas, I. A. N. 1970. Observations on the fine structure of the Cryptophyceae. I. The genus *Cryptomonas. Journal of Phycology* 6: 30–38.

Ludwig, M., and S. P. Gibbs. 1985. DNA is present in the nucleomorph of cryptomonads: Further evidence that the chloroplast evolved from a eukaryotic endosymbiont. *Protoplasma* 127: 9–20.

Lüning, K. 1980. Critical levels of light and temperature regulating the gametogenesis of three *Laminaria* species (Phaeophyceae). *Journal of Phycology* 16: 1–15.

Luttenton, M. R., and R. G. Rada. 1986. Effects of disturbance on epiphytic community architecture. *Journal of Phycology* 22: 320–26.

Madsen, T. V., and S. C. Maberly. 1990. A comparison of air and water as environments for photosynthesis by the intertidal alga *Fucus spiralis* (Phaeophyta). *Journal of Phycology* 26: 24–30.

Maggs, C. A., and C. M. Pueschel. 1989. Morphology and development of *Ahnfeltia plicata* (Rhodophyta): Proposal of Ahnfeltiales ord. nov. *Journal of Phycology* 25: 333–51.

Makarewicz, J. C., and P. Bertram. 1991. Evidence for the restoration of the Lake Erie ecosystem. *BioScience* 41: 216–23.

Manhart, J. R., and J. D. Palmer. 1990. The gain of two chloroplast tRNA introns marks the green algal ancestor of land plants. *Nature* 345: 268–70.

Mann, D. G., and H. J. Marchant. 1989. The origins of the diatom and its life cycle. In J. C. Green, B. S. C. Leadbeater, and W. L. Diver, eds., *The Chromophyte Algae*, 307–23. Clarendon Press.

Mann, K. H. 1973. Seaweeds: Their productivity and strategy for growth. *Science* 182: 975–81.

Mann, K. H. 1982. *Ecology of Coastal Waters*. University of California Press.

Marchant, H. J., K. R. Buck, D. L. Garrison, and H. A. Thomsen. 1989. *Mantoniella* in Antarctic waters, including the description of *M. antarctica* sp. nov. (Prasinophyceae). *Journal of Phycology* 25: 167–74.

Marshall, A. T. 1996. Calcification in hermatypic and ahermatypic corals. *Science* 271: 637–39.

Martin, J. H., K. H. Coale, K. S. Johnson, S. E. Fitzwater, R. M. Gordon, S. J. Tanner. 1994. Testing the iron hypothesis in ecosystems of the equatorial Pacific Ocean. *Nature* 371: 123–29.

Mathieson, A. C. 1979. Vertical distribution and longevity of subtidal seaweeds in northern New England, U.S.A. *Botanica Marina* 22: 511–20.

Mattox, K. R., and K. D. Stewart. 1977. Cell division in the scaly green flagellate *Heteromastix angulata* and its bearing on the origin of the Chlorophyceae. *American Journal of Botany* 64: 931–45.

Mattox, K. R., and K. D. Stewart. 1984. Classification of the green algae: A Concept based on comparative cytology. In D.E.G. Irvine and D. M. John, eds. *Systematics of the Green Algae*. 29–72. Academic Press.

Mazumder, A., W. D. Taylor, D. J. McQueen, and D. R. S. Lean. 1990. Effects of fish and plankton on lake temperature and mixing depth. *Science* 247: 312–15.

McCook, L. J., and A. R. O. Chapman. 1993. Community succession following massive ice-scour on a rocky intertidal shore: Recruitment, competition and predation during early, primary succession. *Marine Biology* 115: 565–75.

McCourt, R. M., R. W. Hoshaw, and J. C. Wang. 1986. Distribution, morphological diversity and evidence for polyploidy in North American Zygnemataceae (Chlorophyta). *Journal of Phycology* 22: 307–13.

McCourt, R. M., K. G. Karol, S. Kaplan, and R. W. Hoshaw. 1995. Using *rbc*L sequences to test hypotheses of chloroplast and thallus evolution in conjugating green algae (Zygnematales, Charophyceae). *Journal of Phycology* 31: 989–995.

McFadden, G. I., P. R. Gilson, and D. R. A. Hill. 1994. *Goniomonas*: rRNA sequences indicate that this phagotrophic flagellate is a close relative of the host component of cryptomonads. *European Journal of Phycology* 29: 29–32.

McFadden, G. I., P. R. Gilson, C. J. B. Hofmann, G. J. Adcock, and U. G. Maier. 1994. Evidence that an amoeba acquired a chloroplast by retaining part of an engulfed eukaryotic alga. *Proceedings of the National Academy of Sciences USA* 91: 3690–94.

McNally, K. L., N. S. Govind, P. E. Thomé, and R. K. Trench. 1994. Small-subunit ribosomal DNA sequence analyses and a reconstruction of the inferred phylogeny among symbiotic dinoflagellates (Pyrrophyta). *Journal of Phycology* 30: 316–29.

McQuoid, M. R., and L. A. Hobson. 1995. Importance of resting stages in diatom seasonal succession. *Journal of Phycology* 31: 44–50.

Medcof, J. C. 1985. Life and death with *Gonyaulax*: An historical perspective. In D. M. Anderson, A. W. White, and D. G. Baden, eds., *Toxic Dinoflagellates*, 1–8. Elsevier.

Medlin, L. K., M. Lange, and M. E. M. Baumann. 1994. Genetic differentiation among three colony-forming species of *Phaeocystis:* Further evidence for the phylogeny of the Prymnesiophyta. *Phycologia* 33: 199–212.

Melkonian, M., and H. Robenek. 1984. The eyespot apparatus of flagellated green algae: A critical review. In F. E. Round and D. J. Chapman, eds., *Progress in Phycological Research*, vol. 3, 193–268. Biopress.

Menge, B. A. 1976. Organization of the New England rocky intertidal community: Role of predation, competition, and environmental heterogeneity. *Ecological Monographs* 46: 355–93.

Mensinger, A. F., and J. F. Case. 1992. Dinoflagellate luminescence increases susceptibility of zooplankton to teleost predation. *Marine Biology* 112: 207–10.

Miller, R. J. 1985. Succession in sea urchin and seaweed abundance in Nova Scotia, Canada. *Marine Biology* 84: 275–86.

Milligan, K. L. D., and E. M. Cosper. 1994. Isolation of a virus capable of lysing the brown tide microalga, *Aureococcus anophagefferens*. *Science* 266: 805–7.

Mitman, G. G., and H. K. Phinney. 1985. The development and reproductive morphology of *Halosaccion americanum* I. K. Lee (Rhodophyta, Palmariales). *Journal of Phycology* 21: 578–84.

Mitman, G. G., and J. P. van der Meer. 1994. Meiosis, blade development, and sex determination in *Porphyra purpurea* (Rhodophyta). *Journal of Phycology* 30: 147–59.

Mizuno, M. 1987. Morphological variation of the attached diatom *Cocconeis scutellum* var. *scutellum* (Bacillariophyceae). *Journal of Phycology* 23: 591–97.

Mizuno, M., and K. Okuda. 1985. Seasonal change in the distribution of cell size of *Cocconeis scutellum* var. *ornata* (Bacillariophyceae) in relation to growth and sexual reproduction. *Journal of Phycology* 21: 547–53.

Moe, R. L., and P. C. Silva. 1977. Antarctic marine flora: Uniquely devoid of kelps. *Science* 196: 1206–8.

Monastersky, R. 1995. Iron versus the greenhouse. *Science News* 148: 220–22.

Morrison, D. 1988. Comparing fish and urchin grazing in shallow and deep coral reef algal communities. *Ecology* 69: 1367–82.

Morse, A. N. C. 1991. How do planktonic larvae know where to settle? *American Scientist* 79: 154–67.

Morse, D., P. Salois, P. Markovic, and J. W. Hastings. 1995. A nuclear-encoded form II RuBisCO in dinoflagellates. *Science* 268: 1622–24.

Müller, D. G. 1988. The role of pheromones in sexual reproduction of brown algae. In A. W. Coleman, L. J. Goff, and J. R. Stein-Taylor, eds., *Algae As Experimental Systems*, 201–13. Alan R. Liss.

Müller, D. G., M. N. Clayton, and I. Germann. 1985. Sexual reproduction and life history of *Perithalia caudata* (Sporochnales, Phaeophyta). *Phycologia* 24: 467–73.

Müller, D. G., I. Maier, and G. Gassmann. 1985. Survey on sexual pheromone specificity in Laminariales (Phaeophyceae). *Phycologia* 24: 475–77.

Müller, D. G., and B. Stache. 1989. Life history studies on *Pilayella littoralis* (L.) Kjellman (Phaeophyceae, Ectocarpales) of different geographical origin. *Botanica Marina* 32: 71–78.

Mumford, T. F., and A. Miura. 1988. *Porphyra* as food: Cultivation and economics. In C. A. Lembi and J. R. Waaland, eds., *Algae and Human Affairs*, 87–117. Cambridge University Press.

Murray, S. N., and M. H. Horn. 1989. Seasonal dynamics of macrophyte populations from an eastern North Pacific rocky-intertidal habitat. *Botanica Marina* 32: 457–73.

Muscatine, L., J. W. Porter, and I. R. Kaplan. 1989. Resource partitioning by reef corals as determined from stable isotope composition. $\delta^{13}C$ of zooxanthellae and animal tissue versus depth. *Marine Biology* 100: 185–93.

Nakanishi, K., M. Nishijima, M. Nishimura, K. Kuwano, and N. Saga. 1996. Bacteria that induce morphogenesis in *Ulva pertusa* (Chlorophyta) grown under axenic conditions. *Journal of Phycology* 32: 479–82.

Norton, T. A. 1986. The zonation of seaweeds on rocky shores. In P. G. Moore and R. Seed, eds., *The Ecology of Rocky Coasts*, 7–21. Columbia University Press.

Norton, T. A. 1992. Dispersal by macroalgae. *British Phycological Journal* 27: 293–301.

Norton, T. A., and A. C. Mathieson. 1983. The biology of unattached seaweeds. In F. E.

Round and D. J. Chapman, eds., *Progress in Phycological Research*, vol. 2, 333–86. Elsevier.

Oates, B. R. 1985. Photosynthesis and amelioration of desiccation in the intertidal saccate alga *Colpomenia peregrina*. *Marine Biology* 89: 109–19.

Oates, B. R. 1986. Components of photosynthesis in the intertidal saccate alga *Halosaccion americanum* (Rhodophyta, Palmariales). *Journal of Phycology* 22: 217–23.

Oates, B. R., and K. M. Cole. 1994. Comparative studies on hair cells of two agarophyte red algae, *Gelidium vagum* (Gelidiales, Rhodophyta) and *Gracilaria pacifica* (Gracilariales, Rhodophyta). *Phycologia* 33: 420–33.

Okada, M. 1992. Recent studies on the composition and the activity of algal pyrenoids. In F. E. Round and D. J. Chapman, eds., *Progress in Phycological Research*, vol. 8, 117–38. Biopress.

O'Kelly, C. J., and G. L. Floyd. 1985. Absolute configuration analysis of the flagellar apparatus in *Giraudyopsis stellifer* (Chrysophyceae, Sarcinochrysidales) zoospores and its significance in the evolution of the Phaeophyceae. *Phycologia* 24: 263–74.

Olaizola, M., P. K. Bienfang, and D. A. Ziemann. 1992. Pigment analysis of phytoplankton during a subarctic spring bloom: Xanthophyll cycling. *Journal of Experimental Marine Biology and Ecology* 158: 59–74.

Oliver, R. L. 1994. Floating and sinking in gas-vacuolate cyanobacteria. *Journal of Phycology* 30: 161–73.

O'Neal, S. W., and C. A. Lembi. 1995. Temperature and irradiance effects on growth of *Pithophora oedogonia* (Chlorophyceae) and *Spirogyra* sp. (Charophyceae). *Journal of Phycology* 31: 720–26.

Osborne, B., F. Doris, A. Cullen, R. McDonald, G. Campbell, and M. Steer. 1991. *Gunnera tinctoria*: An unusual nitrogen-fixing invader. *BioScience* 41: 224–34.

Owens, T. G., J. C. Gallagher, and R. S. Alberte. 1987. Photosynthetic light-harvesting function of violaxanthin in *Nannochloropsis* spp. (Eustigmatophyceae). *Journal of Phycology* 23: 79–85.

Padilla, D. K. 1989. Algal structural defenses: Form and calcification in resistance to tropical limpets. *Ecology* 70: 835–42.

Paerl, H. W. 1994. Spatial segregation of CO_2 fixation in *Trichodesmium* spp.: Linkage to N_2 fixation potential. *Journal of Phycology* 30: 790–99.

Palenik, B., and R. Haselkorn. 1992. Multiple evolutionary origins of prochlorophytes, the chlorophyll *b*-containing prokaryotes. *Nature* 355: 265–67.

Parker, B. C., G. M. Simmons, F. G. Love, R. A. Wharton, and K. G. Seaburg. 1981. Modern stromatolites in Antarctic dry valley lakes. *BioScience* 31: 656–61.

Passow, U. 1991. Species-specific sedimentation and sinking velocities of diatoms. *Marine Biology* 108: 449–55.

Patterson, G. M. L., K. K. Baker, C. L. Baldwin, C. M. Bolis, and F. R. Caplan. 1993. Antiviral activity of cultured blue-green algae (cyanophyta). *Journal of Phycology* 29: 125–30.

Paul, V. J., and K. L. Van Alstyne. 1992. Activation of chemical defenses in the tropical green algae *Halimeda* spp. *Journal of Experimental Marine Biology and Ecology* 160: 191–203.

Pearce, C. M., and R. E. Scheibling. 1991. Induction of metamorphosis of larvae of the green sea urchin, *Strongylocentrotus droebachiensis*, by coralline red algae. *Biological Bulletin* 179: 304–11.

Penny, D., and C. J. O'Kelly. 1991. Seeds of a universal tree. *Nature* 350: 106–7.

Pfiester, L. A. 1975. Sexual reproduction of *Peridinium cinctum* f. *ovoplanum*. *Journal of Phycology* 11: 259–65.

Pfiester, L. A. 1984. Sexual reproduction. In D. L. Spector, ed., *Dinoflagellates*, 181–99. Academic Press.

Phillips, J. A., M. N. Clayton, I. Maier, W. Boland, and D. G. Müller. 1990. Sexual reproduction in *Dictyota diemensis* (Dictyotales, Phaeophyta). *Phycologia* 29: 367–79.

Pickett-Heaps, J. D. 1975. *Green Algae*. Sinauer Associates.

Pickett-Heaps, J. D., D. R. A. Hill, and K. L. Blaze. 1991. Active gliding motility in an araphid marine diatom, *Ardissonea* (formerly *Synedra*) *crystallina*. *Journal of Phycology* 27: 718–25.

Pienaar, R. N., and M. E. Aken. 1985. The ultrastructure of *Pyramimonas pseudoparkeae* sp. nov. (Prasinophyceae) from South Africa. *Journal of Phycology* 21: 428–47.

Polanshek, A. R., and J. A. West. 1977. Culture and hybridization studies on *Gigartina papillata* (Rhodophyta). *Journal of Phycology* 13: 141–49.

Porter, J. W. 1976. Autotrophy, heterotrophy, and resource partitioning in Caribbean reef-building corals. *American Naturalist* 110: 731–42.

Proctor, L. M., and J. A. Fuhrman. 1990. Viral mortality of marine bacteria and cyanobacteria. *Nature* 343: 60–62.

Provasoli, L., and I. J. Pintner. 1980. Bacteria-induced polymorphism in an axenic laboratory strain of *Ulva lactuca* (Chlorophyceae). *Journal of Phycology* 16: 196–201.

Pueschel, C. M. 1987. Absence of cap membranes as a characteristic of pit plugs of some red algal orders. *Journal of Phycology* 23: 150–56.

Pueschel, C. M. 1990. Cell structure. In K. M. Cole and R. G. Sheath, eds., *Biology of the Red Algae*, 7–41. Cambridge University Press.

Pueschel, C. M. 1994. Systematic significance of the absence of pit-plug cap membranes in the Batrachospermales (Rhodophyta). *Journal of Phycology* 30: 310–15.

Pueschel, C. M., and K. M. Cole. 1982. Rhodophycean pit plugs: An ultrastructural survey with taxonomic implications. *American Journal of Botany* 69: 703–20.

Pueschel, C. M., and T. J. Miller. 1996. Reconsidering prey specializations in an algal-limpet grazing mutalism: Epithallial cell development in *Clathromorphum circumscriptum* (Rhodophyta, Corallinales). *Journal of Phycology* 32: 28–36.

Quetin, L. B., and R. M. Ross. 1991. Behavioral and physiological characteristics of the Antarctic krill, *Euphausia superba*. *American Zoologist* 31: 49–63.

Rai, A. N. 1990. *Handbook of Symbiotic Cyanobacteria*. CRC Press.

Ramus, J. 1983. A physiological test of the theory of complementary chromatic adaptation. II. Brown, green and red seaweeds. *Journal of Phycology* 19: 173–78.

Rawitscher-Kunkel, E., and L. Machlis. 1962. The hormonal integration of sexual reproduction in *Oedogonium*. *American Journal of Botany* 49: 177–83.

Reed, R. H., I. R. Davison, J. A. Chudek, and R. Foster. 1985. The osmotic role of mannitol in the Phaeophyta: An appraisal. *Phycologia* 24: 35–47.

Regan, M. A., and K. W. Glombitza. 1986. Phlorotannins, brown algal polyphenols. In F. E. Round and D. J. Chapman, eds., *Progress in Phycological Research*, vol. 4, 129–241. Biopress.

Reynolds, C. S., and A. E. Walsby. 1975. Water-blooms. *Biological Reviews* 50: 437–81.

Ringo, D. L. 1967. Flagellar motion and fine structure of the flagellar apparatus in *Chlamydomonas*. *Journal of Cell Biology* 33: 543–71.

Robinson, D. H., K. R. Arrigo, R. Iturriaga, and C. W. Sullivan. 1995. Microalgal light-harvesting in extreme low-light environments in McMurdo Sound, Antarctica. *Journal of Phycology* 31: 508–20.

Roemer, S. C., K. D. Hoagland, and J. R. Rosowski. 1984. Development of a freshwater periphyton community as influenced by diatom mucilage. *Canadian Journal of Botany* 62: 1799–1813.

Rosemond, A. D. 1993. Interactions among irradiance, nutrients, and herbivores constrain a stream algal community. *Oecologia* 94: 585–94.

Rosemond, A. D., and S. H. Brawley. 1996. Species-specific characteristics explain the persistance of *Stigeoclonium tenue* (Chlorophyta) in a woodland stream. *Journal of Phycology* 32: 54–63.

Rothhaupt, K. O. 1996. Utilization of substitutable carbon and phosphorus sources by the mixotrophic chrysophyte *Ochromonas* sp. *Ecology* 77: 706–15.

Rowan, K. S. 1989. *Photosynthetic Pigments of Algae*. Cambridge University Press.

Rumpf, R., D. Vernon, D. Schreiber, and C. W. Birky. 1996. Evolutionary consequences of the loss of photosynthesis in chlamydomonadaceae: Phylogenetic analysis of *Rrn18* (18S rDNA) in 13 *Polytoma* strains (Chlorophyta). *Journal of Phycology* 32: 119–26.

Rushforth, S. R., L. E. Squires, and C. E. Cushing. 1986. Algal communities of springs and streams in the Mt. St. Helens region, Washington, U.S.A. following the May 1980 eruption. *Journal of Phycology* 22: 129–37.

Saffo, M. B. 1987. New light on seaweeds. *BioScience* 37: 654–64.

Sagert, S., and H. Schubert. 1995. Acclimation of the photosynthetic apparatus of *Palmaria palmata* (Rhodophyta) to light qualities that preferentially excite photosystem I or photosystem II. *Journal of Phycology* 31: 547–54.

Sandgren, C. D. 1981. Characteristics of sexual and asexual resting cyst (statospore) formation in *Dinobryon cylindricum* Imhof (Chrysophyta). *Journal of Phycology* 17: 199–201.

Sathyendranath, S., A. D. Gouveia, S. R. Shetye, P. Ravindran, and T. Platt. 1991. Biological control of surface temperature in the Arabian Sea. *Nature* 349: 54–56.

Satoh, M., T. Hori, K. Tsujimoto, and T. Sasa. 1995. Isolation of eyespots of green algae and analyses of pigments. *Botanica Marina* 38: 467–74.

Scheibling, R. 1986. Increased macroalgal abundance following mass mortalities of sea urchins (*Strongylocentrotus droebachiensis*) along the Atlantic coast of Nova Scotia. *Oecologia* 68: 186–98.

Schimek, C., I. N. Stadnichuk, R. Knaust, and W. Wehrmeyer. 1994. Detection of chlorophyll c_1 and magnesium-2,4-divenyl-phaeoporphyrin A_5 monomethylester in cryptophytes. *Journal of Phycology* 30: 621–27.

Schindler, D. W. 1977. Evolution of phosphorus limitation in lakes. *Science* 195: 260–62.

Schindler, D. W., K. H. Mills, and D. F. Malley. 1985. Longterm ecosystem stress: The effects of years of experimental acidification on a small lake. *Science* 228: 1395–1401.

Schnetter, R., B. Brück, K. Gerke, and G. Seibold. 1990. Notes on heterokaryotic life cycle phases in some Dasycladales and Bryopsidales (Chlorophyta). In W. Wiessner, D. G. Robinson, and R. C. Starr, eds., *Experimental Phycology 1*, 124–33. Springer-Verlag.

Schofield, O., B. M. A. Kroon, and B. B. Prézelin. 1995. Impact of ultraviolet-B radiation on photosystem II activity and its relationship to the inhibition of carbon fixation rates for Antarctic ice algae communities. *Journal of Phycology* 31: 703–15.

Schopf, J. W. 1993. Microfossils of the early archean apex chert: New evidence of the antiquity of life. *Science* 260: 640–46.

Schornstein, K. L., and J. Scott. 1982. Ultrastructure of cell division in the unicellular red alga *Porphyridium purpureum*. *Canadian Journal of Botany* 60: 85–97.

Schulz-Baldes, M., and R. A. Lewin. 1976. Fine structure of *Synechocystis didemni* (Cyanophyta: Chroococcales). *Phycologia* 15: 1–6.

Scott, J., and S. Broadwater. 1990. Cell division. In K. M. Cole and R. G. Sheath, eds., *Biology of the Red Algae*, 123–45. Cambridge University Press.

Scott, J., D. Phillips, and J. Thomas. 1981. Polar rings are persistent organelles in interphase vegetative cells of *Polysiphonia harveyi* Bailey (Rhodophyta, Ceramiales). *Phycologia* 20: 333–37.

Searles, R. B. 1980. The strategy of the red algal life history. *American Naturalist* 115: 113–20.

Sears, J. R., and S. H. Brawley. 1982. *Smithsoniella* gen. nov., a possible evolutionary link between the multicellular and siphonous habits in the Ulvophyceae, Chlorophyta. *American Journal of Botany* 69: 1450–61.

Seliger, H. H., J. H. Carpenter, M. Loftus, and W. D. McElroy. 1970. Mechanisms for accumulation of high concentrations of dinoflagellates in a bioluminescent bay. *Limnology and Oceanography* 15: 234–45.

Serrão, E. A., G. Pearson, L. Kautsky, and S. H. Brawley. 1996. Successful external fertilization in turbulent environments. *Proceedings of the National Academy of Sciences USA* 93: 5286–90.

Shashar, N., and N. Stambler. 1992. Endolithic algae within corals: Life in an extreme environment. *Journal of Experimental Marine Biology and Ecology* 163: 277–86.

Sheath, R. G., J. M. Burkholder, M. O. Morison, A. D. Steinman, and K. L. VanAlstyne. 1986. Effect of tree canopy removal by gypsy moth larvae on the macroalgae of a Rhode Island headwater stream. *Journal of Phycology* 22: 567–70.

Sheath, R. G., and K. M. Cole. 1980. Distribution and salinity adaptations of Bangia atropurpurea (Rhodophyta), a putative migrant into the Laurentian Great Lakes. *Journal of Phycology* 16: 412–20.

Sheath, R. G., and J. A. Hambrook. 1990. Freshwater ecology. In K. M. Cole and R. G. Sheath, eds., *Biology of the Red Algae*, 423–53. Cambridge University Press.

Sheath, R. G., and B. J. Hymes. 1980. A preliminary investigation of the freshwater red algae in streams of southern Ontario, Canada. *Canadian Journal of Botany* 58: 1295–1318.

Sheath, R. G., and M. O. Morison. 1982. Epiphytes on *Cladophora glomerata* in the Great Lakes and St. Lawrence Seaway with particular reference to the red alga *Chroodactylon ramosum* (Asterocystis smargdina). *Journal of Phycology* 18: 385–91.

Shimmel, S. M., and W. M. Darley. 1985. Productivity and density of soil algae in an agricultural system. *Ecology* 66: 1439–47.

Sieburth, J. M., P. W. Johnson, and P. E. Hargraves. 1988. Ultrastructure and ecology of *Aureococcus anophagefferens* gen. et sp. nov. (Chrysophyceae): The dominant picoplankter during a bloom in Narragansett Bay, Rhode Island, summer 1985. *Journal of Phycology* 24: 416–25.

Simons, J., and A. P. van Beem. 1987. Observations on asexual and sexual reproduction in *Stigeoclonium helveticum* Vischer (Chlorophyta) with implications for the life history. *Phycologia* 26: 356–62.

Siver, P. A. 1991. *The Biology of Mallomonas*. Kluwer Academic Publishers.

Siver, P. A., and J. S. Hamer. 1992. Seasonal periodicity of Chrysophyceae and Synurophyceae in a small New England lake: Implications for paleolimnological research. *Journal of Phycology* 28: 186–98.

Sjøtun, K., and K. Gunnarsson. 1995. Seasonal growth pattern of an Icelandic *Laminaria* population (section Simplices, Laminariaceae, Phaeophyta) containing solid- and hollowstiped plants. *European Journal of Phycology* 30: 281–87.

Slankis, T., and S. P. Gibbs. 1972. The fine structure of mitosis and cell division in the chrysophycean alga *Ochromonas danica*. *Journal of Phycology* 8: 243–56.

Smayda, T. J. 1990. Novel and nuisance phytoplankton blooms in the sea: Evidence for a global epidemic. In E. Granéli, B. Sundström, L. Edler, and D. M. Anderson, eds., *Toxic Marine Phytoplankton*, 29–40. Elsevier.

Smetacek, V. S. 1985. Role of sinking in diatom life-history cycles: Ecological, evolutionary and geographical significance. *Marine Biology* 84: 239–51.

Smith, R. C., B. B. Prézelin, K. S. Baker, R. R. Bidigare, and N. P. Boucher. 1992. Ozone depletion: Ultraviolet radiation and phytoplankton biology in Antarctic waters. *Science* 255: 952–59.

Smith, V. H. 1983. Low nitrogen to phosphorus ratios favor dominance by blue-green algae in lake phytoplankton. *Science* 221: 669–71.

Sommer, U. 1994. The impact of light intensity and daylength on silicate and nitrate competition among marine phytoplankton. *Limnology and Oceanography* 39: 1680–88.

Sousa, W. P. 1979. Experimental investigations of disturbance and ecological succession in a rocky intertidal algal community. *Ecological Monographs* 49: 227–54.

Spear-Bernstein, L., and K. R. Miller. 1989. Unique location of the phycobiliprotein light-harvesting pigment in cryptophyceae. *Journal of Phycology* 25: 412–19.

Starr, M., J. H. Himmelman, and J. C. Therriault. 1990. Direct coupling of marine invertebrate spawning with phytoplankton blooms. *Science* 247: 1071–74.

Stebbins, G. L., and G. J. C. Hill. 1980. Did multicellular plants invade the land? *American Naturalist* 115: 342–53.

Steidinger, K. A., J. M. Burkholder, and H. B. Glasgow. 1996. *Pfiesteria piscicida* gen. et sp. nov. (Pfiesteriaceae fam. nov.), a new toxic dinoflagellate with a complex life cycle and behavior. *Journal of Phycology* 32: 157–64.

Steinberg, P. D. 1985. Feeding preferences of *Tegula funebralis* and chemical defenses of marine brown algae. *Ecological Monographs* 55: 333–49.

Steinkötter, J., D. Bhattacharya, I. Semmelroth, C. Bibeau, and M. Melkonian. 1994. Prasinophytes form independent lineages within the Chlorophyta: Evidence from ribosomal RNA sequence comparisons. *Journal of Phycology* 30: 340–45.

Steinman, A. D., and G. A. Lamberti. 1988. Lotic algal communities in the Mt. St. Helens region six years following the eruption. *Journal of Phycology* 24: 482–89.

Steneck, R. S. 1982. A limpet-coralline algal association: Adaptations and defenses between a selective herbivore and its prey. *Ecology* 63: 507–22.

Sterner, R. W. 1986. Herbivores' direct and indirect effects on algal populations. *Science* 231: 605–7.

Stevens, J. E. 1995. The Antarctic pack-ice ecosystem. *BioScience* 45: 128–32.

Strömgren, T. 1983. Temperature-length growth strategies in the littoral alga *Ascophyllum nodosum* (L.) *Limnology and Oceanography* 28: 516–21.

Suda, S., M. M. Watanabe, and I. Inouye. 1989. Evidence for sexual reproduction in the primitive green alga *Nephroselmis olivacea* (Prasinophyceae). *Journal of Phycology* 25: 596–600.

Sumich, J. L. 1984. *An Introduction to the Biology of Marine Life*. 3d ed. Wm. C. Brown Publishers.

Sumich, J. L. 1988. *An Introduction to the Biology of Marine Life*. 4th ed. Wm. C. Brown Publishers.

Sun, W. Q. 1996. Mechanism of survival of algae in desiccated soils. *Phycologia* 35: 270.

Suttle, C. A., A. M. Chan, and M. T. Cottrell. 1990. Infection of phytoplankton by viruses and reduction of primary productivity. *Nature* 347: 467–69.

Suzaki, T., and R. E. Williamson. 1985. Euglenoid movement in *Euglena fusca*: Evidence for sliding between pellicular strips. *Protoplasma* 124: 137–46.

Sze, P. 1980. Aspects of the ecology of macrophytic algae in high rockpools at the Isles of Shoals (USA). *Botanica Marina* 23: 313–18.

Sze, P. 1982. Distributions of macroalgae in tidepools on the New England coast (USA). *Botanica Marina* 25: 269–76.

Takamura, N., and Y. Nojiri. 1994. Picophytoplankton biomass in relation to lake trophic state and the TN:TP ratio of lake water in Japan. *Journal of Phycology* 30: 439–44.

Talbot, M. M. B., G. C. Bate, and E. E. Campbell. 1990. A review of the ecology of surf-zone diatoms, with special reference to *Anaulus australis*. *Oceanography and Marine Biology, Annual Review* 28: 155–75.

Tanner, C. E. 1981. Chlorophyta: Life histories. In C. S. Lobban and M. J. Wynne, eds., *The Biology of Seaweeds*, 218–47. University of California Press.

Takewaki, M., L. Provasoli, and I. J. Pintner. 1983. Morphogenesis of *Monostroma oxyspermum* (Küuz.) Doty (Chlorophyceae) in axenic culture, especially bialgal culture. *Journal of Phycology* 19: 409–16.

Terry, L. A., and B. L. Moss. 1980. The effect of photoperiod on receptacle initiation in *Ascophyllum nodosum* (L.) Le Jol. *British Phycological Journal* 15: 291–301.

Tilman, D. 1977. Resource competition between planktonic algae: An experimental and theoretical approach. *Ecology* 58: 338–48.

Tilman, D., and R. L. Kiesling. 1984. Freshwater algal ecology: Taxonomic trade-offs in the temperature dependence of nutrient competitive abilities. In M. J. Klug and C. A. Reddy, eds., *Current Perspectives in Microbial Ecology*, 314–19. American Society of Microbiology.

Tilman, D., R. Kiesling, R. Sterner, S. S. Kilham, and F. A. Johnson. 1986. Green, blue-green, and diatom algae: Taxonomic differences in competitive ability for phosphorus, silicon, and nitrogen. *Archiv fuer Hydrobologie* 106: 437–85.

Tilman, D., M. Mattson, and S. Langer. 1981. Competition and nutrient kinetics along a temperature gradient: An experimental test of a mechanistic approach to niche theory. *Limnology and Oceanography* 26: 1020–33.

Tomas, R. N., and E. R. Cox. 1973. Observations on the symbiosis of *Peridinium balticum* and its intracellular alga. I. Ultrastructure. *Journal of Phycology* 9: 304–23.

Tranior, F. R., and R. Gladych. 1995. Survival of algae in desiccated soil: A 35-year study. *Phycologia* 34: 191–92.

Turner, S., T. Burger-Wiersma, S. J. Giovannoni, L. R. Mur, and N. R. Pace. 1989. The relationship of a prochlorophyte *Prochlorothrix hollandica* to green chloroplasts. *Nature* 337: 380–82.

Tyler, M. A., and H. H. Seliger. 1978. Annual subsurface transport of a red-tide dinoflagellate to its bloom area: Water circulation patterns and organism distribution in the Chesapeake Bay. *Limnology and Oceanography* 23: 227–46.

Underwood, G. J. C., J. D. Thomas, and J. H. Baker. 1992. An experimental investigation of interactions in snail-macrophyte-epiphyte systems. *Oecologia* 91: 587–95.

Urbach, E., D. L. Robertson, and S. W. Chisholm. 1992. Multiple evolutionary origins of prochlorophytes within the cyanobacterial radiation. *Nature* 355: 267–70.

Vadas, R. L., and R. S. Steneck. 1988. Zonation of deep-water benthic algae in the Gulf of Maine. *Journal of Phycology* 24: 338–46.

Vadas, R. L., W. A. Wright, and S. L. Miller. 1990. Recruitment of *Ascophyllum nodosum*: Wave action as a source of mortality. *Marine Ecology Progress Series* 61: 263–72.

van der Meer, J. P., and E. R. Todd. 1980. The life history of *Palmaria palmata* in culture. A new type for the Rhodophyta. *Canadian Journal of Botany* 58: 1250–56.

van Donk, E. 1989. The role of fungal parasites in phytoplankton succession. In U. Sommer, ed., *Plankton Ecology*, 171–94. Springer-Verlag.

van Donk, E., and S. S. Kilham. 1990. Temperature effects on silicon- and phosphorus-limited growth and competitive interactions among three diatoms. *Journal of Phycology* 26: 40–50.

Vanni, M. J., and D. L. Findlay. 1990. Trophic cascades and phytoplankton community structure. *Ecology* 71: 921–37.

Vannote, R. L., G. W. Minshall, K. W. Cummins, J. R. Sedell, and C. E. Cushing. 1980. The river continuum concept. *Canadian Journal of Fisheries and Aquatic Science* 37: 130–37.

van Tussenbroek, B. I. 1989. Observations on branched *Macrocystis pyrifera* (L.) C. Agardh (Laminariales, Phaeophyta) in the Falkland Islands. *Phycologia* 28: 169–80.

Villareal, T. A., and F. Lipschultz. 1995. Internal nitrate concentrations in single cells of large phytoplankton from the Sargasso Sea. *Journal of Phycology* 31: 689–96.

Vincent, W. F., and S. Roy. 1993. Solar ultraviolet-B radiation and aquatic primary production: Damage, protection, and recovery. *Environmental Reviews* 1: 1–12.

Vymazal, J. 1995. *Algae and Element Cycling in Wetlands*. Lewis Publishers.

Vymazal, J., and C. J. Richardson. 1995. Species composition, biomass, nutrient content of periphyton in the Florida Everglades. *Journal of Phycology* 31: 343–54.

Wagner, G., and F. Grolig. 1992. Algal chloroplast movements. In M. Melkonian, ed., *Algal Cell Motility* 39–72. Chapman and Hall.

Waite, A., P. K. Bienfang, and P. J. Harrison. 1992. Spring bloom sedimentation in a subarctic ecosystem. Nutrient sensitivity. *Marine Biology* 114: 119–29.

Waite, A. M., R. J. Olson, H. G. Dam, and U. Passow. 1995. Sugar-containing compounds on the cell surfaces of marine diatoms measured using concanavalin A and flow cytometry. *Journal of Phycology* 31: 925–33.

Walne, P. L., and H. J. Arnott. 1967. The comparative ultrastructure and possible function of eyespots. *Euglena granulata* and *Chlamydomonas eugametos*. *Planta* 77: 325–53.

Watanabe, M. M., K. Kaya, and N. Takamura. 1992. Fate of the toxic cyclic heptapeptides, the microcystins, from blooms of *Microcystis* (Cyanobacteria) in a hypertrophic lake. *Journal of Phycology* 28: 761–67.

Watanabe, M. M., Y. Takeda, T. Sasa, I. Inouye, S. Suda, T. Sawaguchi, and M. Chihara. 1987. A green dinoflagellate with chlorophylls *A* and *B*: Morphology, fine structure of the chloroplast and chlorophyll composition. *Journal of Phycology* 23: 382–89.

West, J. A., A. R. Polanshek, and D. E. Shelvin. 1978. Field and culture studies on *Gigartina agardhii* (Rhodophyta). *Journal of Phycology* 14: 416–26.

Wetzel, R. G. 1983. Attached algal-substrata interactions: Fact or myth, and when and how? In R. G. Wetzel, ed., *Periphyton of Freshwater Ecosystems*, 207–15. Dr. W. Junk Publishers.

Wharton, R. A., B. C. Parker, and G. M. Simmons. 1983. Distribution, species composition and morphology of algal mats in Antarctic dry valley lakes. *Phycologia* 22: 355–65.

Whittaker, R. H. 1975. *Communities and Ecosystems*. Macmillan.

Wilce, R. T. 1994. Essay: The Arctic subtidal as a habitat for macrophytes. In C. S. Lobban and P. J. Harrison, *Seaweed Ecology and Physiology*, 89–92. Cambridge University Press.

Wilcox, L. W., L. A. Lewis, P. A. Fuerst, and G. L. Floyd. 1992. Assessing the relationships of autosporic and zoosporic chlorococcalean green algae with 18S rDNA sequence data. *Journal of Phycology* 28: 381–86.

Wilcox, L. W., and G. J. Wedemayer. 1985. Dinoflagellate with blue-green chloroplasts derived from an endosymbiotic eukaryote. *Science* 227: 192–94.

Wilcox, L. W., G. J. Wedemayer, and L. E. Graham. 1982. *Amphidinium cryophilum* sp. nov. (Dinophyceae), a new freshwater dinoflagellate. II. Ultrastructure. *Journal of Phycology* 18: 18–30.

Wilhelm, C., I. Lenartz-Weiler, I. Wiedemann, and A. Wild. 1986. The light-harvesting system of a *Micromonas* species (Prasinophyceae): The combination of three different chlorophyll species in one single chlorophyll-protein complex. *Phycologia* 25: 304–12.

Williams, S. L. 1990. Experimental studies of Caribbean seagrass bed development. *Ecological Monographs* 60: 449–69.

Witman, J. D. 1987. Subtidal coexistence: Storms, grazing, mutualism, and the zonation of kelps and mussels. *Ecological Monographs* 57: 167–87.

Wolfe, J. M., and M. M. Harlin. 1988. Tidepools in southern Rhode Island, U.S.A. II. Species diversity and similarity analysis of macroalgal communities. *Botanica Marina* 31: 537–46.

Yokohama, Y., A. Kageyama, T. Ikawa, and S. Shimura. 1977. A carotenoid characteristic of chlorophycean seaweeds living in deep waters. *Botanica Marina* 20: 433–36.

Zechman, F. W., E. C. Theriot, E. A. Zimmer, and R. L. Chapman. 1990. Phylogeny of the Ulvophyceae (Chlorophyta): Cladistic analysis of nuclear-encoded rRNA sequence data. *Journal of Phycology* 26: 700–710.

Zupan, J. R., and J. A. West. 1988. Geographic variation in the life history of *Mastocarpus papillatus* (Rhodophyta). *Journal of Phycology* 24: 223–29.

INDEX

A

Acetabularia, 76, 77
actin, 9
Actinastrum, 66
adenosine triphosphate (ATP), 3, 4, 26, 27
agar, 151
Agardhiella, 158, 159
Agarum, 120, 126, 245
agglutinins, 50–51
aggregations, 55–59
Ahnfeltia, 164–165, 177
akinetes, 25–29
Alaria, 120, 121, 245
Alexandrium, 141, 142
algae
  beneficial aspects, 19
  detrimental aspects, 19
  generally, 1
  humans and, 18–19
alginic acid, 108
allophycocyanin, 22
alternation of generations, 42
ammonia, 26
amoeboid cells, 15
Anabaena, 26, 27, 29, 32, 33
  A. azollae, 34
androspores, 64
anisogametes, 10
Ankistrodesmus, 66
Antarctic
  marine, 200–203
  sea ice, 108
Apatococcus, 61, 62, 221
Aphanizomenon, 29, 32, 33
aplanospores, 56
Arctic, marine, 200–203
Ascophyllum, 124–127, 174, 242, plate 6
asexual reproduction, 10
  Chlamydomonas, 48–49
assimilators, 74, 113
association of cells for colonies, Chlorococcum,
  56–58
astaxanthin, plate 2a
Asterionella, 106, 209

Atlantic rocky shores, 239–248
  intertidal emergent surfaces, 241–244
  spray zone, 240
  subtidal region, 245–248
  tidepools, 244–245
atolls, 232
Audouinella, 152, 154, 164, 174
Aureococcus, 95, 195, 203
autospores, 59
autotrophs, 1
auxiliary cells, 162
auxospores, 103
auxotrophy, 185
Azolla, 34

B

Bacillariophyceae, 98-108
bacteria (excluding cyanobacteria), 1, 2, 7
Bangia, 171, 172
Bangiophyciciae, 153, 171–174
barrier reefs, 232
Basicladia, 215
Batophora, 76, 77
Batrachospermum, 154–156, 158
  Florideophycidae, 167–168
benthic algae, 17
  freshwater, 213–221
benthic cyanobacteria, 31
benthic marine algae, 225–250
  Atlantic rocky shores, 239–248
  coral reefs, 231–237
  morphologic types, 225–230
  Pacific rocky shores, 248–249
  production and food chains, 230–231
  soft-bottom communities, 237–239
Berkeleya, 106
binary fission, 23
blooms, 18, 32
Bonnemaisonia, 164–166, 170
Bostrychia, 238
Botrydiopsis, 96, 97
Brachiomonas, 51
branching, 28–29, 155

brevetoxin, 141
brown algae
    ecology and commercial uses, 126–127
    without free-living haploid phase, 122–125
brown tide, 95, 195, 203
Bryopsis, 74
Bulbochaete, 64, 65

C
calcareousx structures, 160
calcium carbonate, 133
Callithamnion, 155
Callophyllis, plate 7b
Caloglossa, 238
Calothrix, 28, 29, 34, 240
Calvin Cycle, 3, 4, 7, 8, 23, 26
carbon, phytoplankton, 187–188
carbon dioxide, 3, 4
carboxysomes, 23
β-carotene, 4, 5
carotenoids, 3, 4, 5, 6, 8
carpogonium, 162
carpospores, 163
carposporophytes, 154, 162–163, 165, 166
    rhodophyta (red algae), 169–170
carrageenan, 151
Carteria, 51
Catenella, 238
Caulerpa, 74, 234, 235
Caulerpales, 70–77
cell division, 42
cell of filament, 9
cell plates, 41
cellular organization, diversity, 7-13
cellular structure, cyanobacteria, 22–24
cellulose, 9
cell walls, 9
centric diatoms, 100, 101
centrioles, 8
Centritractus, 97
Centroceras, 156
Ceramium, 156
Ceratium, 136, 138
Chaetoceros, 105, 106, plate 3b
Chaetophorales, 47, 59–61
Chara, 81–82, 217
Charales, 81
charophytes, 41, 79–87
    ecology, 86
Chattonella, 98
chemocline, 184
Chesapeake Bay, 206
Chlamydomonas, 47–53, 55, 222
    asexual reproduction, 48–49
    C. mexicana, 221
    life cycle, 51
    palmelloid stage, 48–49
Chlorarachnion, 149
chlorarachniophyta, 130, 149
Chlorella, 60, 66, 67, 218, 219
Chlorococcales, 47, 55–59

chlorophyceae, 55–59
Chlorococcum, 56
    association of cells to form colonies, 56–58
    formation of multinucleate cells, 58–59
    loss of flagellated stages, 59
Chlorogonium, 51
Chlorophyceae
    Chlorococcales, 47, 55–59
    ecology and commercial uses, 64–67
    evolution, 46
    filamentous green algae, 59–61
    flagellated cells, 47–53
    orders, 47
    specialized filaments, 61–64
Chlorophyceae, 40
chlorophyceans, 39, 41, 45–66
chlorophylls, 3
    a, 1, 3, 4
    b, 4
    c, 4, 44
Chlorophyta, 39–88
    chlorophyceans, 45–66
    classes, 40
    prasinophyceae, 43–45
chlorophyte, 39
chloroplast ER (CER), 10, 131, 144
chloroplasts
    origin, 35–36
    types, 11
Chondrus, 158, 159, 164, 166, 242, 243, 245,
        plate 4b, plate 6b
Chorda, 120, 121
Chordaria, 116, 117
Chordariales, 116–117
Chromophyta, 89–127
    Bacillariophyceae, 98–108
    characteristics, 90
    Chrysophyceae, 91–96
    classes, 89, 90
    Eustigmatophyceae, 98
    Phaeophyceae, 108–126
    Raphidophyceae, 98
    Synurophyceae, 91–96
    Tribophyceae, 96–98
Chroomonas, 145
chrysalominarin, 7
Chrysochromulina, 131, 132, 133, 203
Chrysophyceae, 91–96
Chrysophytes, ecology, 94–96
Chrysopyxis, 93, 94
ciguatera, 142
ciguatoxin, 142
Cladophora, 70–72, 106, 214, 215, 217, 219
Cladophorales, 70–77
classification levels, 2
Clathromorphum, 152, 170, 175
Closterium, 86
coastal oceans, marine phytoplankton,
        203–204
coastal upwelling, marine phytoplankton,
        204–207
coccoliths, 133

Cocconeis, 107
Codiolum stages, 69
Codium, 74, 78
coenobium, 53, 58
coenocytes, 58–59
Colacium, 148
Coleochaetales, 79–80
Coleochaete, 79–80, 87
colonies, 55–59
   flagellated cells, 15, 53–54
   green algae, 52
commercial uses
   brown algae, 126–127
   Chlorophyceae, 64–67
   Rhodophyta (red algae), 174–177
   Ulvophytes, 77–79
compensation depth, 180
conceptacles, 123, 164
Conchocelis, 174
conchocelis phase, 174
conjugation, 83
contractile vacuoles, 8
coral bleaching, 232
Corallina, 160, 161, plate 4b
Corallinales, 159–160
cortex, 108
cortication, 81, 155
Coscinodisucs, 100, 105
Cosmarium, 84–85
Crouania, 158
Crucigenia, 66
cryptomonad cells, 144–146
Cryptomonas, 145
cryptophyta, 130, 144–146
   ecology, 146
   typical cell, 144–145
cyanelles, 34–35
Cyanidium, 175
cyanobacteria, 1, 21–36
   cellular structure, 22–24
   filamentous without specialized cells, 25
   heterocystous, 25–29
   nonfilamentous, 24–25
cyanophycin granules, 23
Cyclotella, 100, 209
Cylindrocystis, 85
Cymbella, 101, 107
Cymopolia, 76, 77
cysts, 9
cytoplasm, 9
cytoskeleton, 8–10, 23

D
Darwin, Charles, 248
Dasycladales, 76–77
daughter cells, 10
density stratification and phytoplankton,
     180–184
Derbesia, 71–74
Desmarestia, 117
Desmarestiales, 116–117

Desmidium, 86
desmids, 81
desmokonts, 134
Diadema, 235
Diatoma, 107
diatomaceous earth, 18
diatomite, 18
diatoms, 98–108
   centric, 100, 101
   ecology and diversity, 105–108
   reproduction, 101–104
Dictyoneurum, 121
Dictyosiphon, 117
Dictyosiphonales, 117–118
Dictyosphaerium, 56–58
Dictyota, 95, 113–114, 235
dimethysulfide (DMS), 133, 134
dimictic lakes, 207
Dinobryon, 93, 94
dinoflagellate cells, 135–140
dinoflagellate ecology, 140–143
dinokaryotic nucleus, 137
dinokont cells, 134
Dinophysis, 136, 142
dinophyta, 130, 134–143
   dinoflagellate ecology, 140–143
   typical dinoflagellate cell, 135–140
Diotyocha, 95
diplohaplonts, 10
diploid vegetative phase, 43
diplonts, 10
diversity
   of cellular organization, 7–13
   diatoms, 105–108
   of photosynthetic pigments, 3–6
DNA, 13, 35–36
Draparnaldia, 61, 62
Dumontia, 158, 159, 170
Dunaliella, 18, 66–67, 210
Durvillaea, 125
dystrophic water, 218

E
ecology
   brown algae, 126–127
   Charophytes, 86
   Chlorophyceae, 64–67
   Chrysophytes, 94–96
   Cryptophyta, 146
   diatoms, 105–108
   dinoflagellate, 140–143
   diversity of, 16–18
   Euglenophyta, 148
   Haptophyta, 133–134
   Rhodophyta (red algae), 174–177
   Synurophytes, 91–94
   Ulvophytes, 77–79
Ectocarpales, 109–113
Ectocarpus, 109–113, 115
Egregia, 121, 248
ejectosomes, 144

Elachista, 116
emergent surfaces, 239
Emiliania, 131, 133
encrusting thalli, *Florideophycidae,* 158–159
endolithic algae, 214, 221
endopelic algae, 214
endophytic algae, 214
endosammic algae, 214
endosymbionts, 11
endosymbiotic cyanobacterium, 35–36
Enteromorpha, 69, 77–78, 239, 245
epibionts, 17
epicones, 134
epilignic algae, 214
epilithic algae, 214
epiphytes, 17, 214
epiphytic algae, 214
epipsammic algae, 214
Epipyxis, 93, 94
epitheca, 98
epivalve, 98
epizoic algae, 214
Erythrotrichia, 171, 172
estuaries, marine phytoplankton, 204–207
eubacteria, 7
Euchampia, 105
Eudorina, 52, 53
Euglena, 146–148
Euglenophyta, 130, 146–148
    ecology, 148
eukaryotic algae, 1–3, 7, 10
euphausiids, 192, 193
Eustigmatophyceae, 98
Eutreptia, 146
eutrophication, 209
eutrophy, 184
Everglades, 218
evolution of land plants, 87
exospores, 23
eyespot, 8

F
false branching, 28–29
ferredoxin, 26
fibrils, 9
Fibrocapsa, 98
fibrous roots, 8
filamentous green algae, 68–69
    Chlorophyceae, 59–61
filamentous without specialized cells,
        cyanobacteria, 25
filaments, 9-10, 15
    and pseudoparenchymatous thalli, 109–113
flagellar apparatus, 12
flagellar swelling, 8
flagellated cells
    Chlorophyceae, 47–53
    Volvocales, 47–53
flagellated solitary cells, 14
Florideophycidae
    Batrachospermum, 167–168

Bonnemaisonia, 164–166
    life cycles, 160–169
    palmaria type, 168–169
    Polysiphonia, 164
    Rhodophyta (red algae), 153–171
    thallus structure, 154–160
flotation, phytoplankton, 190–193
flowing water, freshwater benthic algae,
        219–221
food chains, benthic marine algae, 230–231
Fragilaria, 101, 107
freshwater benthic algae, 213–221
    flowing water, 219–221
    standing water, 215–218
    symbiotic algae, 218–219
freshwater phytoplankton, 207–211
    rivers, 210–211
    temperate lakes, 207–209
    tropical and polar lakes, 209–210
fringing reefs, 232
Fritschiella, 61, 62
frustule, 98, 108
fucoxanthin, 3, 5, 6, 109
Fucus, 122–124, 242
    F. distichus, 122
    F. vesiculosus, 122, 123, 124, 239
furrowing (division), 8, 9, 41
    Chlorophyta, 41–42

G
Gambierdiscus, 142
gametes, 10
    Chlamydomonas, 49–50
gametophyte, 10
gas vesicles, 33
Gelidium, 177
genetic systems of cells, 13
Giffordia, 112
Gigartina, 166, 248
girdle bands, 98
Glaucocystis, 35
glaucophytes, 34, 36
Glenodinium, 136
gliding movement, 22
globules, 81
Gloeobacter, 23
Gloeocapsa, 23, 24–25, 34, 221, plate 5a
Gloeocystis, 55
glucans
    α-1,4-linked, 7
    β-1,3-linked, 7
glucose, 3, 4, 7
Gomphonema, 101
gonidium, 54
Gonium, 52, 53
Gonyaulax, 143
Gonyostomum, 98
Gracilaria, 158, 170, 177, plate 4a
grazing, phytoplankton, 193–194
green algae, 39–88
Gunnera, 34, 223
Gymnodinium, 142

H

Haematococcus, 67, plate 2a
Halicystis phase, 73
Halimeda, 78, 234, 235
halocline, 184
Halosaccion, 168
haploid vegetative phase, 43, 122–125
haplonts, 10
haptera, 118
haptophyta, 129–134
  ecology, 133–134
  representative genera, 132–133
Helicodictyon, 191
Heribaudiella, 127
Heterococcus, 96, 97
heterocysts, 25–29
heterokontous, 91
heteromorphic generations, 42
  Phaeophyceae, 114–122
Heterosigma, 98
heterotrichous filaments, 16
heterotrichy, chaetophorales, 60–61
heterotrophic algae, 1
Hildenbrandia, 158, 159, 160, 164, plate 8a
holdfast, 17
hormogonium, 25
humans and algae, 18–19
Hydra, 66, 218
Hydrodictyon, 54, 59
hydrogen sulfide, 30
hypocones, 134
Hypoglossum, 159
hypotheca, 98
hypovalve, 98

I

ice algae, 203
intertidal emergent surfaces, Atlantic rocky
    shores, 241–244
intertidal region, 239
isogametes, 10
isomorphic generations, 42
  Phaeophyceae, 108–114

K

kelps, 118–122, 249
Klebsormidiales, 79
Klebsormidium, 79, 80
krill, 192, 193

L

Laminaria, 110, 120, 122, 126, 127, 245, 247,
    248, 249, plate 7a
  L. digitata, 120
  L. saccharina, 3, 118–119, 120, 121
Laminariales, 3, 118–122
laminarin, 7
Leathesia, 116
Lemanea, 167

lentic environments, 215
Lewin, Ralph, 29
lichens, 223
life cycle
  algae, 10
  Chlamydomonas, 51
light influence, phytoplankton, 180
Littorina, 242, 245, plate 8a
lorica, 51, 53
lotic environments, 219
Lyngbya, 30, 33

M

macrandry, 63, 65
macroalgae, 17
Macrocystis, 122, 126, 127, 248
Mallomonas, 91–92, 93, 94
mangroves, 237–238
Mantoniella, 44
marginal ice zone (MIZ), 201
marine macroalgae, 225–231
marine phytoplankton, 197–206
  coastal oceans, 203–204
  coastal upwelling and estuaries, 204–207
  polar oceans, 200–203
  temperate oceans, 199
  tropical and subtropical oceans, 199–200
mastigonemes, 12
Mastocarpus, 166, 242, 243, 248, plate 6b
medulla, 108
Melosira, 99, 103
Meridion, 107
Merismopedia, 25
meristem, 108
meristoderm, 118
meromictic, 207
Mesodinium, 146
mesotrophy, 207
Micractinium, 66
Micrasterias, 86
Microcystis, 25, 32, 33, 209
Micromonadophyceae, 43–45
Micromonas, 44
microplankton, 179
Microspora, 96
Microsporales, 59–61
microtubular roots, 8
microtubules, 7, 9, 41
mitotic spindle, 8
mixed layer, 182
mixotrophy, 185
molecular phylogeny, 13–14
Monera, 1
Monod equation, 185–186
monomictic, 207
monospores, 164
Monostroma, 69, 70
morphologic diversity, 14–16
Mougeotia, 82, 83, 86, 208
mucilage, 9
multinucleate cells, Chlorococcum, 58–59

multiseriate filaments, 15
mycobiont, 223
Mycosphaerella, 125
Myrionema, 116

N
nannandry, 63, 65
nanoplankton, 179
Navicula, 101, 102, 217
Nemalion, 157, 158, 160, 164
Nereocystis, 121, 126
Netrium, 85
nicotinamide adenine dinuclotide phosphate
        (NADP), 3, 4
nitrogen
    inorganic compounds, 26
    as N_2, 26, 29
    phytoplankton, 188–189
nitrogenase, 26
nitrogen fixation, 26–29
Nitzschia, 106, 217
Nodularia, 32, 33
nonfilamentous cyanobacteria, 24–25
nonflagellated cells, 15, 55–59
Nostoc, 28, 29, 33, 34, 223
nucleomorph, 144
nucule, 81
nutrients and phytoplankton, 184–190

O
Ochromonas, 91–92, 93
Oedogoniales, 47, 61–64
Oedogonium, 214, 216
okadaic acid, 142
oligotrophy, 184
Oocystis, 56–59
oogamy, 10
oogonium, 162
openly branching filaments, *Florideophycidae,*
        154–157
Ophiocytium, 96, 97
Origin of Land Plants (Graham), 87
Ornithocercus, 136
Oscillatoria, 25, 31, 33, 215
Ostreobium, 234

P
Pacific rocky shores, benthic marine algae,
        248–249
Padina, 114, 126
Palmaria, 168, 170, 176
palmaria type, *Florideophycidae,* 168–169
palmelloid aggregations, 15
    Tetrasporales, 54–55
palmelloid stage, *Chlamydomonas,* 48
Pandorina, 52, 53
paraflagella body, 8
Paramecium, 66, 218
paramylon, 7, 146

parenchymatous green algae, 69–70
parenchymatous macrothalli, 117–118
parenchymatous thalli, 16
Pascherina, 52
Pediastrum, 57–58, plate 2b
Pelvetia, 125, 126, 248
pennate diatom, 100–101, 102
peptidoglycan wall, 22
peridinin, 3, 5
Peridinium, 136, 139, 140
periphyton, 213
periplast, 144
periplastidal space, 144
Petalonia, 117
Petrocelis, 166
Pfiesteria, 143
Phacotus, 51, 53
Phaeocystis, 132, 133, 194, 200, 201
Phaeophyceae, 108–126
    heteromorphic generations, 114–122
    isomorphic generations, 108–114
Phormidium, 30, 235
phosphorus, phytoplankton, 189–190
photic zone, 179
photoreceptor, 7
photorespiration, 188
photosynthesis/irradiance (P/I) curve,
        180–181
photosynthetically active radiation (PAR), 3,
        196
photosynthetic pigments, diversity, 3–6
phototaxis, 7, 191
phragmoplasts, 41
phycobilins, 3, 5, 6, 21–22
phycobiliproteins, 3, 5, 6, 22, 26
phycobilisomes, 7, 22
phycobiont, 223
phycocolloids, 18
phycocyanin, 22
phycocyanobilin, 6
Phycodrys, 158, 159, plate 7b
phycoerythrin, 22, 31
phycoerythrobilin, 6
phycoerythrocyanin, 22
phycology, 1–2
phycoplasts, 41
Phyllospadix, 248
phylogeny, 13–14
physodes, 109–110
phytoplankton, 17, 179–211
    defined, 179
    density stratification and, 180–184
    flotation and sinking, 190–193
    freshwater, 207–211
    freshwater phytoplankton, 210–211
    grazing, 193–194
    light influence on, 180
    marine phytoplankton, 197–206
    nutrients and, 184–190
picoplankton, 31, 179
Pilayella, 112
Pithophora, 216–217

pit plugs, 152
   Floridian red algae, 170–171
plakea, 53, 54
planktonic cyanobacteria, 31
Planktoniella, 105
planozygote, 51
Plasmodium, 14
Platymonas, 44, 45
Pleurastrophyceae, 39
Plumaria, 156, 164
plurilocular gametangia, 110
plurilocular sporangia, 110
polar lakes, freshwater phytoplankton,
    209–210
polar oceans, marine phytoplankton, 200–203
Polysiphonia, 152, 157, 162–164, 170
   Florideophycidae, 164
   P. lanosa, 174
polysiphonous cells, 157
Polytoma, 53
Porphyra, 167, 171, 173, 176, 230, 241,
    plate 4b
Porphyridium, 171–172
Postelsia, 121, 122, 248
Prasinophyceae, Chlorophyta, 43–45
Prasiola, 240
procarp, 170
Prochlorococcus, 29, 31, 197
Prochloron, 29
prochlorophytes, 21, 29
Prochlorothrix, 29
production, benthic marine algae, 230–231
prokaryotic algae, 2, 7
propagules, 113
Prorocentrum, 136, 142, 206
Protista, 1
Protogonyaulax, 141, 142
Protosiphon, 58–59
Prototheca, 66
Prymnesium, 133
pseudoflagella, 55
Pseudonitzschia, 102, 106, 204
pseudoparenchymatous, 16, 116–117
pseudoraphe, 99
Pterocladia, 177
pusule, 139
pycnocline, 182
Pyramimonas, 44, 45
pyrenoids, 8
Pyrocystis, 136, 143
Pyrodinium, 143

R
Ralfsia, 112–113, plate 8a
raphe, 99
Raphidophyceae, 98
receptacles, 122
red algae. *See* Rhodophyta (red algae)
red tides, 18, 140
replicate end wall, 83
reproduction

diatoms, 101–104
   sexual and asexual, 10
reserve carbohydrates, 7
Rhizoclonium, 71
Rhizophora, 237
rhizoplast, 8
rhizopodial cells, 15
Rhizosolenia, 34, 105, 108
Rhodochaete, 171
Rhodophyta (red algae), 6, 151–177
   Bangiophycidae, 171–174
   characteristics, 151
   ecology and commercial uses, 174–177
   Florideophycidae, 153–171
   pit plugs, 170–171
   postfertilization events and development of
      carposporophyte, 169–170
Rhoicosphenia, 107
ribosomes, 7, 8, 23, 35–36
ribulose bisphosphate, 4
ribulose bisphosphate carboxylase, 7, 36
Richelia, 34, 108
rivers, freshwater phytoplankton, 210–211
RNA, 13, 35–36
rRNA, 13

S
salt marshes, 30, 237
Sargasso Sea, 126
Sargassum, 125, 126, 127
   S. fluitans, 126
   S. muticum, 127
   S. natans, 126
saxitoxin, 141
scales, 7
Scenedesmus, 57–59, 67, 194
Scytonema, 28, 29
Scytosiphon, 114, 117–118, 230
Scytosiphonales, 117–118
seagrasses, 237–238
semicells, 84
semidiurnal tides, 240
sexual reproduction, 10
sheath, 9
silica deposition vesicle, 101
silicon, 108
   phytoplankton, 190
sinking, phytoplankton, 190–193
siphonaxanthin, 5
siphonous green algae, 70–77
Skeletonema, 105
snow algae, 222, plate 5b
soft-bottom communities, benthic marine algae,
    237–239
   mangrove swamps and seagrass meadows on
      tropical coasts, 237–238
   temperate shorelines, 238–239
somatic meiosis, 168
Southern Ocean. *See* Antarctic
Spartina, 239
specialized filaments, Chlorophyceae, 61–64

sperm packet, 54
Sphacelaria, 113, 127
Sphagnum, 218
spindle, 8
Spirogyra, 82, 83, 86, 214, 215–216
Spirotaenia, 85
Spirulina, 18, 26, 30
sporophylls, 121
sporophyte, 10
spray zone, 239
    Atlantic rocky shores, 240
Spyridia, 156
standing water, freshwater benthic algae,
        215–218
starch, 7
    cyanobacteria, 23
statocyst, 93
Staurastrum, 86
Stauroneis, 101
Stephanodicus, 106
stephanokontous flagellation, 61
Stigeoclonium, 46, 61, 109
stigma, 8
Stigonema, 28, 29
stipe, 118
stratification and phytoplankton, 180–184
Strongylocentrotus, 247
subtidal region, 240–241
    Atlantic rocky shores, 245–248
sulfates, 117
suspension-feeding phytoplankton, 193
Symbiodinium, 143
symbionts, 18, 36
symbiotic algae, 218–219
symbiotic cyanobacteria, 34
Synechococcus, 30, 31
Synechocystis, 31
Synura, 91, 93–94, 95, plate 3a
Synurophyceae, 91–96
Synurophytes, ecology, 91–94

T
temperate lakes, freshwater phytoplankton,
        207–209
temperate oceans, marine phytoplankton, 199
temperate shorelines, soft-bottom communities,
        238–239
terrestrial algae, 221–224
Tetraselmis, 44, 45
Tetraspora, 55
Tetrasporales, 47
    palmelloid aggregations, 54–55
tetrasporophyte, 165–166
Thalassia, 237
thallus, 9
    Florideophycidae, 154–160
theca, 135
thermocline, 181
thermophilic algae, plate 3
thylakoids, 7, 10, 21, 23
tidepools, 239, plate 8b
    Atlantic rocky shores, 244–245

Tolypothrix, 28, 29
toxins, 33
Trachelomonas, 147, 148
Trebouxia, 66, 224
Trentepohlia, 222, 223
Trentepohliales, 79
Tribonema, 96, 97
Tribophyceae, 96–97
trichocysts, 137
Trichodesmium, 31, 32, 199, 200
trichogyne, 162
trichomes, 25, 28, 31
trichothallic meristem, 109
tropical and subtropical oceans, marine
        phytoplankton, 199–200
tropical lakes, freshwater phytoplankton,
        209–210
true branching, 28–29
Turbinaria, 126
turf algae, 234

U
Udotea, 234, 237
Ulothrix, 68, 70, 217
Ulotrichales, 68–69
Ulva, 3, 69, plate 2c
Ulvaa, 77–79
Ulvales, 3, 69–70
Ulvaria, 69
Ulvophyceae, 40, 67–79
ulvophytes, 41, 67–68
    ecology and commercial uses, 77–79
Undaria, 127
unilocular sporangia, 110
uniseriate filaments, 15
upwelling, coastal, 204–207
Uroglena, 93, 94

V
Vacuolaria, 98
Vaucheria, 96, 97, 237, 239
vegetative cell divisions, 10
vitamins, 185
Volvocales, 47
    colonies of flagelated cells, 53–54
Volvox, 52–55

W
water net, 59

X
xanthophyll cycle, 6

Z
zonation (intertidal), plate 7a
zoochlorellae, 218
zoospores, 10
zooxanthellae, 143, 231
Zostera, 96, 239
Zygnemataceae, 81–86
Zygnematales, 81–86
zygotes, 10, 42